LIGHT
Its Interaction with Art and Antiquities

This limited facsimile edition has been issued for the purpose of keeping this title available to the scientific community.

LIGHT
Its Interaction with Art and Antiquities

Thomas B. Brill
University of Delaware
Newark, Delaware

Plenum Press · New York and London

Library of Congress Cataloging in Publication Data

Brill, Thomas B
Light, its interaction with art and antiquities.

Includes index.
1. Light. 2. Optical materials. 3. Photochemistry. 4. Artists' materials – Effect of light on. I. Title.
QC355.2.B74 702'.8'8 80-16975
ISBN 0-306-40416-8

A Division of Plenum Publishing Corporation
227 West 17th Street, New York, N.Y. 10011

Printed in the United States of America

To PATRICIA, BARBARA, and RUSSELL

"There is no science without fancy,
and no art without facts."

Vladimir Nabokov

Preface

Light is an ultimate. It is basic to defining the universe. It lies at the heart of cosmology, physics, and philosophy. Its properties have influenced and, in part, controlled existence. The physicist, philosopher, artist, historian, art conservator, museum scientist, chemist, and engineer intuitively or expressly recognize, although in different ways, the fundamental significance of light. At the very least it provides a common ground where their interests meet.

Our comprehension of the beauty as well as the potential harmfulness of light has not come easily. The attention of natural philosophers and scientists as preeminent as Aristotle, Isaac Newton, and James Clerk Maxwell was directed toward understanding light. More recently, the subject of the interaction of light with matter has been a central issue in science, as attested to by the number of Nobel laureates whose work has touched upon the subject. The names include Becquerel, Bethe, Bohr, de Broglie, Einstein, Heisenberg, Hertz, Lenard, Lorentz, Mulliken, Pauli, Pauling, Planck, Rayleigh, Rutherford, Schrödinger, Van Vleck, and Werner.

The serious artist and craftsman may develop a love–hate relationship with light. Light is the entity which the master manipulates to highlight a work and attract admirers, but in the process light can also chemically and physically alter an object. An art conservator or curator is confronted with the same problems, but receives, in addition, a constant flux of new ideas and materials originating from the work of chemists and physicists. It is formidable, if not impossible, for anyone to possess at the level of a specialist the technical expertise that is needed to effectively deal with all of the problems involving art materials. Fortunately, useful principles about the effect of light usually can be discussed on the practical level without resorting to a highly theoretical exposition.

The subject at hand—*Light: Its Interaction with Art and Antiquities*—concerns the nature of light, its properties, and its good and bad effects on materials. The "good" effects of light are considered to be those that produce color and the desired appearance in works of art. "Bad" effects are those that result in deterioration and color changes. In presenting the material, I have presumed that the reader has a modest background in general and organic chemistry. The presentation is organized with the theme of understanding objects from the point of view of material science rather than specific types of art. Chapters I–V deal with the basic concepts concerning light itself, namely what it is, how it is generated, how it is measured, and how it is affected by materials. Chapter VI gathers these concepts together and focuses them on understanding why certain materials look the way they do. Chapters VII–IX are concerned with the origin of color and photoreactions in organic materials, such as dyes, fibers, and coatings. Chapters X and XI deal with these concepts with regard to inorganic materials, such as pigments and glass. Chapter XII surveys the use of light to make photographic images.

This book is an outgrowth of lectures presented to the students of art conservation in the University of Delaware–Winterthur Museum graduate program. The comments, criticisms, and ideas of the students in the program, now too numerous to mention by name, were invaluable and are most appreciated. Dr. Peter G. Sparks, former Director of the program, provided constant encouragement and support. Without his help, this project would not have been undertaken. I am indebted and give special thanks to Dr. David Onn, Dr. Mark Sharnoff, and Dr. William Weaver, and to Hilton Brown, Joyce Hill Stoner, and my wife, Patricia, for making valuable comments on various parts of the book.

Thomas B. Brill

Contents

I. Light and Its Properties 1
The Description 2
Molecular Response to Radiation 7
Ultraviolet Radiation 10
Visible Radiation 11
Infrared Radiation 12
Bibliography 14

II. Primary Light Sources 15
The Blackbody Radiator 16
Direct Sunlight 18
The Incandescent Lamp 20
The Mercury Lamp and the Sodium Lamp 22
The Fluorescent Lamp 24
The Flashbulb 29
Ultraviolet Content of Some Light Sources 30
Bibliography 32

III. The Measurement of Properties of Light 33
Luminous Intensity of a Source 34
The Solid Angle 35
Luminous Flux 36
Luminous Intensity at a Surface 36
Quantity of Light 38
Light Distribution Diagrams 39
Bibliography 41

IV. The Effect of Materials on Light 43
The Index of Refraction 45
Polarization 47
Reflected Light Intensity 52
Interference of Light 55

Absorption of Light . . . 59
Bibliography . . . 62

V. **Color and Color Mixing** . . . 63
Color . . . 63
Perceived Color . . . 67
Additive Primary Colors . . . 71
Subtractive Colors . . . 72
Bibliography . . . 77

VI. **The Optical Appearance of Materials** . . . 79
Index of Refraction . . . 80
Practical Use of the Index of Refraction . . . 80
Hiding Power . . . 84
Change in Hiding Power with Age . . . 89
Anomalous Dispersion . . . 90
The Effect of Particle Size on Appearance . . . 92
Rayleigh Scattering . . . 93
Pigment Particles . . . 95
A Summary of Opacity . . . 98
The Reflection, Transmission, and Absorption of Light . . . 99
Gloss and Matte Surfaces . . . 99
Surfaces with Transparent Coatings and Films . . . 101
Colored, Transparent Films . . . 106
Metals . . . 108
Glass, Glaze, and Enamel . . . 111
Fibers and Woven Materials . . . 113
Reflection of Ultraviolet Radiation . . . 116
Bibliography . . . 117

VII. **The Origin of Color in Organic Compounds** . . . 119
Atomic Structure . . . 119
Chemical Bonding and Molecular Structure . . . 128
Electronic Transitions in Organic Molecules: Color . . . 144
Bibliography . . . 153

VIII. **Organic Dyes** . . . 155
Method of Application . . . 156
Historical Classification . . . 159
Chemical Structure . . . 166
Bibliography . . . 170

IX. **Decomposition of Organic Materials by Light** . . . 173
Photochemical Principles . . . 174
Rates of Photochemical Reactions . . . 176
Mechanisms of Photochemical Reactions of Dyes . . . 178
Environmental Factors . . . 183
Photoreactions of Paper, Textiles, Wood, Varnish, and Oils . . . 185
Minimization of Photochemical Effects . . . 189

The Future of Natural Organic Materials . . . 191
Bibliography . . . 194

X. **The Origin of Color in Inorganic Compounds** . . . 197
Covalency of Bonds . . . 200
Charge Transfer Transitions . . . 203
Geometry and Electronic Configuration of Ions . . . 205
The "*d–d*" Transition . . . 210
Miscellaneous Sources of Color . . . 213
Effects of Crystal Structure Variation on Color . . . 213
Effects of Particle Size and Concentration on Color: The "Alexandrite Effect" . . . 214
Bibliography . . . 216

XI. **Pigments and Colored Glass** . . . 217
Artist's Pigments . . . 217
White Pigments . . . 219
Blue Pigments . . . 221
Green Pigments . . . 224
Yellow, Orange, and Brown Pigments . . . 227
Red and Violet Pigments . . . 230
Black Pigments . . . 231
Metallic Paints and Coatings . . . 232
The Mixing of Pigments . . . 233
Decomposition of Inorganic Pigments by Light . . . 235
Color in Glass and Glaze . . . 237
Iron . . . 238
Manganese . . . 239
Copper . . . 239
Cobalt . . . 239
Chromium . . . 240
Lanthanides and Actinides . . . 240
Nonmetallic Coloring Agents . . . 241
Metals in Glass . . . 241
Ultraviolet Absorption by Glass . . . 243
Photochemical Effects . . . 243
Bibliography . . . 244

XII. **Light-Sensitive Materials: The Photographic Process** . . . 247
Historical Milestones . . . 248
Photographic Image Materials . . . 250
The Action of Light on Silver Halide Grains . . . 256
Spectral Sensitivity . . . 261
Development and Fixation of the Image . . . 263
Reducers, Intensifiers, and Toners . . . 266
Degradation of Photographic Materials . . . 267
Color Photography . . . 268
Bibliography . . . 273

Index . . . 275

I

Light and Its Properties

In order to understand how light produces color, texture, and surface appearance as well as damage to an object, we need to consider what light is, how it is generated, how it is measured, and how it interacts with matter. Such an exposition can be laid out in various levels of complexity with concomitant degrees of rigor. The subject, if covered in full detail, would generate an incomprehensibly long treatise. In most cases in order to work on a practical problem, the nonspecialist does not need the thorough and sophisticated approach sought by the physicist. In this chapter the nature of light is introduced as it pertains to producing the appearance and alteration of materials of artistic and historical importance. The properties and characteristics of light are defined in the energy regions of interest. The coupling of electromagnetic radiation with atomic and molecular components and events is then briefly considered. Attention is finally focused in more detail on the middle energies of the spectrum of electromagnetic radiation, namely the ultraviolet, visible, and infrared energies.

It might be expected that we should begin by saying what light really is, but this is not possible, because light is more fundamental than any of the terms used to describe it. In fact, the efforts made to explain light have produced the stormiest debates in the most recent centuries of physics. The nature of light can be described by discussing its properties and couching them in terms of the simplest principles. To be cast correctly, these properties and theories must be put in pure mathematical form. But in doing this, a practitioner not concerned with subtle theory often loses the practical significance of the properties of light and the powers of analogy. As long as the limitations of the pictorial and general discussion here are kept in mind, it is better for the non-physical scientist to have an imperfect but descriptive tool rather than the pure truth about light in a mathematical form.

THE DESCRIPTION

Since the onset of the modern theory of light, there has been a conflict between two very different models. On the one hand, light is pictured as a form of wave motion, and on the other as fast-moving particles.

The meaning of wave motion and its defining parameters can be best understood by considering the most common type of wave encountered in nature. This is the sine wave depicted in Figure I-1. The sine wave involves simple harmonic motion. The amplitude a of the wave corresponds to the extreme vertical excursions of the wave from the average. The horizontal rate of passage of the wave is described by the wave velocity c. The peak-to-peak distance in the wave determines the *wavelength* λ. If a point p is viewed on the wave, then p will oscillate up and down on a line whose length is determined by the amplitude. The number of times per second that p completes the oscillation is known as the *frequency* ν of the wave. The velocity, wavelength, and frequency are related to one another by

$$c = \nu\lambda \tag{I-1}$$

This type of wave motion is characteristic of light waves, but, as will be noted shortly, it does not give a complete description of the particular properties of light.

In the nineteenth century, by virtue of a remarkable series of experimental and theoretical results, the wave theory of light gained universal acceptance. James Clerk Maxwell, who was the director of the Cavendish Laboratory at Cambridge, brought together the various aspects of the wave concept of light in 1864 in his paper, "A Dynamical Theory of the Electromagnetic Field." Maxwell proposed that light waves were electromagnetic waves and, apart from wavelength, were identical with all waves which

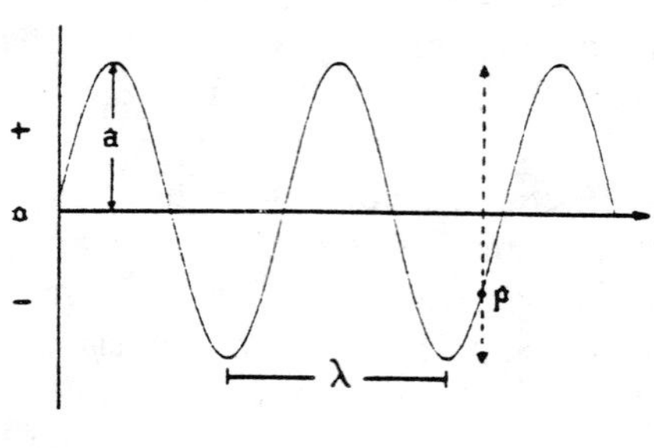

Figure I-1. A wave describing simple harmonic motion.

could be generated by radiation from electrical circuits. He proved to be correct, and his momentous discovery is the basis of modern electromagnetic wave theory. Maxwell's synthesis was built upon the work of such predecessors as Robert Hooke (1665), Francesco Grimaldi (1665), and Christian Huygens (1678), whose opinions represent in various degrees the first modern foreshadowings of the wave theory; Thomas Young (1801), who showed that light waves are subject to interference effects in the same way that water waves are; Etienne Malus (1808), who showed that light could be polarized upon reflection; Augustin Jean Fresnel (1817), who showed how optical phenomena could be explained in terms of the propagation of transverse wave vibrations; Michael Faraday (1830–1850), who developed the theory and experiments to show that light had both electric and magnetic components; Lord Kelvin (William Thomson), who drew an analogy between the flow of heat and the conduction of current; and J. L. Lagrange, whose mathematical methods Maxwell used. Ludwig V. Lorentz in 1867 independently developed a similar wave theory of light. In 1888, Heinrich Hertz crowned the theory by experimentally showing that electromagnetic waves generated by electrical circuits obey all the laws of reflection, refraction, polarization, etc., that optical waves do. The wave properties of light were thus established and light was proven to be electromagnetic in nature.

This satisfying and complacent view was severely shaken at the beginning of the twentieth century when properties of light were discovered which could be explained only by the *particle* or *corpuscular* theory. Isaac Newton in the late seventeenth century had envisioned light as a stream of particles, but with the advent of the wave theory of light his theory lost credence. Albert Einstein in 1905, considering the difficulties encountered by Maxwell's theory in explaining experiments on blackbody radiation, proposed that light was in the form of small energy "quanta," and that this concept could also account for Lenard's photoelectric experiments. The nature of these theorized quanta of light, later called *photons*, is more complex than first imagined, but they were thought of as possessing energy and momentum in discrete packets. Max Planck in 1900 had established that the energy E of emitted radiation is proportional to its frequency ν according to the following equation, where h is a constant:

$$E = h\nu \tag{I-2}$$

These two models, one involving waves and the other particles, and each partially successful in explaining the behavior of light, presented a formidable dilemma to anyone wishing to "understand" light. Indeed, the dilemma produced a great upheaval in physics. It is now generally conceded that one must accept the inadequacy of any simple model in explaining something so basic as light. The modern theory of light can only be stated in mathematical form. It combines the electromagnetic theory of light from Maxwell with the concepts of photons and relativity from Einstein. Rather than trying to go further in "explaining" light, it is more useful to concentrate on its practical properties.

In the phenomena that suggest the existence of photons, the property most readily measured is the energy associated with them. The different colors of light, for example, are thought of as representing photons of different energy. The link between the corpuscular theory and the wave theory lies in Planck's fundamental postulate of quantum theory given by equation (I-2); that is, the energy of the photon is directly proportional to its wave frequency ν. Planck's constant h is the link. If c is set equal to the velocity of light in a vacuum (3×10^{10} cm/sec), then another useful form of Planck's equation is

$$E = \frac{hc}{\lambda} \tag{I-3}$$

because of the relationship between ν and λ shown in equation (I-1). Frequency, which measures the number of vibrations of the wave per second, fixes the wavelength λ, so that different energies also can be expressed in terms of different wavelengths. Equation (I-1) tells us that the larger ν is, the smaller λ must be proportionately, because c is constant. From Planck's equation [equation (I-2) or (I-3)] the energy of light quanta increases with increasing frequency or decreasing wavelength, because the energy of the photon is directly proportional to frequency, but inversely proportional to wavelength. One need not be bothered about the appropriateness of using the photon description of light in one case and the wave description in another. It is convenient for later discussion of interference and polarization to use the wave model, while for photochemistry the photon concept is more often used.

Pictorially, we can envision a single light wave as having the appearance shown in Figure I-2a according to Maxwell's electromagnetic theory. In a simple electromagnetic wave there are electric and magnetic waves which are perpendicular to, but in phase with, one another. They vibrate in a direction perpendicular to the z axis which is called the wave propagation vector. The velocity of light refers to the rate of passage of light in the direction of propagation (z direction). The electric and magnetic waves are also often referred to as vectors. The electric field vector of the wave interacts with electric fields in atoms and therefore is most important to discussions taken up in this book.

The frequency ν represents the number of cycles of oscillation the wave completes per second. The unit of ν is reciprocal seconds (sec^{-1}) or more

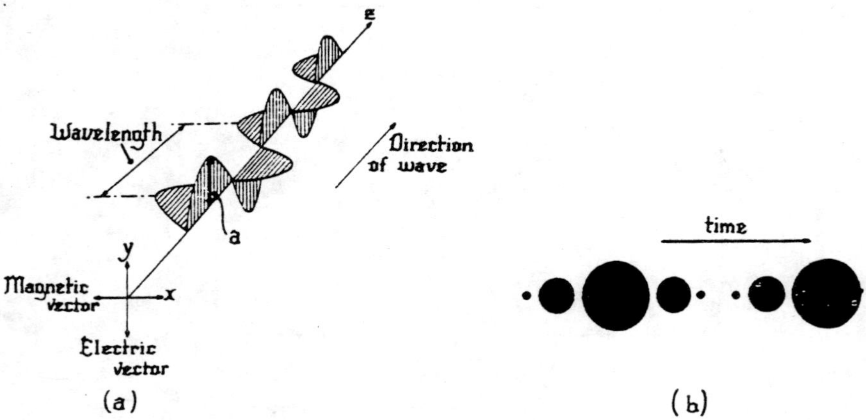

Figure I-2. (a) A modular picture of a single electromagnetic wave, and (b) the view along the z axis of an in-phase packet of many waves oscillating with the passage of time.

properly Hertz (Hz), which is equivalent to sec^{-1}. The value of h in equation (I-2), which relates energy to frequency, is $h = 6.62 \times 10^{-27}$ erg sec. Hence E takes on the unit ergs. An erg is the unit of energy in the centimeter-gram-second system of units. A useful unit which is related to energy is the wavenumber $\bar{\nu}$. The wavenumber is the number of wave peaks per centimeter. It is equal to $1/\lambda$ and has units of reciprocal centimeters (cm^{-1}). Wavenumbers are related to energy but are not equal to energy unless $\bar{\nu}$ is multiplied by the constant h/c, as can be seen from equations (I-1) and (I-3). Wavenumbers are frequently used in spectroscopy because they are directly related to energy, whereas wavelength is reciprocally related, as shown in Figure I-3.

Table I-1 shows comparative numerical values of wavelength, wavenumber, and energy in kilocalories (kcal) for electromagnetic radiation in the visible and ultraviolet ranges. These numbers can be obtained directly from equation (I-2) using the appropriate conversion factors.

The wavelength can be measured in a variety of ways, but the most common units are shown in Table I-2. The various units can be interconverted easily using this table. All wavelength values used in this book will be given in nanometers (nm). The nanometer is an acceptable, internationally used unit and is equivalent to the millimicron.

Several further comments need to be made about the electromagnetic wave depicted in Figure I-2. First of all, according to the wave model, the intensity of the beam of light radiation is measured by the square of the amplitude a of the electric vector,

$$I = ka^2 \tag{I-4}$$

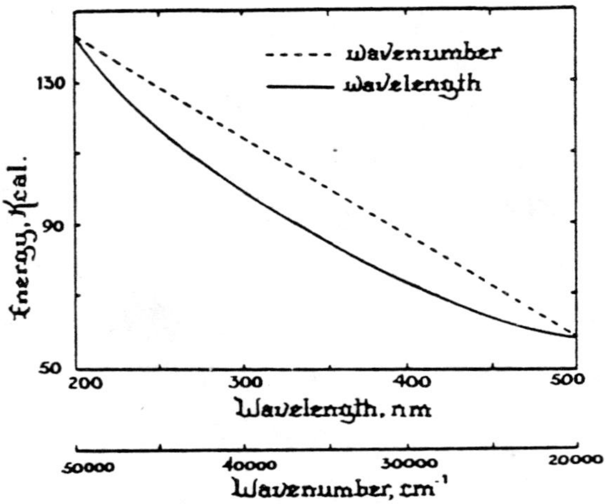

Figure I-3. The relationship between wavelength λ or wavenumber $\bar{\nu}$ and energy E.

where k is a constant. Therefore, the larger the amplitude of the wave, the more intense the radiation is. In the particle model of light, however, amplitude has no meaning, because the model centers on the concept of photons. Hence an alternate way of describing light intensity is needed. In the particle model the intensity of light is proportional to the number of photons per unit volume of the beam. This is a form of "photon density." Both the density and the amplitude concepts of intensity can be shown to be consistent with one another and equation (I-4) holds no matter which model is being used. One can speak of the intensity of light either as the flux

Table I-1. The Energy, Wavelength, and Wavenumber Relationships in the Ultraviolet and Visible Regions

E, kcal	λ, nm	$\bar{\nu}$, cm^{-1}
143.0	200	50,000
95.3	300	33,300
81.7	350	28,570
71.5	400	25,000
57.2	500	20,000
47.7	600	16,670
40.8	700	14,290

Table I-2. Conversion Factors for Units of Length[a]

Multiply number of → by / To obtain ↓ number of	Angstroms	Nanometers	Micrometers	Millimeters	Centimeters	Meters
Angstroms	1	10	10^4	10^7	10^8	10^{10}
Nanometers	10^{-1}	1	10^3	10^6	10^7	10^9
Micrometers	10^{-4}	10^{-3}	1	10^3	10^4	10^6
Millimeters	10^{-7}	10^{-6}	10^{-3}	1	10	10^3
Centimeters	10^{-8}	10^{-7}	10^{-4}	0.1	1	10^2
Meters	10^{-10}	10^{-9}	10^{-6}	10^{-3}	10^{-2}	1

[a] Note that micrometers are sometimes referred to as microns, and nanometers as millimicrons.

of photons or as the square of the amplitude of the wave. Both have their use, depending upon the application.

Second, the magnetic vector of electromagnetic radiation is not of as much interest here as the electric vector, because it is the electric vector which can interact with the electrons and the electric fields in an atom or molecule. This electric vector interaction produces refraction, reflection, and transmission of the wave, in addition to color, chemical reactions, and heating effects in most materials. All of these are subjects to be taken up subsequently. The magnetic vector is important in other applications, such as spectroscopy.

Third, the symbolism "$h\nu$" is often used in writing a chemical reaction to indicate that a photon of electromagnetic radiation is necessary to produce the reaction. For example, a reaction which is important in human vision involves the light-induced isomerization of the vitamin A derivative retinal,

$$cis\text{-retinal} + h\nu \rightarrow trans\text{-retinal}$$

The symbol "$h\nu$" represents light energy and does not disrupt the mass balance of the chemical reaction.

MOLECULAR RESPONSE TO RADIATION

Before turning specifically to the infrared, visible, and ultraviolet regions of radiation, which are of most interest here, it is important to realize that molecules and atoms are affected in one way or another by an encounter with practically all types of radiation. Table I-3 summarizes the events that occur in molecules upon exposure to the various electromagnetic

Table I-3. Molecular Events Induced by Various Wavelengths of Radiation

Wavelength	Frequency, Hz	Region name	Response of atoms and molecules
(3×10^8)–(3×10^5) m	1–10^3	Power	None
(3×10^5)–300 m	10^3–10^6	Audio	None
300–3 m	10^6–10^8	Radiowave	Molecular translations, nuclear reorientations
3–10^{-4} m	10^8–(3×10^{12})	Microwave	Molecular rotations, electron reorientations
10^5–700 nm	(3×10^{12})–(4×10^{14})	Infrared	Molecular vibrations and direct heat effects
700–400 nm	(4×10^{14})–(7×10^{14})	Visible	Low-energy electronic transitions in valence shell
400 nm–10 nm	(7×10^{14})–(3×10^{16})	Ultraviolet	High-energy electronic transitions in valence shell
10 nm–0.03 nm	(3×10^{16})–10^{19}	X-ray	Electronic transitions in the inner shells; diffraction by atoms
0.03–(3×10^{-4}) nm	10^{19}–10^{23}	Gamma-ray	Nuclear transitions

wavelengths. Celestial events are so dramatic and complex that all energies of electromagnetic radiation are produced and bombard the earth at once. In terrestrial situations, however, several energy sources are needed if the intention is to produce radiation over a broad energy range. That is, phenomena that produce X-ray energies do not also produce radio waves and vice versa. In point of fact the events listed in Table I-3 as molecular responses to exposure of matter to the various energy ranges are often conveniently used to *produce* that energy in the first place. That is, visible light will *cause* low-energy valence-shell electronic excitations in atoms to occur, but visible light can be *produced* by valence-shell electronic de-excitations from higher levels back to the ground state.

The lowest energy forms of electromagnetic waves occur in generators used for the production of electric current. Commercial alternating electric current is standardized to a frequency of 60 cycles/sec (60 Hz) in the United States. Such a frequency produces a wavelength of 5×10^6 m. The so-called audio region of electromagnetic radiation is used in microphones, telephones, and telegraphs.

The lowest energy electromagnetic waves that can have a direct effect on individual atoms are the radiowaves. The energies of these waves are nevertheless so small that they can only move whole molecules a short distance in space (called translation) and reorient certain nuclei relative to other nuclei in molecules. This latter effect is the basis of the spectroscopic technique of nuclear magnetic resonance. Microwave energies cause gas molecules to rotate about their center of mass and also cause electrons to reorient relative to other electrons. The former effect is the basis of microwave spectroscopy used for studying molecular rotations, and the latter is the basis of electron spin resonance spectroscopy for studying unpaired electrons in chemical systems. Infrared energies resonate with vibrations of atoms in chemical bonds. This effect is put to use in infrared spectroscopy. Visible and ultraviolet energies can cause excitation of electrons from lower energies to higher energies in atoms and molecules. As the energy of the rays increases, the excited electrons come from progressively more stable energy levels. Visible absorption spectroscopy deals with excitation of electrons in the outermost shells of atoms and molecules, while ultraviolet absorption spectroscopy deals with both outer- and inner-shell electron excitations of higher energy. X-radiation (X-rays) causes electron excitations in the innermost electron shells and, because it has wavelengths which are similar to the dimensions of the atoms themselves, X-rays can be diffracted by atoms. The former effect is the basis of X-ray fluorescence and X-ray photoelectron spectroscopy (ESCA), while the latter is used for crystal lattice identification and crystal structure determinations. Gamma-rays are the highest energy form of usable electromagnetic radiation. They cause excitations of nuclei from low-energy states to high-energy states and are the basis of Mössbauer spectroscopy.

Most of the energies of electromagnetic radiation have important physical, chemical, and biological applications. However, in reference to art and historical materials, the middle energies (infrared–visible–ultraviolet) have the greatest importance because materials are routinely exposed to them in the course of display. Consequently, the infrared–visible–ultraviolet regions are the focus of this monograph. If we expand the infrared–visible–ultraviolet region as shown in Figure I-4, additional subclassifications can be made.

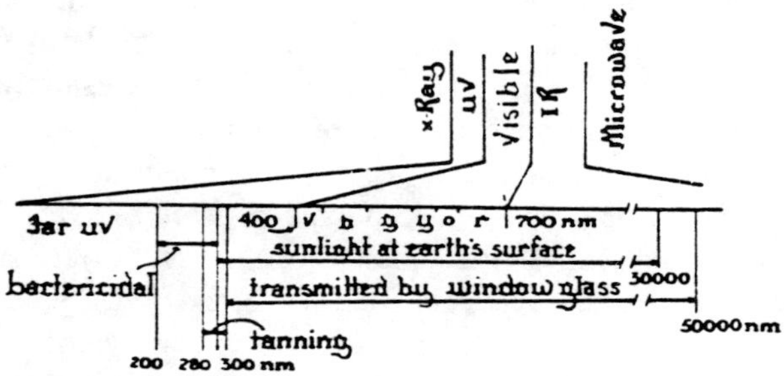

Figure I-4. Infrared, visible, and ultraviolet wavelengths expanded for further detail.

ULTRAVIOLET RADIATION

The existence of ultraviolet radiation was discovered by J. W. Ritter in 1801, using sunlight, a glass prism, and a silver chloride-coated plate. The silver halides are sensitive to uv radiation (Chapter XII). Ritter found that the plate darkened beyond the violet end first, and then in the violet, and finally in the blue, which demonstrated that radiation existed at wavelengths shorter than violet. These wavelengths were invisible to the eye and were called "ultraviolet." We now define the ultraviolet range as roughly the 10–400 nm wavelength region. This region is sometimes further subdivided for convenience. The 10–180 nm range is known as the vacuum ultraviolet because the radiation can be transmitted only by a vacuum. The 180–280 nm range is called the short- or far-ultraviolet. Quartz and photographic gelatin transmit radiation in this range. The 280–300 nm wavelengths are known as the middle-ultraviolet and are responsible for the tanning of skin. The 300–400 nm region is called the long- or near-ultraviolet region. Ordinary glass will transmit radiation in this range. Excepting the sun and the mercury discharge tube, vacuum-ultraviolet energies are not generated by sources commonly used to produce visible light. Because of this and because vacuum-ultraviolet radiation does not travel any appreciable distance in the absence of a vacuum, it will not be considered further. On the other hand, radiation in the 200–400 nm wavelength range is prevalent and causes photochemical reactions and bond rupture in many organic compounds.

However, there is a positive side to these photochemical reactions. Painters have known for years that exposure of a freshly painted object to daylight promotes drying and oxidation of the oils which is necessary before varnishing. Ultraviolet radiation is also of benefit to a conservator or museum scientist. It can be used in the examination of paint films and varnishes for evidence of retouching. Organic compounds frequently fluoresce in different ways from one another when exposed to ultraviolet radiation. For example, mastic and dammar in old varnish fluoresce yellow-green, but this fluorescence may vary in intensity with age. Fresh synthetic varnish does not fluoresce. Wax fluoresces a brilliant white and shellac fluoresces orange. The fluorescence of paint vehicles often tends to increase with age. Recent retouches on paintings usually appear purple or black in ultraviolet illumination but become grayer with age, while unvarnished areas of dark paint are a deep purple-brown. "Foxing" in paper deterioration is evident under ultraviolet illumination, as are alterations and erasures on old paper. Materials such as teeth, bones, and minerals fluoresce on exposure to ultraviolet radiation. Artificial gems which look the same as genuine ones in visible light may appear quite different in ultraviolet illumination. On the negative side, ultraviolet radiation is very damaging to many materials of artistic and historical interest. These concerns will be encountered in much greater detail later in the book.

VISIBLE RADIATION

Most species in the animal kingdom have the ability to "see" in some way. The human eye responds to only a tiny portion of the range of electromagnetic radiation. We refer to this range as the visible region. For the human eye the visible wavelength range is about 400–700 nm. Not all eyes in the animal kingdom would consider this to be the visible region, however. Bees, for example, can "see" in the near-ultraviolet region, which is invisible to humans. This allows bees to sense differences in flowers which humans cannot.

The response of the human eye and brain to different wavelengths and intensities of light differs in the 400–700 nm range and gives us the sensations we call color, texture, transparency, etc. White light might be called a mixture of all colors, although as far as the human eye is concerned, there are situations which give the impression that white light exists though it may not be spectrally pure.

The subject of color and its origin has captured the imagination of many of the great natural philosophers, but it was not until Isaac Newton that a sound basis for a theory of color was developed. In 1672, Newton showed experimentally that a beam of white light, when passed through a glass prism, is decomposed into a spectrum consisting of a large number of colors, from the red to the violet, which gradually merge into each other, and that these colors are components, not modifications, of white light. Figure I-5 shows this well-known property of transparent materials and light. The explanation for Newton's experimental observations with the prism lies in the fact that light of all wavelengths travels at the same velocity in empty space, but light in any other medium does not travel with the same velocity for all wavelengths. Hence, a separation of wavelengths can result. The separation of white light into colors, or, equivalently, wavelengths, by a medium is called *dispersion*. It is thereby convenient to subdivide the visible range according to the various color responses induced in the human eye. Six intervals ranging from the longest to the shortest wavelength are red, orange, yellow, green, blue, and violet.

The case of normal dispersion is shown in Figure I-5 and it is observed for colorless transparent media. It is "normal" because red light (longest wavelength) has the greatest velocity and least dispersion, whereas the violet light (shortest wavelength) has the least velocity and the greatest

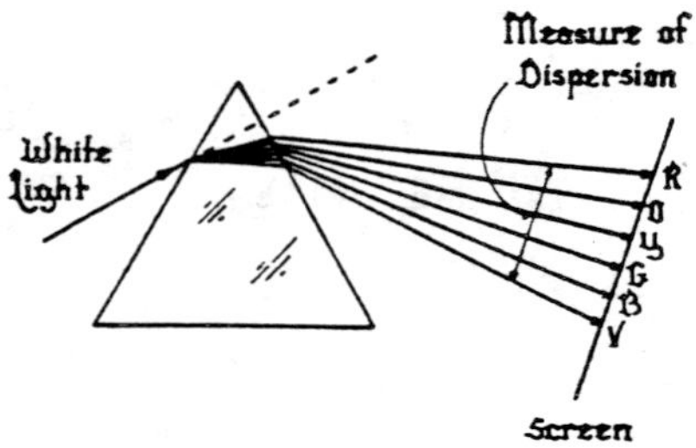

Figure I-5. The decomposition of visible white light into its component wavelength regions (colors) by a glass prism.

dispersion. The other colors are divided sequentially between red and violet. In a more specific way, dispersion of visible light varies with wavelength approximately as $1/\lambda^3$. For this reason the shorter wavelengths show the greatest dispersion ($1/\lambda^3$ is larger) and also show a much greater rate of change of dispersion for small changes in λ ($1/\lambda^3$ is nonlinear in λ) than do longer wavelengths. It should be noted that another type of wavelength separation of light called *anomalous dispersion* is observed in colored media. With anomalous dispersion the longer wavelengths have greater dispersion than shorter wavelengths in the region of the spectrum where the absorption of light takes place. Hence the progression of colors in Figure I-5 is not observed. Anomalous dispersion is taken up in more detail in Chapter VI, where its application to art materials is considered.

Many chemical reactions can be made to occur with visible light, but the higher energies from 400 to 500 nm are generally considered to be most worrisome because they can produce photochemical reactions.

INFRARED RADIATION

The existence of radiation beyond the red was discovered in 1800 by William Herschel. He found that a blackened thermometer placed in the spectrum of sunlight registered a strong temperature rise even beyond the red. This demonstrated that invisible wavelengths longer than red exist, and this radiation became known as infrared. The effects of infrared radiation have, of course, been known for a very long time. The infrared radiation produced by fire has been one of the most influential phenomena in human advancement.

Infrared energies occur in a broad range beginning at the low-energy side of the visible region. Energies corresponding to the 700–1400 nm wavelengths are capable of penetrating filters that are usually used to block out ultraviolet radiation, whereas longer wavelengths are nonpenetrating. Hence the terms "penetrating" and "nonpenetrating" are sometimes applied to infrared energies in these ranges.

Infrared radiation produces heating effects which can mechanically and chemically alter materials, although seldom by photochemical mechanisms. Mechanical changes, such as shrinking, cracking, and drying, can occur in wood, glass, and ceramics with exposure to infrared radiation. The dreadful damage infrared radiation might induce in wax objects goes without mention. Chemical changes, if they occur, are usually an indirect result of infrared radiation. If a chemical reaction is already occurring, then no matter how slow or fast the reaction is, heat from the infrared exposure will always speed up the reaction. Yellowing in natural varnish

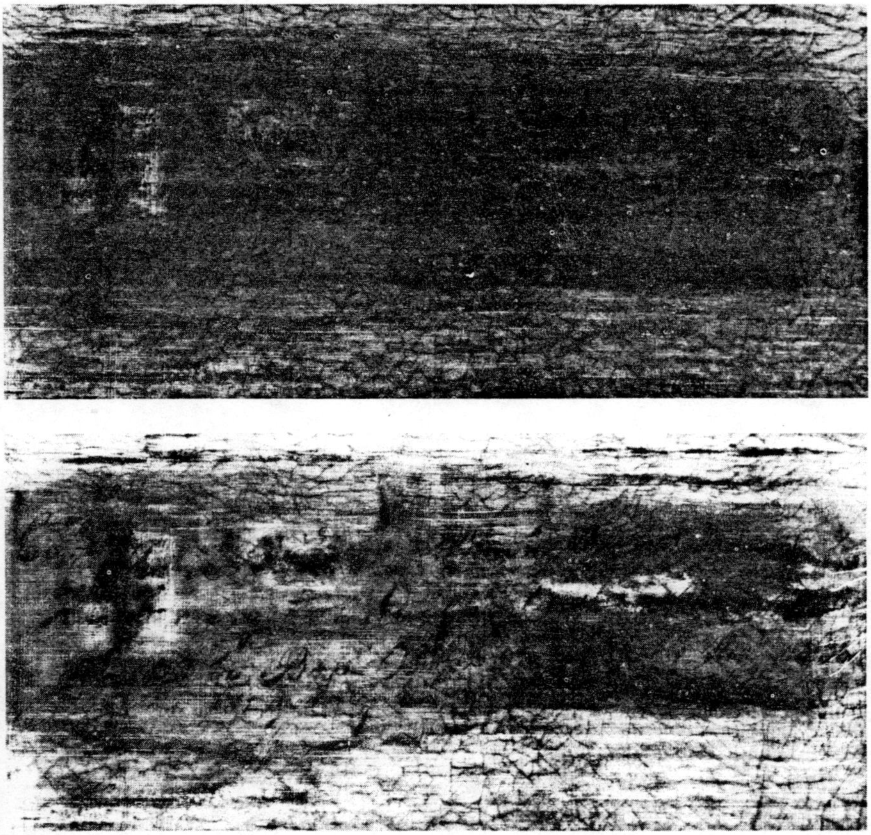

Figure I-6. A visible light (top) and infrared (bottom) photograph of an obscured inscription on the back of a painting (Courtesy, Winterthur Museum–University of Delaware Art Conservation Program).

films may be a direct result of infrared energies. However, synthetic varnish films are usually not sensitive to infrared radiation.

Infrared radiation, through the use of infrared photography, can be important tool in the museum. Infrared rays can often penetrate visibly opaque varnishes and thin paint films to reveal the underpaint, drawings, retouched areas, and changes in composition, using a viewer gun or infrared photography. The reason infrared radiation can be used to "see through" opaque films stems from the fact that infrared radiation has a longer wavelength than does visible radiation. Hence infrared radiation is less efficiently scattered by small particles (Chapter VI) in the varnish film than is visible light. The infrared rays are therefore able to penetrate and overcome the opacity of the upper layers. It becomes possible to observe detail in the paint layer which has become obscured by old varnish and grime. Forgeries can sometimes be detected in this way because the underpainting may be different from what is on the surface. Figure I-6 shows an obscured inscription on the back of a painting. Through the use of infrared photography, the inscription became legible.

BIBLIOGRAPHY

Bouma, P. J., *Physical Aspects of Color*, Macmillan, New York (1971).

Bridgman, C. F., and Gibson, H. L., Infrared luminescence in the photographic examination of paintings and other art objects, *Stud. Conserv.* **8**, 77 (1963).

Clulow, F. W., *Color: Its Principles and Application*, Fountain Press, London (1972).

Dibdin, F. J. H., *Essentials of Light*, Cleaver-Hume Press, London (1961).

Ditchburn, R. W., *Light*, Interscience, London (1963).

Jaffe, H., and Orchin, M., *Theory and Application of Ultraviolet Spectroscopy*, Wiley, New York (1962).

Koller, L. R., *Ultraviolet Radiation*, 2nd ed., Wiley, New York (1965).

Ronchi, V., *The Nature of Light*, Harvard University Press, Cambridge, Massachusetts (1970).

Weisskopf, V. F., How light interacts with matter, *Sci. Am.* **219**(3), 60 (1968).

II

Primary Light Sources

Before one can appreciate how materials interact with light, it is necessary to become familiar with the properties of some of the common primary sources of light. The nature of the light source used in illumination is critically important in terms of color rendering and also in terms of its effects on the properties of the material. For example, to give good overall color rendering, a light source should produce all of the wavelengths of light that are characteristic of the true colors of the object. In a purely physical sense, visible and ultraviolet radiation are generated by an incredibly rapid oscillation of electrons in atoms. All sources of light in this energy range have this fact in common, but they differ in the types of atoms, the geometry of the radiator, and other controllable features.

The blackbody radiation source is considered first in this chapter even though it is not a source of light per se. It serves rather as a standard for comparing the color of the light produced by other sources. The commonly encountered light sources of sunlight, incandescent light, fluorescent light, sodium and mercury discharge lamps, and the flashbulb are discussed here. The ultraviolet content of commonly encountered light sources is taken up at the conclusion of this chapter.

Radiant energy from an illuminating source rather than reflected light is the subject of this chapter. Reflected light certainly is important in determining the appearance of surfaces, but this and other points about reflected light require discussion of the particular reflecting body. At this juncture the most commonly encountered *primary* illuminating sources are outlined. Reflected light is more appropriately taken up as part of Chapter VI.

There are many specialized light sources which are not included as subjects of this chapter because they are likely to find little use in a museum setting of the current day. For example, the tallow lamp and the candle are of considerable historical interest. Later, kerosene and gas lamps were used for lighting. However, apart from recognizing that some works may have been smudged by combustion products, and that eyes were no doubt strained by their low luminous intensity, we will not devote any attention to these light sources. Instead, attention is paid to the light sources commonly used in industrialized cultures. Except for the sun, all of these sources require electrical energy to produce light.

THE BLACKBODY RADIATOR

If a material is heated to a sufficient temperature, it will either be destroyed or it will begin to emit light as a result of excitation of the electrons in the atoms or molecules comprising the material. If light is emitted, then the material is a thermal radiator. When a thermal radiator is heated, it emits radiant energy of a continuous spectral energy distribution which depends upon both the nature of the material and its temperature. A *blackbody thermal radiator* can be considered as a source emitting spectral energy whose intensity and wavelength depend upon the temperature of the body alone and not on the material from which it is made. This is to say that all wavelengths of light are absorbed or emitted with equal efficiency. The radiator is therefore perfectly "black." The cavity inside an enclosed body which has been heated to a uniform high temperature T approximates a blackbody and will contain radiant energy characteristic only of the temperature. If a small hole is made in the wall of the body, the emitted radiant energy can be measured. Figure II-1 shows the electromagnetic radiation distribution as a function of wavelength from a blackbody heated to various temperatures. The visible energy region is the 400–700 nm wavelength region. As the temperature of the body is increased, the principal wavelengths emitted by the body become shorter. At every temperature most of the energy released is in the infrared region. Visible light begins to appear above 1000°K. Only when the temperature is rather high (>2000°K) does ultraviolet radiation begin to appear.

The blackbody radiator is important because the temperature of the body is indicative of its spectral energy distribution. Its temperature in degrees Kelvin (°K) is the absolute temperature of the body. Note that the vertical scale shown on Figure II-1 does not begin at 0 W/cm^2 of radiation, but rather at a much higher radiated energy. Only the top of each temperature curve is shown. Even at 300°K, which is slightly above room temperature, the blackbody radiates energy. Since virtually all of the radiated energy is in the infrared at this temperature, the body is "warm." No

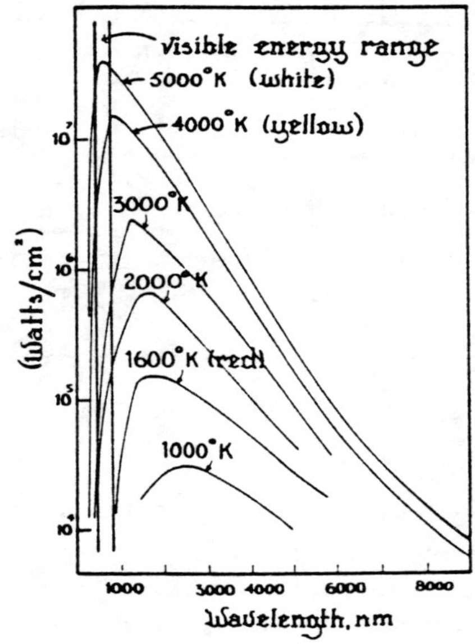

Figure II-1. The spectral energy distribution of a blackbody thermal radiator as a function of its temperature.

visible energy is radiated, of course. As the body is heated to well above 1000°K, it begins to glow by producing the lowest visible light energies (a weak red color). At 2000°K the intensity of the red glow is increased and at 3000°K the body has a bright orange-yellow cast. At 4000°K it is bright yellow. At 5000°K the energy emitted is quite uniformly distributed over the visible range and the body emits white light ("white heat"). At still higher temperatures the blackbody becomes blue, due to the dominance of shorter, blue wavelengths in the emitted radiation.

The application of blackbody radiation to the subject of illumination comes from the fact that the color and saturation (the strength of color) of any light source can be matched to the color of a blackbody radiator operating at a particular temperature. The temperature in °K of the blackbody that gives a color match to the light source of interest is called the *color temperature* of the source of interest. (The term *correlated color temperature* is also in use when referring to fluorescent lamps and other discharge sources.)

In order to carry out this matching in the visible region, it is useful to have a greatly expanded version of Figure II-1 which shows the visible region in detail. Figure II-2 shows this expansion for the 1000–100,000°K temperature range. Unlike Figure II-1, however, the temperature curves in the visible energy range have been moved up or down as necessary so that the intensity at 560 nm occurs at a common point. This permits direct comparisons to be made of the colors for each temperature. The trend in

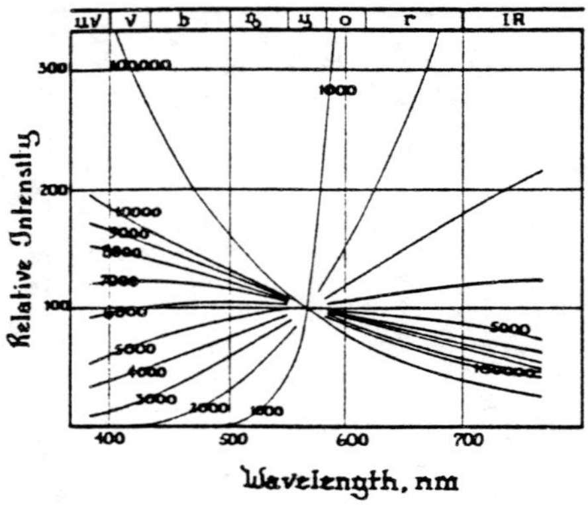

Figure II-2. Spectral energy distributions of a blackbody radiator in the visible region at different temperatures from 1000°K to 100,000°K. The lines are normalized at 560 nm for purposes of comparison of the colors in the wide range of temperature.

colors with temperature increase follows the energy ideas discussed in Chapter I. For example, at 1000–2000°K reds and yellows (lower energies) are emphasized. White heat is achieved at 5000–6000°K because all energies in the visible are about equally represented. At 10,000°K and above, blue colors (higher energy) appear. It is very important to understand that the light source of interest to which a color temperature has been assigned is not necessarily heated to that temperature. The color temperature is obtained by matching the color of the source of interest to the color of one of the blackbody temperature curves shown in Figure II-1 or II-2. Fluorescent lights, for example, operate at reasonably low temperatures, yet their color temperatures may be 3000–7000°K (yellow to white light). These temperatures mean that the spectral color of fluorescent lamps approximately matches the color of a blackbody operating at 3000–7000°K.

The color temperature of a source is not a highly accurate specification of the color-*rendering* properties of a source. The color temperature specifies only the color and saturation of the light and there are many spectral energy distributions which yield the same color temperature. Color temperatures are most meaningful when used to compare sources having similar spectral energy distributions. For example, color temperatures of various fluorescent lamps can be compared satisfactorily, but the color temperatures of a fluorescent lamp and an incandescent lamp should not be compared to one another for the purpose of analyzing their relative color-rendering capabilities.

DIRECT SUNLIGHT

The most ubiquitous light source of all is, of course, direct sunlight. The spectral energy distribution of sunlight depends upon a wide variety

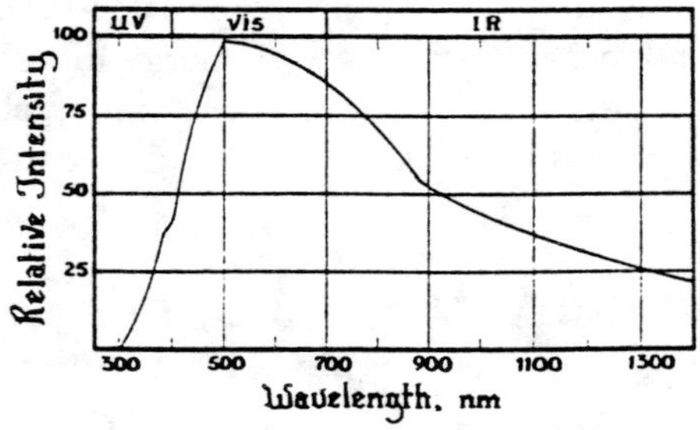

Figure II-3. Solar energy distribution of mid-summer noonday sun in Washington, D.C.

of factors, such as latitude, time of day, season, altitude, and local atmospheric conditions. Its general characteristics are shown in Figure II-3 for Washington, D.C.

Roughly half the energy of the sun reaching the surface of the earth is in the visible region. The other half is in the ultraviolet and infrared, but with only about 3% of the total being in the ultraviolet region. Of most interest for chemical and visual effects is the radiation in the 300–800 nm region. Figure II-4 shows an expanded version of this region. The top curve in Figure II-4 is the distribution of light at 200 miles altitude above the earth and the lower curve is the distribution at sea level. Across the visible region, the sun's energy measured at sea level is relatively uniform in wavelength distribution. At the visible–infrared edge, water and ozone, O_3, in the atmosphere have several absorption bands, and at the ultraviolet–visible edge, Fraunhofer lines appear. Fraunhofer lines result from the absorption of energy by gases in the sun's atmosphere. The rapid dropoff in intensity through the ultraviolet is mainly a result of the absorption of the ultraviolet radiation by ozone in the earth's atmosphere. The intensity of the sunlight is, of course, somewhat dependent upon the amount of atmosphere through which the light has to pass. For example, the noon sunlight passes through less atmosphere than does the evening sunlight and thus noon sunlight is more intense. Light that is reflected from the earth has lower energy than the incident rays because much of the energy is converted to heat after it encounters the earth.

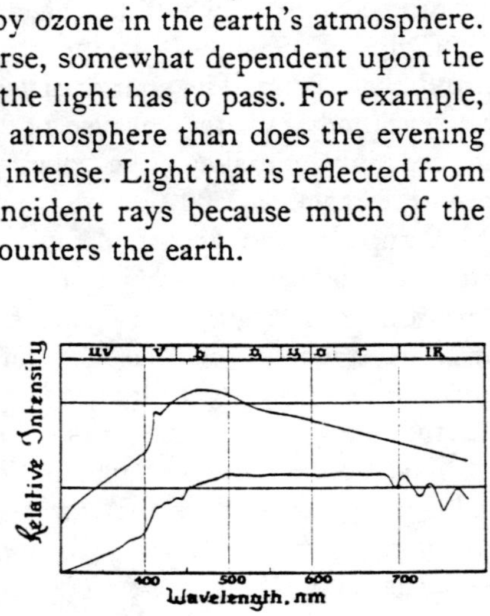

Figure II-4. Spectral distribution of direct sunlight. Upper curve: above atmosphere; lower curve: at sea level.

Sunlight has excellent color-rendering capability because of its relatively uniform spectral energy distribution in the visible range. Direct sunlight has a color temperature of about 5000°K. North sky light is considerably bluer and has a color temperature of 7500°K or higher.

THE INCANDESCENT LAMP

Incandescence is the shining of hot bodies because they are hot. When an electric current flows through a wire, there is a temperature rise which is greatest for small wires of high electrical resistance. The heating increases the resistance further and as the voltage is increased, the wire glows and becomes a light source. Thomas Edison put this principle to use in the design of the incandescent lamp. Initially he used a carbon filament inside an evacuated bulb, but carbon had the disadvantage of having a high evaporation rate, and thus a short life, when it was operated at a desirable level of illumination. Carbon was replaced with tungsten filaments, which are still in use today. Tungsten was chosen because it has the highest melting point of all metals and has a low evaporation rate in an evacuated bulb even near its melting point. About 1913, the procedure was introduced of placing a nonreactive gas, such as argon or nitrogen, in the evacuated bulb to further reduce the evaporation rate of tungsten and increase the life of the lamp. The pressure of the gas was made about 80% of atmospheric pressure, but it is approximately equal to the atmospheric pressure when in use. The wire was also coiled to increase its operating temperature. The gray or black deposit on the inside of a spent bulb is evaporated tungsten which has been deposited on the glass wall.

Figure II-5 shows the spectral energy distribution of a 500-W incandescent lamp. The energy distribution of light closely resembles that of a blackbody radiator because both work on the principle of thermally induced light emission. The color temperatures for incandescent lamps are in the range of 2700–3200°K.

Lamp *efficiency* is defined as the amount of visible light produced compared to the total amount of energy radiated. Seventy to eighty-five percent of the energy of a tungsten filament lamp is released as heat in the infrared region, which makes it rather uneconomical as a visible light source. The incandescent lamp thus has low efficiency. Absorption of the infrared energy by an object results in heating of the object.

The spectral energy distribution of an incandescent lamp in the visible range depends somewhat upon the wattage rating. The visible energy emission is shown in Figure II-6 for several wattages. A strong emphasis on the longer wavelengths exists in these lamps and gives the

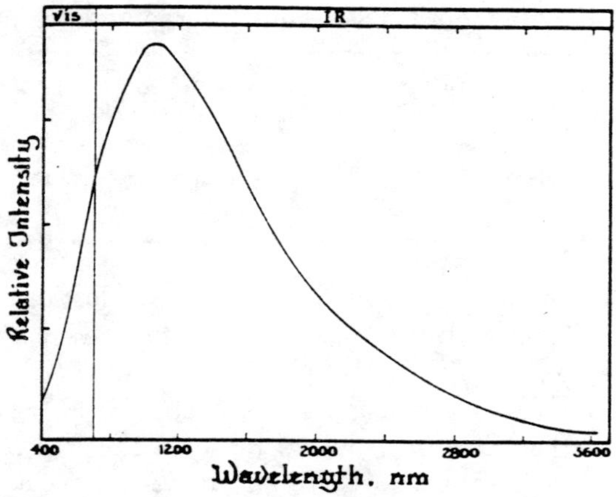

Figure II-5. Spectral energy distribution of a 500-W incandescent lamp.

orange-yellow tinge to areas being illuminated with incandescent light. Because of the low level of blue light intensity, it can be hard to distinguish slightly different blue shades with incandescent illumination. Because of the large amount of yellow light, it is difficult to see light yellow lines on a white background using this source. These color-rendering concepts will be taken up further in Chapter V.

Table II-1 summarizes some of the important features of the incandescent lamp. The *luminous efficacy* of a lamp is the total flux of the

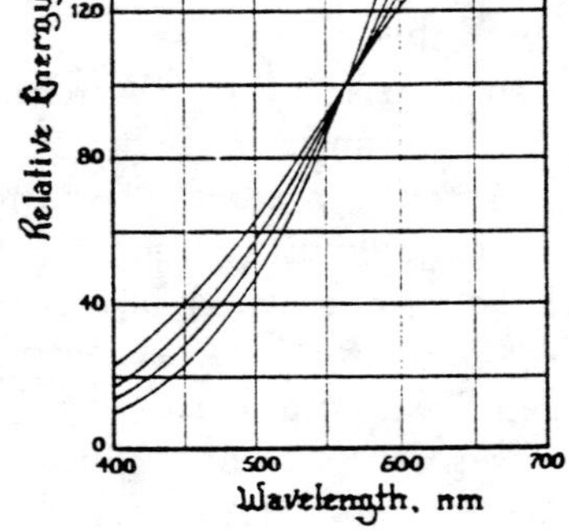

Figure II-6. The distribution of energy radiated by tungsten filament incandescent lamps. (a) 25-W, (b) 50-W, (c) 500-W, (d) 1000-W; all curves are normalized to 560 nm.

Table II-1. Incandescent Lamp Characteristics

Efficacy, lumens per watt	Low
Lamp appearance on neutral surfaces	Yellowish white
Effect on surroundings	Heat
Colors strengthened	Red, orange, yellow
Colors grayed	Blue

emitted light per unit of energy supplied to the lamp. Most of this energy is lost in the form of heat with the incandescent lamp. The efficacy of a tungsten filament lamp is 12–25 lm/W (lumens per watt).

Quartz lamps often found in conservation laboratories are tungsten filament incandescent lamps with a quartz rather than glass bulb. They give good color-rendering properties because they run at a higher color temperature than do normal incandescent lamps. The quartz envelope can withstand the higher temperature much better than can glass. For this reason the lamps can be made small with highly efficient light-reflecting components surrounding them. Often a small amount of iodine is placed in the bulb to extend the life of the lamp and reduce the tendency for the lamp envelope to blacken. These lamps are called tungsten–halogen lamps. It should be noted that quartz is more transparent to ultraviolet radiation than is ordinary glass. As a result, extended use of quartz lamps around highly light-sensitive materials would be inadvisable.

THE MERCURY LAMP AND THE SODIUM LAMP

The principle behind the mercury and sodium lamps is different from that of the incandescent lamp. They are sometimes called discharge lamps and have been used since the early 1940s. In these light sources the volatile metal mercury or sodium is excited by electrons which have been accelerated between two electrodes as a result of an applied voltage. Collisions between the excited gaseous metal atoms produce luminescence, which yields a great deal of visible light. The luminous efficacy of a sodium lamp is 60–90 lm/W. The emission spectrum of sodium and mercury when heated in the gas phase contains some very intense, sharp emission lines rather than a broad, continuous spectrum. This can be seen in the spectral energy distribution of a mercury lamp shown in Figure II-7. The blocklike representation of the emission line regions is meant to show that a large number of intense, sharp emission lines occur over a narrow energy range. The light produced is silver-green in appearance. The sodium lamp produces yellow-orange light characteristic of sodium vapor emission and has a color temperature of about 2100°K. The yellow-orange color comes

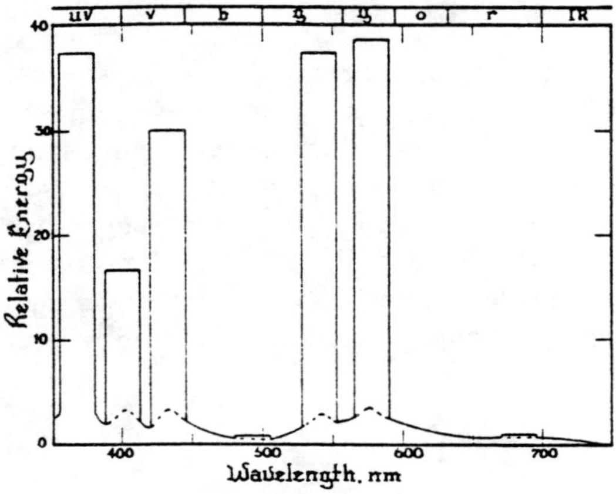

Figure II-7. The spectral energy distribution of a mercury discharge lamp.

from intense sodium emission lines in the 580–600 nm range. These sodium emission lines are also partly responsible for the yellow-orange color of flames in burning wood.

Because the wavelength distribution of light from metal-atom electric discharge sources is very discontinuous in the visible range, these sources render color very poorly. Table II-2 indicates the color reproduction qualities which can be obtained from the sodium and mercury light sources. The use of mercury and sodium lamps for exhibition lighting in a museum is not generally advisable, although they are common in lighting of highways and public areas, such as a museum parking lot. The mercury lamp also contains intense ultraviolet emission lines.

Table II-2. Color-Rendering Characteristics of Sodium and Mercury Lamps

As seen in daylight	As seen under sodium vapor (low pressure)	As seen under mercury vapor
Blue	Dark brown or black	Deep violet
White	Light yellow	Bluish white
Green	Brownish yellow	Deeper green
Yellow	Yellow	Greenish yellow
Black	Black	Black
Orange	Brown	Brown
Light red	Yellowish brown	Brown
Brown	Brown	Gray
Red	Brown	Dark brown or black

Consequently, exposure of light-sensitive materials to this light source should be avoided.

As a side point, light from a mercury lamp and light from the moon have some similarities in their appearance to the eye. In reality, moonlight has a spectral energy distribution similar to, but slightly more yellow than, that of daylight. The reason that moonlight is very poor for color reproduction is not because it has an irregular spectral energy distribution, but because it is not a radiant source of light. Moonlight is reflected light and its intensity is thus very low, being only 10^{-6} or less as intense as bright sunlight. There simply is not enough light to produce good color contrast.

THE FLUORESCENT LAMP

Experiments with fluorescence were carried out as early as 1867 by Henri Becquerel, but it was not until the 1930s that a commercial fluorescent lamp became available. Fluorescence is an effect whereby a substance can absorb radiation at one wavelength and reradiate it at different, usually longer, wavelengths. If ultraviolet radiation in addition to visible light falls on such a substance, some of the ultraviolet radiation is reemitted as visible light. If the emission occurs immediately, it is known as *fluorescence*. If the emission is delayed and the energy is emitted over a period of time, it is known as *phosphorescence*. Fluorescent lamps are electric discharge sources similar to the sodium and mercury lamps, but with the major difference that a fluorescent material called a *phosphor* coats the inside of the glass envelope. A common design of this type of lamp is shown in Figure II-8. The tube often contains a small amount of an inert gas, such as neon, to carry the electric discharge until the mercury droplet is vaporized. When vaporized, the mercury atoms emit energy in sharply defined lines (Figure II-7). About 50% of the emission occurs at 253.7 nm, in the ultraviolet. This ultraviolet radiation excites the phosphors coated on the inside of the tube envelope. The phosphors fluoresce, giving off considerable light in the visible region. Some typical fluorescent materials used are listed in Table II-3 along with the wavelength of the principal emission line of each. Calcium halophosphates are among the most commonly used fluorescent materials, but research for new and better

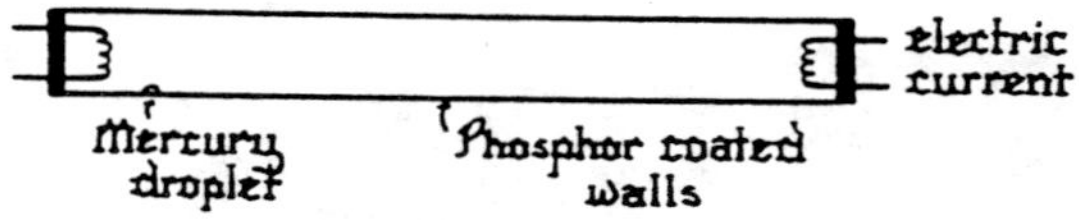

Figure II-8. The basic design of one type of fluorescent lamp.

Table II-3. Some Phosphors Used in Fluorescent Lamps

Phosphor	λ_{max}, nm
Calcium tungstate	440
Magnesium tungstate	480
Zinc silicate	540
Calcium halophosphate	590
Calcium silicate	610
Cadmium borate	615
Calcium–strontium phosphate	640
Magnesium arsenate	660

phosphors continues vigorously. One of the most recent advances involves the use of rare earth compounds as fluorescent materials to obtain better color reproduction.

The emission spectrum of the fluorescent lamp is characteristic of the phosphors and the mercury atom. Figures II-9 to II-13 show the spectral energy distribution of several types of 40-W lamps. The broad background in each results from the phosphor emission, while the square-topped columns are emission lines produced by the mercury atoms.

Fluorescent lamps can be classified according to different phosphors as "standard" or "deluxe." With the standard lamp the lack of a deep red causes bright red objects to appear dull red or brownish. Strong lines in the yellow combined with a weak red distort red toward greenish. The strong blue-violet lines and the lack of a pure blue and green-blue make

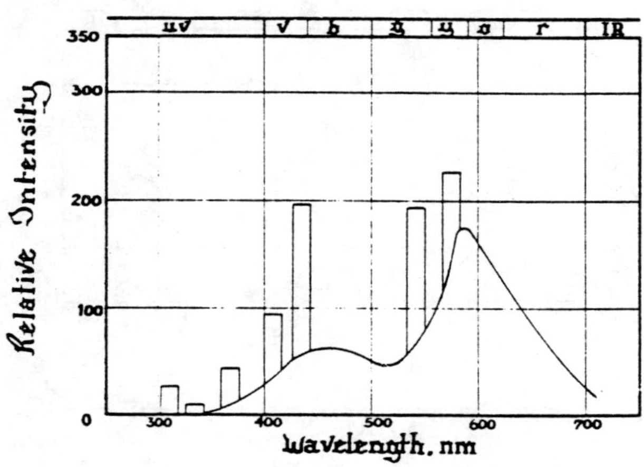

Figure II-9. The Cool White fluorescent lamp.

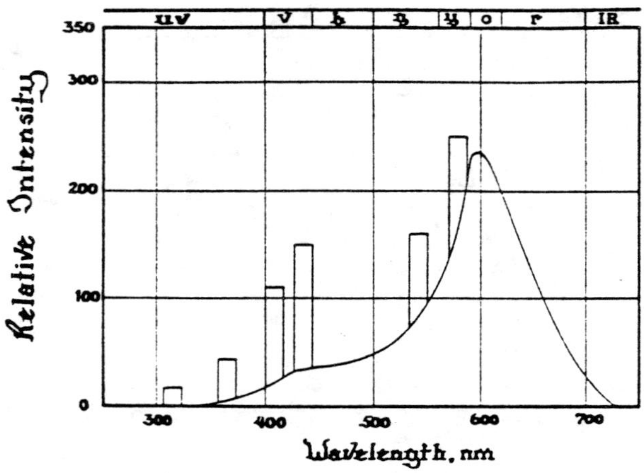

Figure II-10. The Warm White fluorescent lamp.

blue objects appear violet. With the deluxe lamp, color reproduction is better than with standard lamps, but the efficacy is about 30% less than the standard lamp. The luminous efficacy of fluorescent lamps is typically 30–80 lm/W.

Within the standard and deluxe classes, the lamps are further separated into the "warm" and "cool" groups. The term warm refers to the presence of redder color and relatively more infrared radiation (heat), whereas cool lamps produce less red and relatively less infrared energy. The warm fluorescent lamps operate at a lower color temperature than do the cool lamps and thus emit radiation of relatively lower energy. Warm

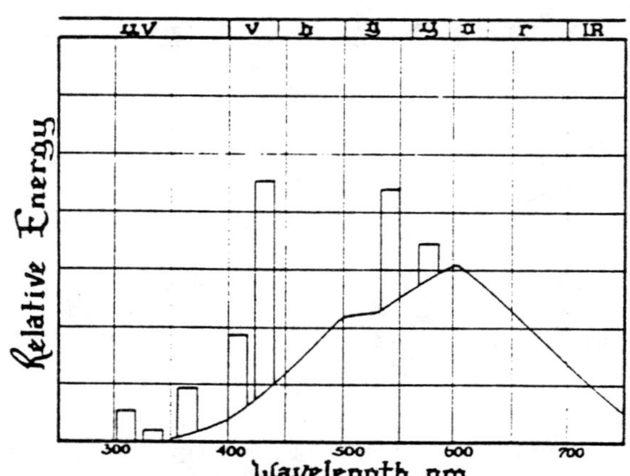

Figure II-11. The Deluxe Cool White fluorescent lamp.

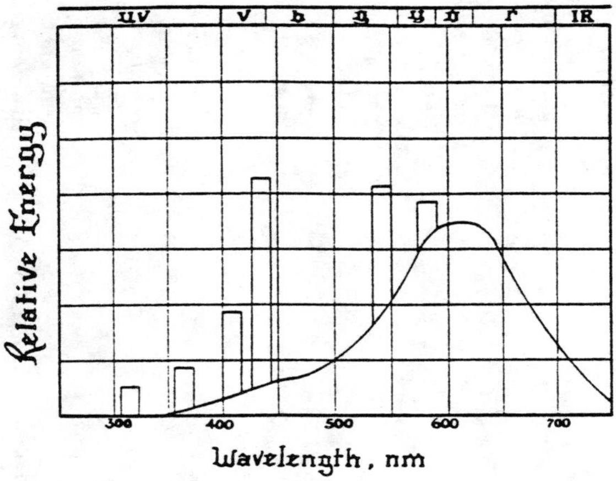

Figure II-12. The Deluxe Warm White fluorescent lamp.

fluorescent lamps are less destructive in terms of photodegradation than are the cool lamps. Table II-4 summarizes a number of the important characteristics of the fluorescent lamps whose spectral energy distributions are shown in Figures II-9 to II-13.

Fluorescent lamps are more efficient and economical than incandescent lamps because they operate at lower temperatures, produce more light per watt, and are longer lived. Since the light is produced by the surface of a rather large tube, there are no bright spots and fewer harsh shadows. If

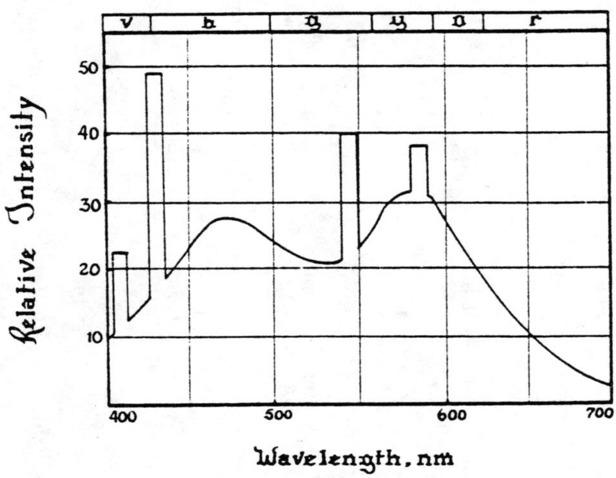

Figure II-13. The Daylight fluorescent lamp.

Table II-4. Some Characteristics of Fluorescent Lamps Shown in Figures II-9 to II-13[a]

Lamp names	Efficacy, lumens/watt	Lamp appearance effect on neutral surfaces	Effect on "atmosphere"	Colors strengthened	Colors grayed	Color temperature, °K	Subjective remarks
Cool white	High	White	Neutral to moderately cool	Orange, yellow, blue	Red	4200	Blends with natural daylight; good color acceptance
Deluxe cool white	Medium	White	Neutral to moderately cool	All nearly equal	None appreciably	4200	Best overall color rendition; simulates natural daylight
Warm white	High	Yellowish white	Warm	Orange, yellow	Red, green, blue	3000	Blends with incandescent light; poor color acceptance
Deluxe warm white	Medium	Yellowish white	Warm	Red, orange, yellow, green	Blue	3000	Good color rendition; simulates incandescent light
Daylight	Medium high	Bluish white	Very cool	Green, blue	Red, orange	7000	Usually replaceable with cool white

[a] From General Electric Lamp Division, TP-119.

accidentally broken, the lamp has the disadvantage of releasing toxic mercury metal and mercury vapor.

THE FLASHBULB

Flashbulb light sources operate on the principle of combustion of a "wool" of reactive metal (magnesium, aluminum, zirconium, etc.) to produce a bright flash upon a discharge of electricity in the bulb. The bulb contains oxygen to enhance the combustion rate and temperature. The bulbs often have a blue coating to absorb and thereby reduce the intensity of the longer wavelengths. There are many types of flashbulbs and several examples are shown in Figure II-14 and II-15.

The use of flashbulbs is often restricted in museums because ultraviolet radiation is present in the emitted light. The exposure time from a flashbulb is very short, but the time is offset to a considerable degree by the very great intensity of the flash. Exposure of materials to 25,000 flashes has caused no significant deterioration in light-stable materials, but deterioration can occur in watercolors and fabrics where the sensitivity to ultraviolet radiation is high.

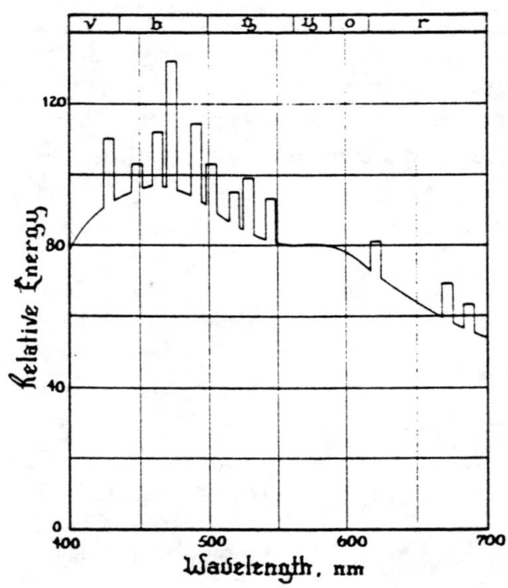

Figure II-14. Spectral energy distribution of a Kodatron Type II bulb.

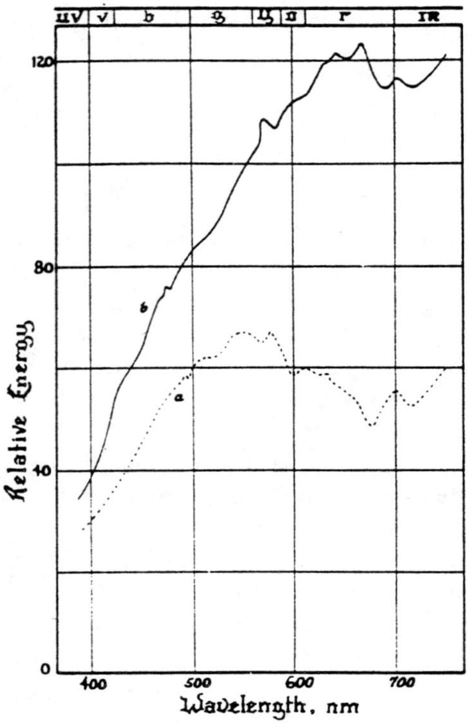

Figure II-15. Spectral energy distribution of a zirconium-filled bulb (a) with and (b) without a blue filter.

ULTRAVIOLET CONTENT OF SOME LIGHT SOURCES

When thinking of the color-rendering properties of a light source it is tempting to consider only the visible components of the energy emission. Yet of considerable concern is the ultraviolet content of the light source, because on a comparative basis ultraviolet radiation can cause much more

Table II-5. Relative Amounts of Ultraviolet Radiation in Various Sources

Source	Ultraviolet content
Sunlight	50
Cloudy daylight	20–35
Fluorescent lamp	2–15
Incandescent lamp	3–5

Table II-6. Radiation from Incandescent Lamps[a]

Lamp	Color temperature °K	280–320 nm		320–400 nm		Total UV		Total visible	
		Watts	Percent of lamp Watts	Watts	Percent of lamp Watts	Watts	Percent of lamp Watts	Watts	Percent of lamp Watts
40-W Standard	2560	—	—	0.027	0.068	0.027	0.068	2.65	6.6
100-W Standard	2890	—	—	0.117	0.117	0.117	0.117	9.32	9.32
500-W Standard	3000	0.005	0.001	0.88	0.175	0.89	0.176	54.84	11.0
1000-W Standard	3050	0.012	0.001	1.98	0.198	1.99	0.199	115.5	11.6
1000-W Photoflood	3360	0.123	0.012	4.40	0.44	4.52	0.45	156.3	15.6
250-W Purple X	3100	0.009	0.004	0.482	0.19	0.49	0.19	2.79	1.11

[a] General Electric Lamp Bulletin LD-1 (1956).

chemical change in materials than can visible light. *Ultraviolet content* is defined as the ultraviolet energy divided by the visible energy multiplied by a luminosity function. It gives an indication of the relative amount of ultraviolet radiation in the various sources at the same level of illumination.

Table II-5 shows the ultraviolet content of several of the light sources as measured through $\frac{1}{4}$-in. glass. The data indicate the relative desirability of some important primary sources of illumination from the standpoint of potential damage from ultraviolet radiation. The natural light sources have the greatest potential for producing photochemical degradation reactions. Fluorescent lamps are the next most potentially hazardous, with the "cool" fluorescent bulbs being more worrisome than "warm" bulbs. The ultraviolet hazard of incandescent lamps is comparatively small. Table II-6 shows the relative amounts of energy produced in the ultraviolet and visible regions by various incandescent sources. As the wattage increases, the ultraviolet content gradually increases because the color temperature increases.

One must also be concerned with the amount of ultraviolet radiation reflected from walls. Since this radiation is not originating from a primary light source, we defer its discussion until Chapter VI, where reflection of light by materials is taken up in more detail.

BIBLIOGRAPHY

Billmeyer, F. W., and Saltzman, M., *Principles of Color Technology*, Interscience, New York (1966).

Bouma, P. J., *Physical Aspects of Color*, Macmillan, New York (1971).

Evans, R. M., *An Introduction to Color*, Wiley, New York (1948).

Genard, J., Extreme ultraviolet radiation from tubular fluorescent lamps and its effect on museum lighting, *Mus. J.* **52**, 5 (1952).

Henderson, S. T., *Daylight and Its Spectrum*, American Elsevier, New York (1970).

Judd, D. B., and Wyszecki, S. G., *Color in Business, Science and Technology*, Wiley, New York (1975).

Koller, L. R., *Ultraviolet Radiation*, 2nd ed., Wiley, New York (1965).

Kraehenbuehl, J. O., *Electric Illumination*, Wiley, New York (1951).

Rawlins, F. I. G., and Rawson-Bottom, W. E., Ultraviolet radiation from fluorescent lamps, *Mus. J.* **52**, 84 (1952).

Thomson, G., Visible and ultraviolet radiation, *Mus. J.* **57**, 27 (1957).

Wyszecki, S. G., and Stiles, W. S., *Color Science*, Wiley, New York (1967).

III

The Measurement of Properties of Light

Up to this point we have concentrated on the description of light itself and on the generation of visible light in particular by natural and artificial sources. Now attention is focused on the measurement of useful parameters of light. One of the most important and fundamental quantities of interest is the velocity of light. As important as it is to physical science, the measurement of the absolute velocity of light has little direct importance in the examination of art materials. The rate of passage of light is constant in the absence of matter and gravitational fields. When matter is present the velocity is changed, however. This fact is very important with regard to the index of refraction to be discussed later, but, fortunately, direct measurement of the absolute velocity of light in materials is unnecessary. In this chapter attention is aimed at another parameter of light, namely its intensity, since the amount of light striking an object affects degradation as well as color rendering.

The first attempts to measure the intensity of light were made by Pierre Bouguer in 1729, but he engaged in little basic work on mathematical descriptions or definitions. These came in 1760, when Johann Lambert developed a system of concepts which is still in use today. Most of these concepts and relationships could not be experimentally tested in Lambert's time because no practical methods of measurement were available. It was not until the late nineteenth century that measurements could be made which showed Lambert's ideas to be correct.

The subject of light intensity can be confusing to the beginning or sporadic user because it is an area riddled with duplicate-sounding terms. The word "intensity" is necessarily used in several senses. A lamp designer, for example, concerned only with the source itself defines intensity differently from a museum employee who is concerned with the

Table III-1. Units of Measurement of Light Intensity

Parameter	Units
Intensity of a light source	Candlepower (cp), candela (cd)
Amount of light (luminous flux)	Lumen (lm)
Level of illumination	Foot-candle (fc), lux (lx), phot

intensity of light at the surface of an object. There are, in fact, three standards for the measurement of light intensity, depending upon the particular application. These are the measurements of (1) the total light power emitted from a source in all directions (luminous or radiant intensity); (2) the light power incident on a surface at some distance from the source (illuminance or irradiance); (3) the light power emitted from the source at the surface of the source (luminance or, where the eye is involved, brightness).

In addition to these measurement standards, there are two standards for the source itself. These are (1) the radiant standard, which is used where the human eye is not involved, and which is applicable to any part of the spectrum; (2) the illumination standard, which is used in the visible range for applications involving the eye. The radiant standard uses radiometric units such as watts or ergs per second. The illumination standard uses photometric units such as lumens. The illumination standard will be employed here. Table III-1 lists some of the common units encountered in lamp work.

LUMINOUS INTENSITY OF A SOURCE

The common sources of visible light were discussed in the preceding chapter. Light sources emit energy in the form of electromagnetic waves which spread out in all directions. The intensity I of this light is proportional to the square of the amplitude of its electric vector [equation (I-4)]. A more practical means of measuring intensity is needed since wave amplitudes seem an unnecessarily basic formalism. Instead, we can use the amount of energy radiated per unit time (power), expressed in terms of watts. For discussions about specific light characteristics, the watt still has limited value. The commerical method of rating the relative intensities of lamps is the luminosity factor, lumens per watt.

For many years the standard against which intensity was measured was the *candle* (c). One *candlepower* (cp) represented the luminous intensity of a flame of a certain make of candle, the constituents of which were specified by international agreement. The standardized international unit of intensity now used is the *candela* (cd). One candela is 1/60 of the luminous intensity emitted by 1 cm^2 of a surface of a blackbody at the temperature of solidifying platinum (2042°K) when measured in a direction perpendicular to that surface.

The intensity of the light from any source in a particular direction is expressed by a number of candlepower or candela. For example, a spotlight may have a candlepower of 10,000 in a certain part of its beam.

THE SOLID ANGLE

In order to define the intensity of light in space *away from the source*, it will be necessary to digress for a moment to solid geometry and define the *solid angle*. If a light source is envisioned as a point in space from which light is emitted, one can imagine a sphere of illumination around the source with the source at the center. One is usually concerned with measuring the light intensity on the surface of the imaginary sphere. This requires an understanding of what is meant by a solid angle in the sphere.

Consider a sphere of radius r measured from the apex of a solid angle ω. The part of the surface of the sphere s that is enclosed by the conical boundary surface of the solid angle is proportional to the solid angle subtended by s. Figure III-1 pictorializes this statement. When the size of the portion of the spherical surface s is equal to r^2, then the solid angle equals one steradian (sr; the unit of the solid angle). In general, then,

$$\omega = \frac{s}{r^2}$$

For example, if a solid angle subtends an area of 16 ft^2 on the surface of a sphere having a radius of 3 ft, then the solid angle is $\omega = s/r^2 = 16\ ft^2/(3\ ft)^2 = 1.78$ sr. It is to be noted that the solid angle in solid geometry is

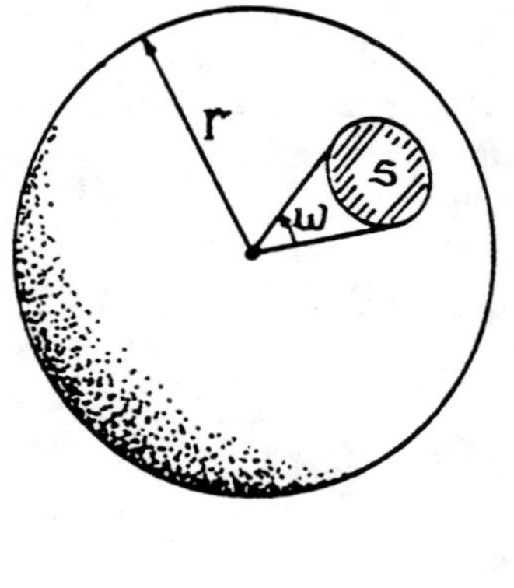

Figure III-1. The solid angle ω defines an area s on the surface of a sphere.

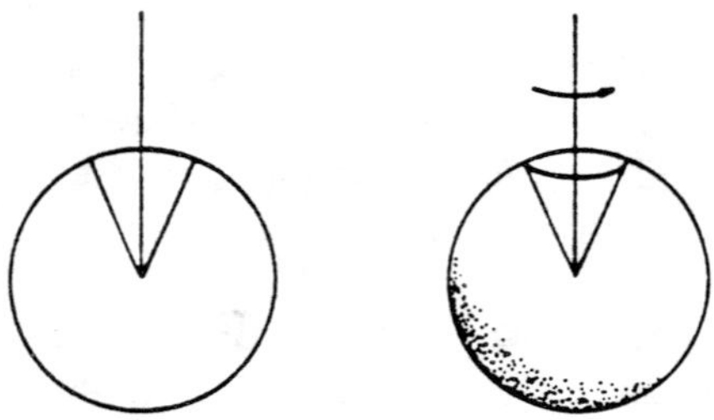

Figure III-2. The generation of a sphere and a cone angle by rotation of a circle containing a plane angle.

simply the counterpart of the plane angle in plane geometry. The plane angle is defined according to some number of degrees out of 360° or in terms of the amount (in radians) of the circle subtended by that angle. In turn, the solid angle is measured by the area of a spherical surface transected by a cone. Figure III-2 illustrates this point. There are 2π radians about a point in a complete circle because its circumference is $2\pi r$. There are 4π steradians about a point in a complete sphere because the surface area of a sphere is $4\pi r^2$.

LUMINOUS FLUX

The rate of flow of radiated visible energy from a light source evaluated on the basis of the impression of light which it induces in the eye is called the *luminous flux*. It can be likened to the amount of light energy. The symbol for luminous flux is Φ. Its magnitude depends upon the wavelength of the light, since the eye has different sensitivity for different wavelengths of light (Chapter V). The distribution of flux varies directly with the solid angle ω and the luminous intensity I according to

$$\Phi = I\omega \tag{III-1}$$

The photometric unit of flux is the lumen (lm). One lumen is the luminous flux (the amount of light) radiated by a homogeneous point light source of 1 cd intensity through a solid angle of 1 sr. The total luminous flux in all directions is $\Phi = 4\pi I$, so that a 1-cd source produces a total of 4π or 12.566 lm. The luminous flux of typical light sources is as follows: an incandescent lamp produces 10–23 lm/W; a flashbulb gives 1.2×10^6 lm; fluorescent lamps give 30–80 lm/W.

LUMINOUS INTENSITY AT A SURFACE

The flux of light from a source may not be the same in all directions or, more importantly, in the solid angle of interest. The design of the lamp itself, the lamp housing, and other factors can produce an irregular distribution of the light in space. Therefore, a term for intensity is needed to express the flux or level of illumination in an area separated from the source. Such a term has great utility where the display of art objects is concerned.

The units for measuring light intensity on a surface some distance from the source are *flux per unit area*. Some commonly encountered units are the foot-candle (1 fc = 1 lm/ft^2), the lux (1 lx = 1 lm/m^2), and the phot (1 phot = 1 lm/cm^2). For instance, an illumination of 1 fc is produced on a flat surface when 1 lm falls on each square foot of the surface. Table III-2 gives the numerical relationships between the various units. To gain a feeling for magnitudes, bright moonlight gives about 0.01 fc of illumination, while full summer sunlight gives about 10^4 fc. The minimum level of illumination for reading without eyestrain is about 10–30 fc. In fine work, such as surgical operations and the repair of some art objects, 500–1000 fc of direct light is recommended.

For illumination of objects, one is interested in the level of illumination at the object as a function of distance from the source. The relationship between the extent of illumination at a surface and the distance that surface is from the light source is given by the *inverse square law*

$$I = \frac{I_0}{d^2} \qquad \text{(III-2)}$$

I_0 is the light intensity of the source as viewed at unit distance and d is the distance between the source and the object being illuminated. The level of illumination is directly proportional to the luminous intensity of the source, but is inversely proportional to the square of the distance between the source and the object. That is, if a surface is illuminated with a single light source directly in front of it, and then the distance between them is doubled, the illumination at any point on the surface will

Table III-2. Comparative Magnitudes of Units of Illumination

Multiply number of → by To obtain ↓ number of ↘	Lux	Phot	Foot-candle
Lux	1	10,000	10.76
Phot	0.0001	1	0.00108
Foot-candle	0.0929	929	1

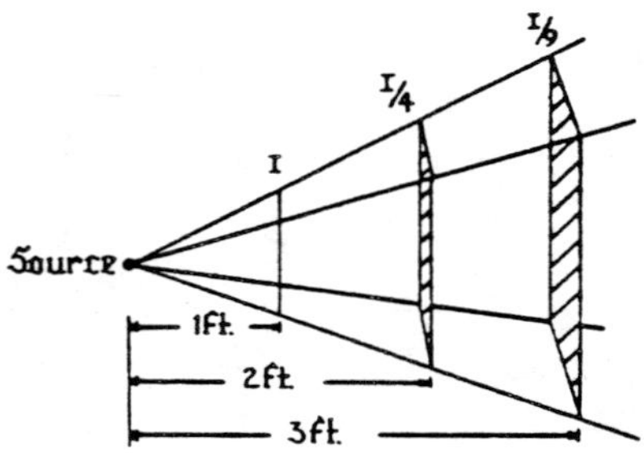

Figure III-3. A pictorial representation of the inverse square law.

be only one-fourth as great. This relationship is depicted in Figure III-3. The inverse square law only strictly applies (1) to the center of each square shown, because the distance *d* is measured to the center of the square, and (2) when the dimension of the light source is insignificant compared to *d*. If the illuminated plane happens to be tilted from that shown in Figure III-3, then the illumination level at points on the plane depends upon the angle of tilt according to

$$I = \frac{I_0 \cos \theta}{d^2} \qquad \text{(III-3)}$$

If 15 lx illuminates a point on a plane and the plane is then tipped by 30°, then 15 cos(30°) or 13 lx falls on the same point.

Many different kinds of direct-reading instruments for measuring luminous intensity on a surface are available. Most employ a photoelectric cell. Light falling on the photocell generates an electric current, which in turn activates a highly sensitive ammeter. The scale of the ammeter is calibrated in foot-candles, lux, or some other unit of luminous intensity. By turning the photometer at various angles to the direction of the incident light, one notes variations in the light intensity. Equation (III-3) is being put into practice by this procedure.

QUANTITY OF LIGHT

The quantity of light $\underset{\sim}{Q}$ is the amount of luminous flux emitted by a source per unit of time. The following equation defines the quantity of light if Φ is constant with time *t*:

$$\underset{\sim}{Q} = \Phi t \qquad \text{(III-4)}$$

Time can be measured in seconds, minutes, hours, days, etc. The luminous intensity *I* is sometimes used in place of flux. If *t* is in seconds, the unit of $\underset{\sim}{Q}$ is lumen-seconds.

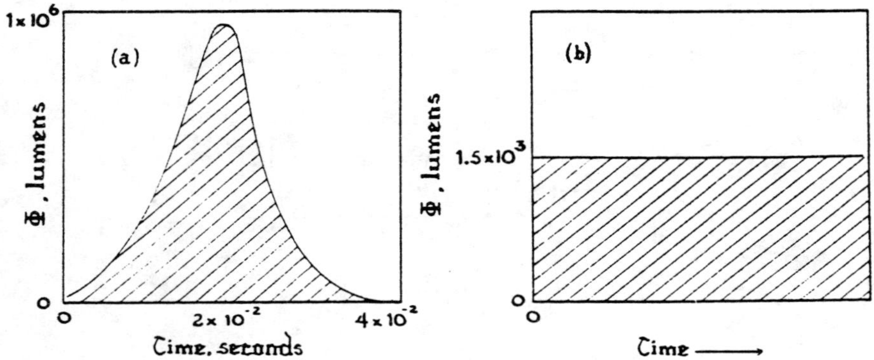

Figure III-4. The quantity of light from (a) a flashbulb and (b) a 100-W incandescent lamp. The flux is not drawn to the same scale in (a) and (b).

The quantity of light has several important applications. For example, the amount of light from and the operational characteristics of a flashbulb and an incandescent lamp are different. They cannot be easily compared using Φ or I alone. The flashbulb produces a large flux for a very short time, whereas a lamp can produce a small flux for a long time. A plot of the flux vs time for these two sources is shown in Figure III-4. The area under the curve or line is the quantity of interest and Q contains this information. The areas can be calculated to obtain a comparison of the total exposure to light that an object receives. According to equation (III-4), an object exposed to 50 fc for 50 days receives about the same quantity of light as if the object were exposed to 25 fc for 100 days. The quantity of light striking an artifact is a very important factor, particularly if the object is light-sensitive.

A second application of the quantity of light comes from the fact that all sources of artificial light experience a decrease in luminous flux as they age (i.e., Φ is not constant with time). The magnitude of the decrease depends upon the lamp make and type. In order to compare illumination characteritics among lamps with the same initial flux rating, it is useful to know the quantity of light produced during the life of the lamp rather than just its initial flux rating. In turn, equation (III-4) must be modified to account for the variation of Φ with time.

LIGHT DISTRIBUTION DIAGRAMS

The field of lighting engineering uses intensity diagrams to depict the distribution of light in space around various lamp fixtures and housings. Several examples of these diagrams are shown in Figure III-5. The top

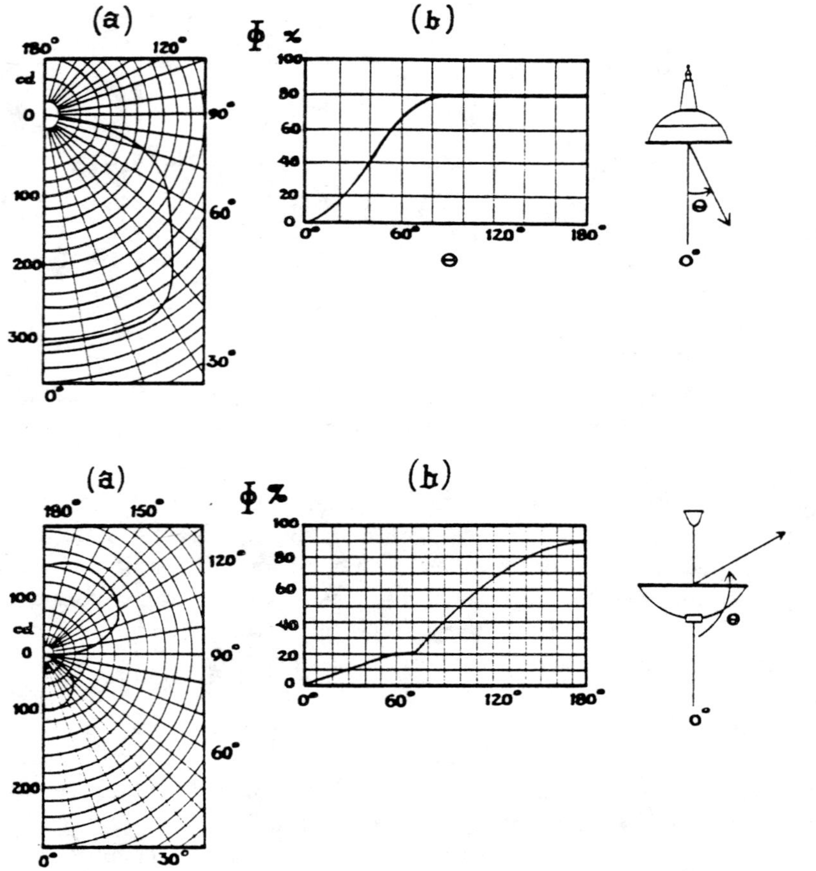

Figure III-5. (a) Polar light distribution diagram and (b) the zonal flux diagram for a direct lighting housing made of white, enameled metal (top) and an indirect lighting housing made of translucent plastic open on the upper side (bottom).

diagram shows a housing used in direct lighting. The polar plots labeled (a) show the flux of light from the luminaire (the complete lighting fixture) at angles from zero to 180° about the housing. In the top example, the intensity of light from a 1000 lm source is rather constant at 308 cd up to 30°. Between 30° and 60° the intensity drops from 308 to 200 cd as a result of interference by the housing. At 90° no light is emitted because of blockage by the edge of the housing. In the lower example in Figure III-5, 100 cd is produced at the base of the luminaire, but up to 150 cd is reached at the top.

The plots labeled (b) show the percentage of the total flux available from the luminaire as a function of the angle θ. In the top example between 0° and 80°, 80% of the light of the luminaire is radiated. The total efficiency of the housing for projecting light is about 80%. In the bottom example between 0° and 90° about 20% of the light is radiated, and between 90° and 180° about 70% of the light is radiated. The total efficiency of the housing from 0° to 180° is 90%. These diagrams are useful for determining quantitatively the intensity of light at various points around a particular light fixture.

BIBLIOGRAPHY

Kaufman, J. E., and Christensen, J. F., eds., *IES Lighting Handbook*, 5th ed., Waverly Press, Baltimore, Maryland (1972).

Keitz, H. A. E., *Light Calculations and Measurements*, Macmillan, New York (1971).

Lape, P. B., and Ryer, D. V., How to Measure Light, in *1968 IEEE Northeast Electronic Research and Engineering Meeting* (November 1968). *Principles of Light Measurement*, Bureau Central de la CIE, Paris, France (1970).

Roche, R., Environment, Level of Illumination and Protection Against Harmful Radiations and Their Influence, *Public Archives of Canada*, No. 3 (April 1971).

Walsh, J. W. T., *Photometry*, 3rd ed., Dover, New York (1958).

IV

The Effect of Materials on Light

This chapter could also be titled Physical Optics. To understand why an object appears the way it does, it is important to fully appreciate how light behaves once it strikes the surface of the object. Appearances would be very monotonous if all that could happen to light were simple reflection and absorption. Fortunately, many other phenomena also can occur which enhance the variety of appearances observed in art objects and in nature.

In this chapter some of the most important modifications caused by matter in the wavelength and intensity composition of electromagnetic radiation are taken up. The meaning of the index of refraction of a material is presented. The index of refraction allows one to show quantitatively how light is refracted in a material. It also allows consideration of an important reason for the various degrees of opacity found in materials. The index plays a role in controlling the amount of reflectance of light at an interface. The subjects of the polarization and interference of the electric vectors of electromagnetic waves as a result of reflection and transmission of light are presented. Finally, the absorption of light is introduced. Selective absorption of light by an object can lead to the appearance of color on the object. The actual molecular processes involved in color production by absorption are subjects of Chapter VII and X, however. In this chapter, only the physical considerations of light absorption are attended to.

Once light encounters an interface of some kind, its fate can be separated into three general categories. These are: (1) reflection and/or scattering of the light from the surface; (2) transmission and/or refraction of light into and through the object; (3) partial or total absorption of light by the object. Kirchhoff's law (a law of conservation of energy), formulated by Gustav Robert Kirchhoff in the nineteenth century, requires that the sum of the intensity of the light involved with events 1, 2, and 3 must equal the intensity of the incident light. This fact is sometimes known as the RAT law (Reflected intensity + Absorbed intensity + Transmitted intensity = Incident intensity). Processes 1 and 2 depend upon the surface characteristics, the composition of light, and the nature of the material. Process 3 is intimately related to the opacity and color of the object. The laws governing these phenomena at surfaces of homogeneous, non-conducting materials will be presented here. The optical effects that occur at other surfaces, composite materials, and specific art objects are the subject of the remaining chapters.

Figure IV-1 summarizes how reflection, refraction, and transmission interrelate in a transparent glass plate. If the glass is colored, a fraction of the light will also be absorbed; this is not represented. Several features merit mention. First, note that the angle of incidence i equals the angle i'. That is, reflected rays occur at the same angle as the incident rays. This is one of the laws of reflection. All reflected light obeys this law, although two types of reflection are usually defined. *Specular reflection* is characteristic of light coming from a smooth, polished surface, and *diffuse* or *irregular reflection* occurs from a roughened surface. In both specular and diffuse reflection $i = i'$. (Reflection and scattering of light also can occur from particles dispersed in a medium. However, the laws controlling the interaction of light with particles depend upon the size of the particles. This fact is discussed in Chapter VI.)

Second, note that the angle i equals the angle i''. That is, transmitted rays leave the glass surface at the same angle as the incident rays entered, except that the exit point is displaced from the entrance point by an amount which depends upon the thickness and refractive index of the glass and the

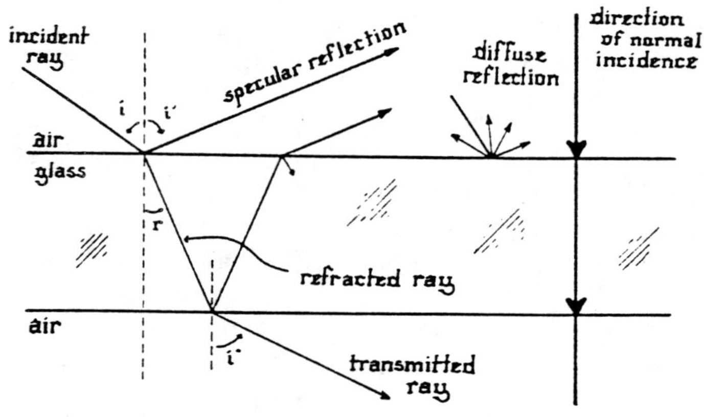

Figure IV-1. A pictorial summary of reflection, refraction, and transmission of light by a transparent material such as glass.

incidence angle i. This is why an object covered by a flat plate of glass appears to be displaced to one side if the glass is viewed at any angle other than perpendicular to the surface. If the angle of incidence is 0° as shown on the right side of Figure IV-1, then no displacement occurs. A ***transparent*** object transmits virtually all light, while a ***translucent*** object only partially transmits light. An ***opaque*** object is one which is incapable of transmitting light.

Third, because light rays are refracted upon entering the glass, the angle i does not equal the angle of refraction r.

Fourth, internal reflection of light can occur within the glass, which can affect the appearance as perceived by the eye. We now consider the basis for these and other facts which determine how light interacts with matter.

THE INDEX OF REFRACTION

The index of refraction of a material is an important parameter in qualitative discussions of how the material affects light. It is one of the few constants of a bulk material which can be defined by ordinary light. The concept of the refractive index was introduced by Newton in the late seventeenth century, and the development of the theory was essentially complete by the end of the nineteenth century. In Chapter I, it was noted that Newton split visible light into its component wavelengths (colors) by using a homogeneous glass prism. This dispersion occurs because light of different wavelengths has different velocities in the glass.

Refraction refers to the change in direction of light resulting from the change in the velocity as light passes from one medium into another. Only when the light beam is traveling perpendicular to the surface of the material will a change in velocity occur without a change in the direction of the light beam. Every medium produces a characteristic velocity for a specified wavelength of light, except that in crystals, as discussed in Chapter VI, this velocity may depend upon the orientation of the molecules in the lattice relative to the direction of the light source. In this latter case, one, two, or three characteristic velocities of light can be defined, depending upon the crystal structure.

The refractive index η of a medium refers to the ratio of the velocity of light in a vacuum c to its velocity in the medium of interest. That is,

$$\eta_{\text{medium}} = \frac{c_{\text{vacuum}}}{c_{\text{medium}}} \qquad \text{(IV-1)}$$

Equation (IV-1) tells us that, because water has a refractive index of 1.33, visible light will travel 1.33 times slower in water than in a vacuum because the refractive index for a perfect vacuum is exactly unity by definition. The refractive index for air is 1.003. Henceforth, air, instead of vacuum, will be used as the reference medium because η for the two differ insignificantly.

The velocity changes of light in materials depend upon temperature (in air this is responsible for the shimmering effect of an object when it is viewed through air rising from a hot surface and mixing with cold air), the atomic number of the constituent atoms and the density of the medium (iodides, for example, have greater indices of refraction and atomic number than do chlorides, and conjugated aromatic hydrocarbons have greater density and thus higher indices of refraction than do straight-chain aliphatic hydrocarbons), and the wavelength of light (η increases with decreasing wavelength, which produces normal dispersion). Most of these dependences can be correlated to differences in the density of materials. In general, as the density of a material becomes greater, the value of η becomes larger, and, as noted shortly, the dispersion it is capable of producing becomes larger.

Figure IV-2 shows some characteristics of η as a function of the wavelength of light for several colorless transparent materials. Three generalities about the changes in index of refraction with wavelength of light can be noted. First, the index of refraction of these materials decreases with increasing wavelength of light. Second, the degree of curvature in η is not the same over the wavelength region shown. Indices of refraction change much more rapidly for shorter wavelengths than for longer wavelengths. In fact, an approximate $1/\lambda^3$ dependence of the index exists, which accounts for this curvature. Third, the curvature of the lines in the visible region becomes more pronounced as the refractive index of the material becomes greater. Hence the larger the value of η for a material, the more

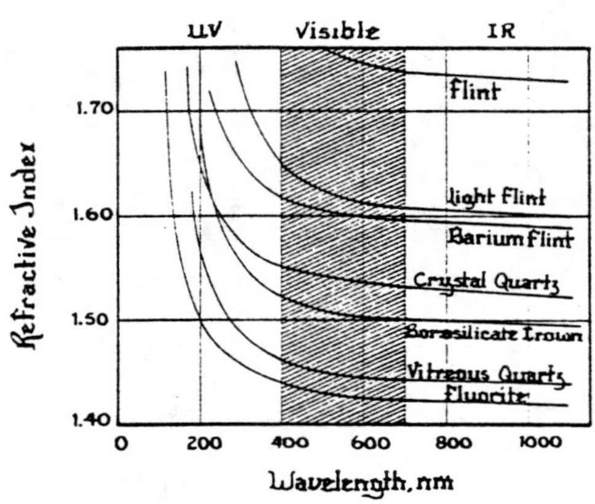

Figure IV-2. The dispersion curves for several types of colorless transparent materials.

dispersion it produces. A good example of this fact is the diamond, which has a very large η value. The striking beauty of a cut diamond is partly a result of the high degree of the color-producing dispersion and partly a result of the high degree of reflection at the facets.

To obtain the index of refraction of a medium, it would be quite formidable to be faced with the task of having always to measure the velocity of light, particularly considering the short distances the light usually travels in the medium. Fortunately there is a very simple means of obtaining η in materials. The index of refraction can be obtained from the ratio of the sine of the angle i to the sine of the angle r shown in Figure IV-1. *Snell's law*, formulated in 1621 by Willebrord Snell, gives the relationship:

$$\frac{\eta_2}{\eta_1} = \frac{\sin i}{\sin r} \qquad \text{(IV-2)}$$

η_1 is the index of refraction of the less dense medium, and η_2 is the refractive index of the more dense medium. If air is one of the media, then $\eta_1 = 1.00$, and Snell's law reduces to

$$\eta_2 = \frac{\sin i}{\sin r} \qquad \text{(IV-3)}$$

Snell's observation predated Newton's work on refraction. Therefore, Snell did not use the term "index of refraction," but merely noted that $(\sin i)/(\sin r)$ is a constant for a given material.

A situation exists where Snell's law of refraction is not followed. In Chapter VI, the concept of the ordinary and extraordinary beams of light in uniaxial and biaxial crystals is developed. The extraordinary ray does not follow equation (IV-3), but the ordinary ray does.

POLARIZATION

A beam of light incident at a surface can be reflected according to the law of reflection, but the reflected beam rarely has the same distribution of wave oscillations as the incident beam. This difference is called *polarization*, and it manifests itself in intensity differences which depend upon the angle of incidence and the nature of the material. Polarization is general to all electromagnetic radiation in that waves of all energy from X-ray to radio waves can be polarized in one way or another. We concentrate here on visible energies, but the same discussion can be equally well applied to all energies. To comprehend what is meant by polarization of light, imagine a pencil of light containing many waves traveling through space in a direction labeled *z*. This pencil of light is made up of many oscillating waves, each characterized by different amplitudes and orientations of vibrations in all possible *x* and *y* directions. Figure IV-3 pictures this complicated light beam in terms of its component wave characteristics.

Linearly polarized light corresponds to electric vector oscillations which are located in a single plane. It is often called *plane polarized* light for this reason. An end-on view of the light wave oscillations resolved in this mode would appear as a line. *Circularly polarized* light corresponds to waves which pass in a "corkscrew" motion, turning either to the left or to the right. An end view would show a circle with the beam rotating to the left or to the right. *Elliptically polarized* light describes the motion of the electric vectors traveling in an ellipse. The motion can be rotation and oscillation in opposite directions in space as shown in Figure IV-3. Elliptically polarized light can be thought of as a composite of plane and circularly polarized light.

A beam of light passing through space has little polarization if it has not encountered a reflecting or scattering surface. The beam is continually changing from one class of polarization to another such that all three types of polarization are equally represented. If the light encounters a reflecting or

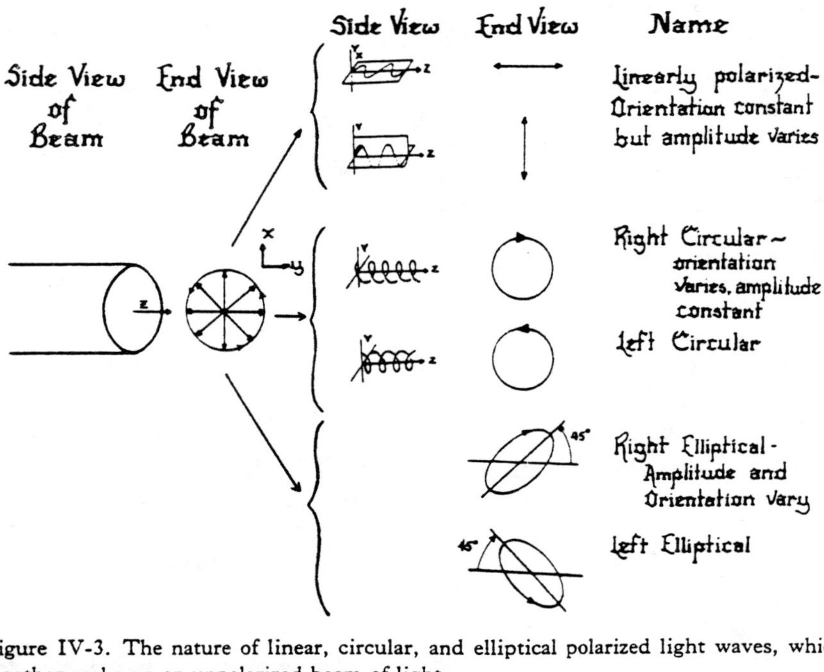

Figure IV-3. The nature of linear, circular, and elliptical polarized light waves, which together make up an unpolarized beam of light.

scattering surface, some of the oscillating motions are suppressed and the reflected beam becomes polarized. Light can be partially or completely polarized, depending upon the degree of suppression that occurs.

Many discussions of polarized light focus only on plane polarized light. In part, this is because plane polarization is conceptually more easily visualized than are the other types. There are spectroscopic methods which rely on circular and elliptical polarization of light, but they are not the subject of interest here. Plane polarization will be exemplified for simplicity.

A pencil of light is made up of many waves with random electric vector oscillations. In the course of polarization of the light, the oscillation in one direction becomes quenched relative to the perpendicular direction. Two ways of representing a light beam (as a "light beam" and in "two-vector" notation) are shown in Figure IV-4. Three degrees of plane polarization of light (unpolarized, partially polarized, completely polarized) are shown in the two representations. Partial and complete polarization occur when light is specularly reflected from a transparent material such as glass or water. A lesser amount of polarization will occur in light reflected from opaque but highly reflecting materials, such as metals.

The polarization of certain electric vector oscillations in one direction relative to those in another upon reflection is illustrated in Figure IV-5. An unpolarized beam of light containing electric vectors a_1 and a_2 oriented at right angles to one another in the two-vector notation of Figure IV-4 strikes the glass surface. The electric vector a_1, which is perpendicular to the surface, is suppressed relative to a_2 in the course of the reflection. The reflected beam containing a_2 can be completely polarized or just partially polarized, but it is shown to be completely polarized in the figure. The reflected light a_2 is polarized in the plane of the reflecting surface as shown.

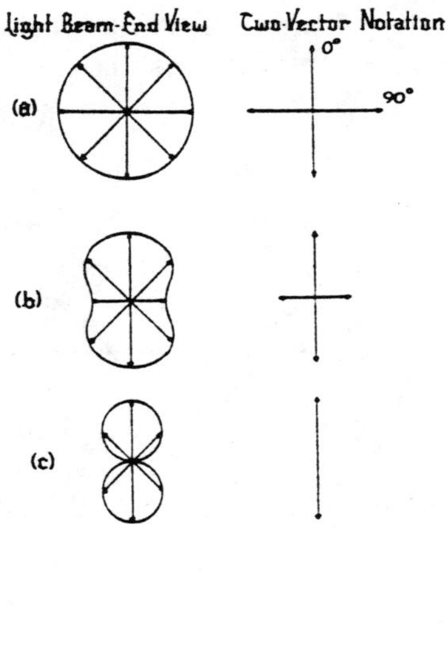

Figure IV-4. An end-on view of (a) unpolarized, (b) partially plane polarized, and (c) completely plane polarized pencils of light. The two-vector notation at the right is often used for simplicity.

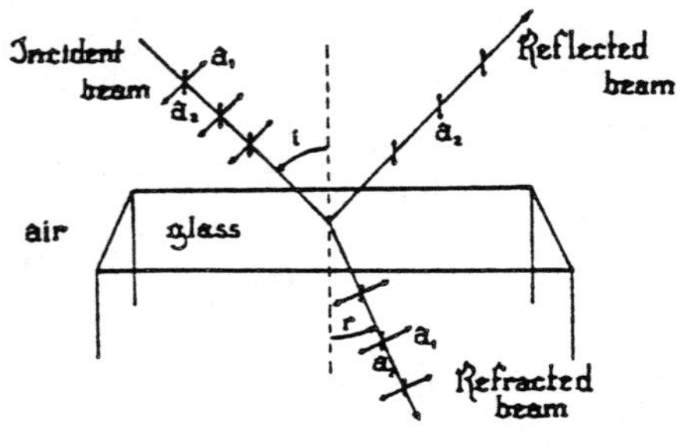

Figure IV-5. Polarization of reflected light from a transparent glass surface. The reflected beam is completely polarized in the plane parallel to the surface (only a_2 oscillations occur). The refracted beam is partially polarized (all a_1 oscillations and some a_2 oscillations).

The refracted beam will contain all a_1 oscillations and the fraction of a_2 that was not reflected. Hence the refracted beam is only partially polarized in all cases and has its main polarization in the plane perpendicular to the surface.

The degree to which polarization takes place in a transparent medium depends upon the angle of incidence i of the light. The mathematical treatment of the relationship between incidence and reflection and the degree of polarization involves Fresnel's equations and Brewster's equations. Augustin Jean Fresnel derived the following equations in 1820 to relate the magnitude of a_1 and a_2 in the reflected light to the angles i and r shown in Figure IV-5 (the value of a is the amplitude of the unpolarized incident beam):

$$a_1 = a\,\frac{\tan(i - r)}{\tan(i + r)}, \qquad a_2 = a\,\frac{\sin(i - r)}{\sin(i + r)} \tag{IV-4}$$

The ratio of a_1 and a_2 is the degree of polarization. David Brewster noted that if $i + r = 90°$, then $a_1 = 0$ because the tangent of 90° is infinitely large. Hence at this angle relationship, the reflected light only has a_2 amplitude because $\sin 90° = 1$. This is to say that when $i + r = 90°$, the reflected beam is completely polarized. It can also be shown by simple geometric relationships that this is equivalent to having a 90° angle between the reflected and the refracted beams as pictured in Figure IV-6. The

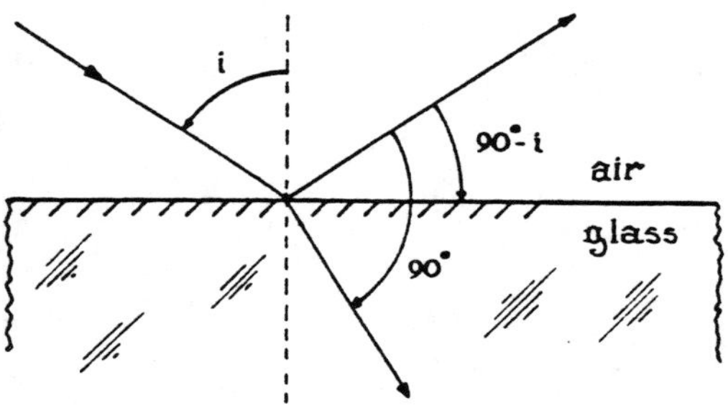

Figure IV-6. Some angular relationships when the reflected light is completely polarized. The angle i is the Brewster angle.

particular value of the angle i such that $i + r = 90°$ is known as the *Brewster angle* i_p at which all reflected light is completely polarized. Complete polarization can only occur in transparent media, however. The Brewster angle is related to the index of refraction of the material through the following equation and accordingly differs from one material to the next:

$$\tan i_p = \frac{\eta_{\text{material}}}{\eta_{\text{air}}} \simeq \eta_{\text{material}} \qquad \text{(IV-5)}$$

For silica glass, $\eta = 1.5$, so that $i_p = 57°$. Therefore, light striking a silica glass surface at 57° from the normal produces completely polarized reflected light. For water, the Brewster angle is 53° because $\eta = 1.33$.

A pictorial representation of Fresnel's equations (IV-4) is shown in Figure IV-7, where the percent reflection of the a_1 and a_2 electric vectors is plotted as a function of the angle of incidence of the incoming light. Ordinary silica glass ($\eta = 1.5$) is used as the surface in this illustration. It can be seen that the degree of polarization in the reflected light increases as i increases from 0° to 57°. At 57° the reflected light is completely polarized because $a_1 = 0$ at this angle. From $i = 57°$ to $i = 90°$, the percentage of reflected light for the two vectors rises to 100% and some polarization exists until this maximum value is reached.

Most natural light exhibits some degree of polarization. Blue sky light traveling roughly at right angles to the sun's rays can reach 70–80% polarization as a result of Rayleigh scattering (Chapter VI).

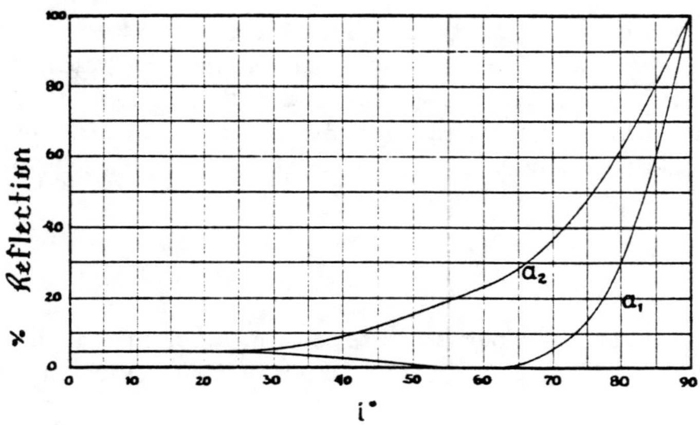

Figure IV-7. A plot of Fresnel's equations. The variation with angle of incidence of the percent reflection for the two electric vector components a_1 and a_2 of the light beam is shown. a_1 is perpendicular to the plane of incidence and a_2 is in the plane.

Even though humans are unaware of the polarization, this polarized light is of great importance to other species. Insects such as bees, ants, beetles, and flies detect the polarization of sunlight and navigate by it. Polarized light is inconspicuous to humans because the human eye is unable to distinguish it from unpolarized light without the aid of some device. Such a device is a polarization filter. Materials can be made which transmit light with oscillations in one direction but absorb oscillations in the other. An example is "Polaroid" material. Polaroid consists of a thin layer of small crystals of a salt of iodine and quinine called *herapathite* all aligned with their axes parallel. These crystals absorb light oscillations in one direction but not in the other. The same principle is used by the photographer to remove unwanted glare and reflection when photographing an object. A polarization filter is simply placed in front of the camera lens to remove polarized reflected light.

Polarized light has important analytical uses. In microscopy, certain pigments can be distinguished by their unique appearance in polarized light. Strain in materials becomes visible in polarized light. For example, strain lines in glass can be made apparent by placing the glass between two polarizing filters. Polarized X-rays have been used to reveal strains in opaque materials.

REFLECTED LIGHT INTENSITY

In Figure IV-7, the intensity of light reflected from a glass surface is seen to be 4% of the incident light intensity for angles of incidence less than 15°. This intensity value can be calculated in the following way from the fact that the intensity of the reflected light I is related to the electric vector amplitude a according to equation (I-4). When light beams traveling perpendicular to a surface strike the surface, the angles i and r are zero and the law of reflection [equation (IV-3)] reduces to $\eta = i/r$. The ratio of the reflected beam amplitude to the incident beam amplitude is expressed by

$$\frac{a_{\text{reflected}}}{a_{\text{incident}}} = \frac{\eta_2 - \eta_1}{\eta_2 + \eta_1} \qquad \text{(IV-6)}$$

η_2 is the index of refraction of the more dense medium and η_1 is the index of the less dense medium. If one medium is air, then $\eta_1 = 1.00$, and equation (IV-6) becomes

$$\frac{a_{\text{reflected}}}{a_{\text{incident}}} = \frac{\eta - 1}{\eta + 1} \qquad \text{(IV-7)}$$

The intensity is proportional to the square of the electric vector amplitude [equation (I-4)]. Therefore, the intensity of the reflected light compared to the incident light can be obtained from

$$\frac{I_{\text{reflected}}}{I_{\text{incident}}} = \left(\frac{\eta - 1}{\eta + 1}\right)^2 \tag{IV-8}$$

For silica glass having $\eta = 1.5$, the reflected intensity relative to the incident intensity of light in percent is

$$\frac{I_{\text{reflected}}}{I_{\text{incident}}} \text{(in percent)} = \left(\frac{1.5 - 1}{1.5 + 1}\right)^2 100 = 4\%$$

This percent reflection occurs at the top and bottom surfaces of the glass, so that the total amount of light reflected, assuming the faces are parallel to each other, is 8% of the incident light. It is evident from equation (IV-8) that the larger the value of η for the material, the more the light is reflected when the angle of incidence is 0°. At $\eta = 1.9$, the reflection at each surface is 10%. Therefore, materials with large indices of refraction can be expected to appear more opaque with respect to light incident at a surface that is at a right angle to the direction of the light rays.

It is often found that strong reflection of light incident perpendicular to the surface is accompanied by strong absorption of the light by the medium. Metals are an extreme example of this fact. Metals both absorb and reflect a great deal of energy. Polished nonmetallic materials having large refractive indices reflect a high percentage of light from their surface, but also allow less light to escape once it enters the material. As a result, materials with larger indices of refraction tend to be more opaque than those with smaller indices. In transparent or partly transparent media, the reason for this is that the *critical angle of incidence* of light rays depends upon the value of η. A light beam entering a transparent or semitransparent material which has an index of refraction higher than the medium from which the light came follows the laws of reflection when it is inside the material. A feature of the picture not yet mentioned is the fact that if this light, once in the material, strikes the surface at angles greater than a particular angle called the critical angle of incidence i_c, it is reflected back into the material and does not escape. This is called *total internal reflection.* Light that is not transmitted across the interface is not available for viewing. Hence one perceives the object as physically opaque insofar as this light is concerned.

Figure IV-8 shows the effect on *interior* light beams of a glass plate as a function of the angle of incidence. In case (a), the light beam exits the glass after undergoing refraction at the glass–air interface or is reflected. In

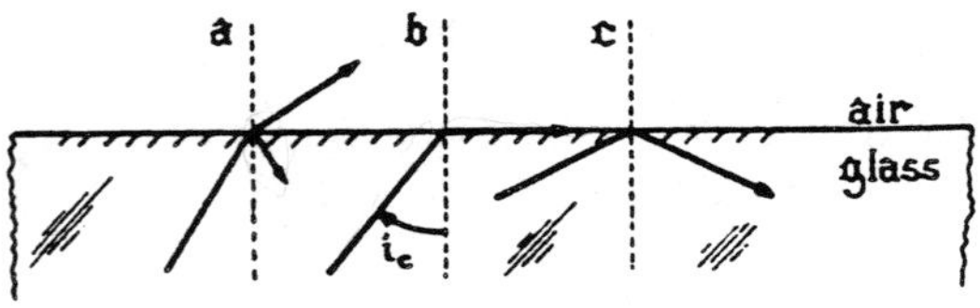

Figure IV-8. The behavior of a beam of light in a glass layer, depending upon the angle of incidence.

case (b), the light beam strikes the surface at an angle equal to the critical angle and passes along the glass–air interface. In case (c), the light beam is reflected back into the material because it strikes the surface at an angle greater than the critical angle of incidence. The magnitude of the critical angle i_c in case (b) is related to the indices of refraction of the two media forming the interface. This relationship is given by

$$\sin i_c = \frac{\eta_1}{\eta_2} \qquad \text{(IV-9)}$$

η_1 is the refractive index of the less dense medium and η_2 is the value for the more dense medium. For a glass–air interface shown in Figure IV-8,

$$\sin i_c = \frac{1.0}{1.5}$$

so that $i_c = 41°$. This means that light in the glass which is attempting to exit at any angle greater than 41° to the normal will be reflected from the surface back into the glass. The cross-hatched region in Figure IV-9 is the region in which light beams remain in the glass.

According to equation (IV-9), the higher the value of η for the material, the smaller i_c becomes, and the less the amount of light that can escape from the material. This fact provides another reason for the apparent opacity of materials having high indices of refraction. Earlier

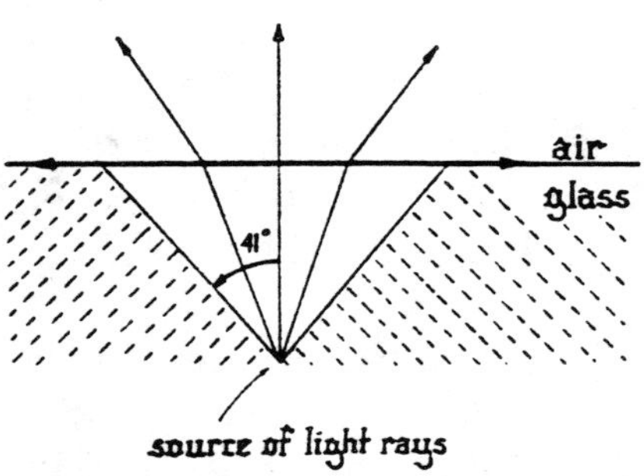

Figure IV-9. The internal reflectance of light in glass. Light beams in the shaded region do not leave the material, because of reflection at the interface.

it was noted that light striking at right angles to the surface of a material is increasingly subject to reflection as η increased. Less light penetrates the surface, making it appear opaque. Now we know that, in addition, as η increases, an object becomes more opaque because less light penetrates the surface boundary and leaves *once it is in the material.*

The principle of the critical angle of incidence is put to use extensively in commercial and industrial problems, where light-carrying fibers have found application. As an example, the gastrofibrescope is a medical tool used for examining patients' stomachs. It has application in art conservation for examining interiors of hollow objects which have an opening. The instrument contains many thousands of glass fibers bundled together. Each fiber has a core of transparent glass with a high index of refraction and is coated with a sheath of transparent material with a low index of refraction. Light in the fibers undergoes a high degree of internal reflectance at the interface and hence travels along the interior of the fibers with little loss through the walls. By viewing the end of the fiber bundle with an eyepiece, one can use light projected into the interior to reveal features otherwise hidden from view.

INTERFERENCE OF LIGHT

Interference is responsible for a number of important optical effects. The color and iridescence of soap bubbles, oil films on water, oyster shells, peacock feathers, certain beetles and butterflies, and to some extent mother-of-pearl can be attributed to the effects of interference in visible light caused by thin films on a reflecting surface. The lusterwares of Tiffany and others were created by using the phenomenon of interference by thin films painted onto a glass surface. Ancient glass, such as the famous antique glass of Cyprus, often possesses iridescent qualities due to the fact that surface layering has developed. Hydration and leaching of the surface expand and contract the surface of the glass and create thin, loosely adhered layers. Interference of light reflected from the successive layers leads to iridescence. Another instance is observed in polished steel which has been heated in air. The surface is oxidized to varying depths, depending upon the temperature. This sheet of metal oxide reflects light as a thin film and affords the metallurgist a valuable aid in judging the proper temperature for tempering, as well as giving a degree of ornateness to the object. Newton's rings in lenses also result from interference patterns. A well-known example related to Newton's rings is camera lens "blooming." Camera lenses tend to look blue or yellow to the eye as a result of light interference patterns.

To determine how interference produces color, a careful examination of the wave properties of reflected light needs to be made. Two light waves are capable of adding or subtracting their amplitudes as they meet and superimpose. It is common knowledge that water waves do this, and light behaves in a similar manner. The resulting superimposition of two waves redistributes the intensity and is known as *interference*. Interference can be either *constructive*, where the wave amplitudes add and become more intense, or *destructive*, where they subtract and become less intense. No light is destroyed in the process; there is merely a redistribution of intensity. For a transparent interference surface, such as a glass coated with a film of oil or a soap solution, the colors of the light reflected are absent in the transmitted light, and vice versa. At any point on the film, the color by reflected light is complementary to its color by transmitted light. This is in keeping with the law of the conservation of energy.

One means of producing interference is reflection from two or more very closely spaced, parallel or approximately parallel surfaces. Double reflection can occur. Figure IV-10 illustrates the simplest kind of interference. The incident light waves strike a surface. Some of the light is reflected and some of it is transmitted into the transparent film, where it is reflected from the lower boundary. Upon reaching the film–air boundary again, this reflected light can be transmitted across the interface into the air and viewed along with the incident rays that were directly reflected initially. However, an important difference may exist between the two beams. This is that their electric vector oscillations may or may not be in phase (i.e., they may or may not be superimposed) with one another. If they are in phase, the amplitudes of the two add and the light is more intense (constructive interference). If they are completely out of phase, the peak of one wave coincides with the trough of the other and they suppress each other (destructive interference). All of the intermediate cases also can occur. The out-of-phase wave relationship is shown in Figure IV-10.

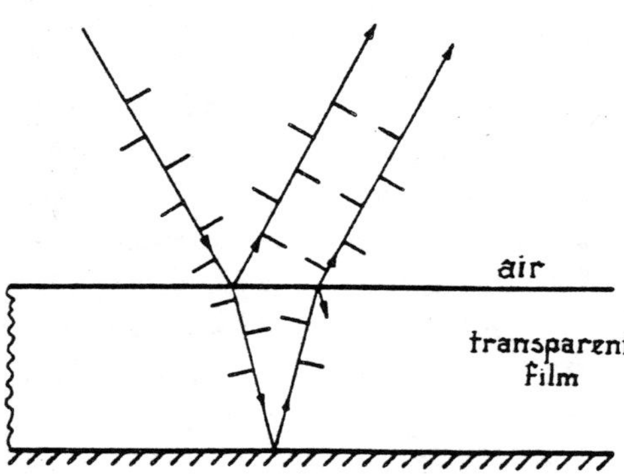

Figure IV-10. Destructive interference between light waves reflected from the upper and lower surfaces of a thin film. The peak maxima (represented by lines) of one wave coincide with the troughs of the other.

There are several important variables to consider in determining whether constructive or destructive interference occurs. The most obvious one is the thickness of the layer. Care must be taken in explaining this point, because phase changes can occur during reflection of light. A phase change represents a displacement of the wave along the propagation vector (Chapter I). There is no change in the magnitude of λ during a phase change. When light strikes a boundary from the side of higher velocity (e.g., light in air striking a soap film), there is a change of phase upon reflection corresponding to half a wavelength. This type of boundary is known as a rare-to-dense interface. Air is rare (the less dense medium) and the water-soap is dense (the more dense medium). A phase change of half a wavelength (or π) occurs during reflection from a rare-to-dense interface. On the other hand, there is no phase change for light reflected from a dense-to-rare boundary. That is, light inside the soap–water film being reflected from the boundary with air undergoes no phase change. *Whether or not light is striking a rare-to-dense or a dense-to-rare boundary plays a critical role in whether constructive or destructive interference will occur in a transparent film of a given thickness.* The refractive indices are useful for determining the type of boundary present. This fact will be illustrated with several examples. Before doing so, the second variable of interest in interference patterns needs to be mentioned. Since the film serves to reinforce and suppress certain wavelengths of light, it follows that the wavelength of light is important. Total attenuation of light occurs only for certain wavelengths, while for others partial interference occurs.

As just mentioned, one must consider the refractive indices at the different boundaries to determine whether the boundary is rare-to-dense or dense-to-rare. Figure IV-11 summarizes the two most commonly encountered types of interference films. Case (a), where $\eta_1 < \eta_2$, occurs with soap bubbles, oil on water or on silica glass, and Tiffanyware. There is a phase change of $\lambda/2$ for ray 1 (rare-to-dense reflection), but no phase change for ray 2 (dense-to-rare reflection). Ray 2 has traveled $2(\lambda/2)$ farther than ray 1, however, and therefore the two rays are exactly out of phase. They will destructively interfere. A film thickness of about 260 nm would destructively interfere with green light of 520 nm wavelength. Case (b), where $\eta_1 < \eta_2 < \eta_3$, occurs in the nonreflecting glass discussed below and in tempered steel. There is a phase change of $\lambda/2$ for ray 1 (rare-to-dense reflection) and also for ray 2 (rare-to-dense reflection because $\eta_2 < \eta_3$). If the film thickness now corresponds to one-quarter of the wavelength of the incident light, then the light beam reflected from the top layer and the beam reflect from the bottom layer destructively interfere. This is because the wave in the film travels a distance of $\lambda/4$ down and $\lambda/4$ up, so the emergent ray is exactly out of phase ($\lambda/2$) with the ray initially

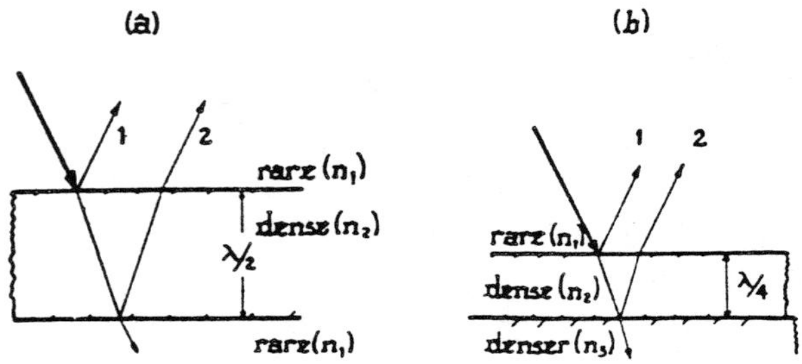

Figure IV-11. Destructive interference of doubly reflected rays from (a) a rare–dense–rare film surface and (b) a rare–dense–denser film surface.

reflected from the surface. Now it will require a film thickness of 130 nm or 390 nm to attenuate green light of wavelength about 520 nm during double reflection. Such films only partially give rise to interference with visible light at other wavelengths. The film would have a purple hue because red and violet can pass.

A surface which acquires its color by creating interference patterns can be expected to change colors with changes in the angle of viewing or illumination. This is because a different thickness of the film layer is observed, depending upon the angle. As the thickness varies, the wavelengths of light that constructively or destructively interfere with one another become different and therefore the color is different. A peacock feather held at different angles to the light shows this effect.

"Nonreflecting" glass is important in display work. Certain types of this glass function on the principle of double reflection to produce destructive optical interference. In one type a thin film of a transparent material, such as MgF_2 or CaF_2, is deposited onto the surface of normal glass. The index of refraction of the film is chosen to be intermediate between that of glass and air (i.e., between 1.0 and about 1.5). The thickness of the film is regulated at one-quarter wavelength, assuming normal incidence. The light reflected from the film–air interface will be largely out of phase* with the light reflected from the film–glass interface, with the net result that, rather than 4% of the light being reflected, less than 1% is reflected. Of course the thickness can be one-quarter wavelength for only one wavelength at a time. The visible region ranges from 400 to 700 nm. The yellow-green region is usually chosen to determine the thickness of the film because this is the middle of the visible region, where the eye is most sensitive. Some nonreflecting glass may appear slightly purple in color

* In order for the light to be exactly out of phase, the thickness d of the interference layer must be equal to $\lambda/4\eta$, where η is the refractive index of the interference film.

in reflected light as a result of this choice. Another way of making nonreflecting glass is to slightly roughen the surface. An object viewed through such glass is clear only if it is placed in tight contact with the glass.

ABSORPTION OF LIGHT

So far two mechanisms that can produce color in an object have been discussed. These are refraction and double reflection. Refraction of light by transparent materials may separate white light by dispersion into its component wavelengths, thus producing color. Double reflection of light by thin films can cause constructive and destructive optical interference, which redirects the passage of certain wavelengths of white light, again producing color. Single reflection, however, causes no change in the wavelength composition of light. If one color is incident, then that same color is reflected. A third means of color production, which will be discussed in Chapter VI, is Rayleigh scattering by very small particles.

By far the most important mechanism for the production of color by materials is the selective removal of certain wavelengths of light from the spectrum by absorption. The light must penetrate the material and encounter energy-absorbing species for this to occur. The absorbed wavelengths are absent from the reflected light and lead to the sensation we perceive as color. The molecular events involved in light absorption by organic and inorganic materials are the subjects of Chapters VII and X. Here we discuss the physical laws and the perceptual and conceptual basis of light adsorption and color.

The loss of light energy by adsorption alone can be evaluated without difficulty once certain constants of the absorbing medium are known. The absorption spectrum of vapors and gases at low pressure may appear in discrete, precisely defined bands as observed in the emission spectra of several light sources shown earlier. The spectral absorption by liquids and solids is usually blurred into broad bands due to the disturbing influence of molecules in close proximity to each other. However, once the absorption for a given thickness or concentration is known for these bulk media, the absorption can, in general, be found for other thicknesses and concentrations. There are two laws, Lambert's law and Beer's law, by which calculations can be made.

Lambert's law deals with the relationship of the absorbing medium to the absorption of radiant energy. It applies to light incident onto the surface at $i = 0°$, to monochromatic light, and to pure, homogeneous materials. These restrictions may sound unduly limiting, but they are necessary to

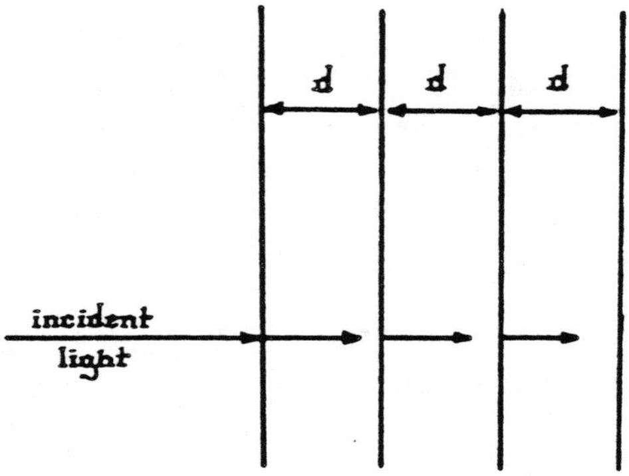

Figure IV-12. Equivalent layers, each of thickness *d*, of a light-absorbing material. The length of the arrows shows the decreasing intensity of transmitted light.

avoid complications which arise otherwise. If n layers of an absorbing medium, each with a thickness of d, are present as illustrated in Figure IV-12, then the fraction of incident light absorbed in each layer is A. The transmittance through each layer is $(1 - A)$ times the incident intensity. As the number of layers increases, the amount of absorbance increases and the amount of transmittance decreases.

For n layers, the absorbance in all layers will be

$$A + A(1 - A) + A(1 - A)^2 + \cdots + A(1 - A)^{n-1} \quad \text{(IV-10)}$$

The first term in equation (IV-10) represents the absorption by the first layer. The second term represents the absorption by the second layer. The second layer absorbs a fractional amount A but it only has an amount of light equal to $1 - A$ (the amount of light transmitted through the first layer) incident onto it. The total amount of light absorbed by all layers together is given by the last term. The total amount of light transmitted through n layers is $(1 - A)^n$. Lambert's law states that the fraction of light which is absorbed is independent of the intensity of the incident light. For a given wavelength of light, each successive layer of the medium absorbs an equal fraction of light passing through it, but the amount of light it receives is successively decreasing. Hence, the intensity of light passing through a medium will exponentially decrease with an increase in the thickness of the medium as shown in Figure IV-13.

Lambert's law is mathematically expressed as

$$I = I_0 e^{-\alpha d} \quad \text{(IV-11)}$$

I is the intensity of the transmitted light, I_0 is the intensity of the incident light, d is the thickness of the medium, and α is the absorption coefficient of the medium. The absorption coefficient is a constant for a homogeneous medium and is called the *extinction coefficient* when $\log_{10}$ is used.

Beer's law states that the absorption of light is directly proportional to the number of molecules in the absorbing substance through which the light passes. It extends Lambert's ideas to systems having a concentration factor. It is applicable only to monochromatic light and often is violated in strongly absorbing media. The mathematical expression of Beer's law, sometimes called the Beer–Lambert law, is

$$\log \frac{I}{I_0} = -Acd \qquad \text{(IV-12)}$$

A is the absorption or extinction coefficient of the medium, depending upon whether natural or common logarithms are used, and c is the concentration of the absorbing species. Thus Beer's law explains the effect of variations in colorant concentration on the transmittance levels. It states that increasing the concentration of the adsorbing species will cause a corresponding logarithmic decrease in the amount of light transmitted by the medium.

If two or more different light-absorbing agents that do not chemically interact with one another are present in the absorbing medium, then Beer's law can be extended as follows to treat such a system:

$$\log \frac{I}{I_0} = -(A_1c_1 + A_2c_2 + A_3c_3 + \cdots)d \qquad \text{(IV-13)}$$

Beer's law and Lambert's law are only qualitatively useful when they are applied to most art and historical materials. Most art materials are not normally able to satisfy the strict requirements demanded by the laws for

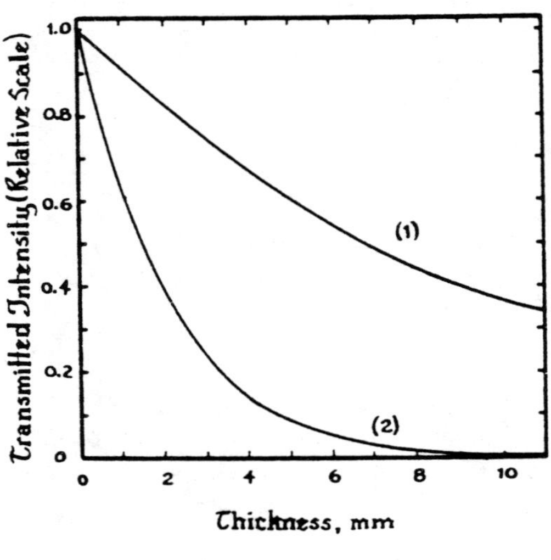

Figure IV-13. The intensity of light transmitted through an absorbing material as a function of thickness d in mm for (1) 10% absorbance and (2) 50% absorbance per mm.

quantitative use. Transparent glass and glazes are an exception, but as their transparency decreases, the Beer–Lambert concept becomes less applicable. The laws can be qualitatively informative. For example, Beer's law states that the amount of light transmission into an absorbing material decreases rapidly with the distance from the surface, particularly when the absorbance is high. Hence, photodegradation mostly occurs at or very close to the surface. The interior regions of a material cannot be directly affected by light unless the absorbance value is low.

BIBLIOGRAPHY

Baumeister, P., and Pincus, G., Optical interference coatings, *Sci. Am.* **223**(6), 58 (1970).

Billmeyer, F. W., and Saltzman, M., *Principles of Color Technology*, Interscience, New York (1966).

Brill, R. H., Ancient glass, *Sci. Am.* **209**(5), 120 (1963).

Evans, R. M., *An Introduction to Color*, Wiley, New York (1948).

Gabbert, G., and Lieb, W., On the use of the gastrofibrescope for reading inscriptions on the inside of Japanese sculptures, *Stud. Conserv.* **15**, 294 (1970).

Jenkins, F. A., and White, H. E., *Fundamentals of Optics*, McGraw-Hill, New York (1957).

Johnson, R. M., and Feller, R. O., Optics of Paint Films: Glazes and Chalking, in *Applications of Science in the Examination of Art Works*, Boston Museum of Fine Arts, Boston, Massachusetts (1968), p. 86.

Shurcliff, W. A., *Polarized Light*, Harvard University Press, Cambridge, Massachusetts (1962).

Shurcliff, W. A., and Ballard, S. S., *Polarized Light*, Van Nostrand, Princeton, New Jersey (1964).

Wright, W. D., *The Measurement of Color*, Macmillan, New York (1969).

V

Color and Color Mixing

A great many references describe in detail the very important area of science dealing with the definition, measurement, designation, and combination of colors. A brief introduction to this vast subject is what is intended in this chapter. The depth is only that needed to understand the topics covered later in the book that require some knowledge of color theory. Attention is first focused on the concept of complementary color relationships. The colors actually perceived by the eye are then analyzed. Finally, color mixing by additive and subtractive methods is discussed.

COLOR

Colored drawings have been found in caves dating from the last Ice Age 15,000 years ago. The Greeks were first to record a theory of color. Democritus in 450 B.C. stated, "In appearance there are sweetness, bitterness, heat and cold, and color. In truth, there are atoms and a vacuum." He distinguished four primary colors: white, black, red, and pale yellow-green. The other colors were supposed to be a mixture of these (e.g., blue coming from black plus yellow-green). Aristotle thought that color was potentially present in all bodies and was made actual by the effect of light. In the early fourteenth century the Dominican, Dietrich of Freiburg and the Arabic scientist, Kamal al-din al-Farisi both proposed that the colors of the rainbow could be explained by the reflection and refraction of light by water droplets. In the late seventeenth century Newton proposed that colors were in a sense already present in white light itself. Subsequent developments, as summarized in Chapter I, have led to our understanding of light as having both wave character and particle character. Today we know with certainty that color is the sensation

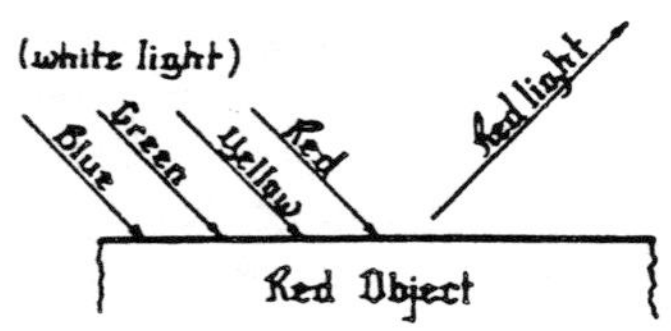

Figure V-1. The relationship of absorption and reflection of light by a red object. Yellow, green, and blue are absorbed. Red is reflected or scattered.

produced in the eye and brain by different wavelengths and intensities of visible light, or, equivalently, by different energies of photons to which the eye responds. The experienced eye of an artist or dyer is capable of distinguishing over ten thousand hues.

We wish to build on the principle that color can be produced by the absorption of selected wavelengths of light by an object, an idea coming from the nineteenth century. The visual effect that we call color originates from the impression created in the eye by the removal or decrease in intensity of certain visible wavelengths relative to others. Objects can be thought of as absorbing all colors except the colors of their appearance, as illustrated in Figure V-1.

An occasional misconception is that the color of an object results from the emission of light by the object. In the case of a light source or a fluorescent or phosphorescent material, color can be due to the selective generation and emission of light by the object. However, most materials are not light sources in and of themselves. They are merely *reflectors* of light. The colors these materials produce are a result of selective reflection of the light incident on their surface and not the emission of light. For example, a red object illuminated by white light absorbs most of the wavelengths except that corresponding to red, which it reflects. The same object illuminated only by red light would, of course, appear a strong red. If it were illuminated by moderate levels of light of any color except red, it would appear dull orange, brown, or black. Hence, in order to see a plethora of colors in their true balance, it is necessary to have illumination with all visible wavelengths adjusted to the same intensity. Individual colors are otherwise strengthened or weakened. In Table V-1 the relationship between absorbed and reflected or transmitted colors and the wavelength of the absorbed light is given. The borderline values of wavelength for the various colors are somewhat arbitrary and may vary from reference to reference by as much as 10–20 nm. Also shown mainly for use later is the energy in kilocalories corresponding to these particular wavelength ranges.

It is convenient to visualize the relationship between absorbed and transmitted colors shown in Table V-1 using the eight-color rosette shown in Figure V-2. We owe this concept primarily to Isaac Newton. The colors opposite one another in the rosette are said to be *complementary* because,

Table V-1. The Color Absorbed and Observed for Various Wavelengths of Light

Color observed	Complementary color absorbed	Wavelength adsorbed, nm	Photon energy, kcal
None (unless fluorescence)	Ultraviolet	<400	>71.5
Greenish-yellow	Violet	400–430	71.5–66
Yellow-orange	Blue	430–490	66–59
Red	Blue-green to green	490–510	59–56
Purple-red	Green	510–530	56–53
Violet	Greenish-yellow	530–560	53–51
Blue	Yellow	560–590	51–49
Greenish-blue	Orange	590–610	49–47
Blue-green to green	Red	610–700	47–42
None	Infrared	>700	<42

ideally, when added together in transmitted light they will give white. The dominant color of an object seen by the eye is approximately diametrically related in this rosette to the dominant color the object absorbs. For example, an object viewed as violet in color absorbs yellow-green light.

The color rosette provides a quick means of relating the color absorbed by an object to that reflected or transmitted. Yet it has several defects which need to be mentioned. First, it implies incorrectly that there are equal intervals among the colors shown. Rather, the wavelength regions spanned by the colors differ in size, depending upon the color. Second, the rosette implies that two complementary colors combined in equal quantities will always produce white. In practice, however, it is not possible to combine a dark violet and a bright yellow-green and have it come out

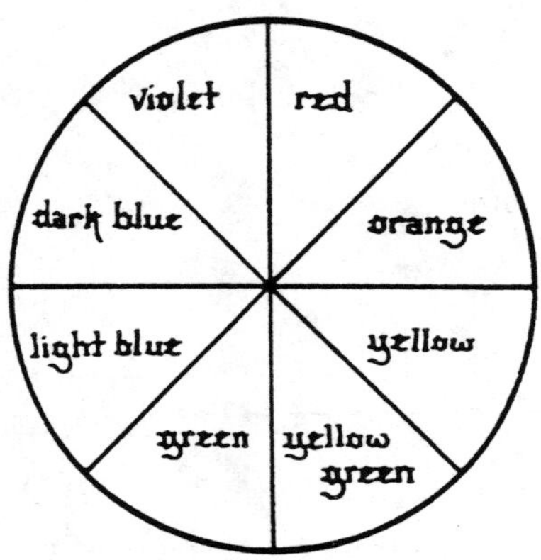

Figure V-2. The color rosette.

white. To achieve white, the brightness levels of two complementary colors must be adjusted to equal one another.

One can more accurately illustrate the relationship between the color available for viewing (i.e., that transmitted or reflected) and the absorption of color by the object by going back to the RAT law mentioned in Chapter IV. The sum of the reflected, absorbed, and transmitted intensities must equal the incident intensity. Pictorially, the RAT law for a transparent blue object in terms of its spectral composition is represented by Figure V-3. Absorption of all colors except blue light leaves only blue to be reflected or transmitted by this object. This is specifically what occurs and what could be qualitatively reasoned from the color rosette. However, the actual spectral absorption curve in Figure V-3 gives much more detailed information than is provided by the rosette. It shows that about 55% of the incident light is transmitted, 35% is absorbed, and 10% is reflected at the color-producing wavelength maximum of about 460 nm. At all wavelengths, at least 4% of the light is reflected, which gives an appearance of some glossiness to this object. The absorption takes place at all wavelengths except those corresponding to blue color. Reasoned from the color rosette, yellow and orange can be seen to be a good but not perfect description of the complementary absorbed wavelengths.

Basic colors can easily be analyzed by their *spectral reflectance* patterns. The reflectance spectrum is essentially the same as the transmittance spectrum, but it is the inverse of the absorbance spectrum. Shown in Figure V-4 are the reflectance spectra of four colors and two achromatics

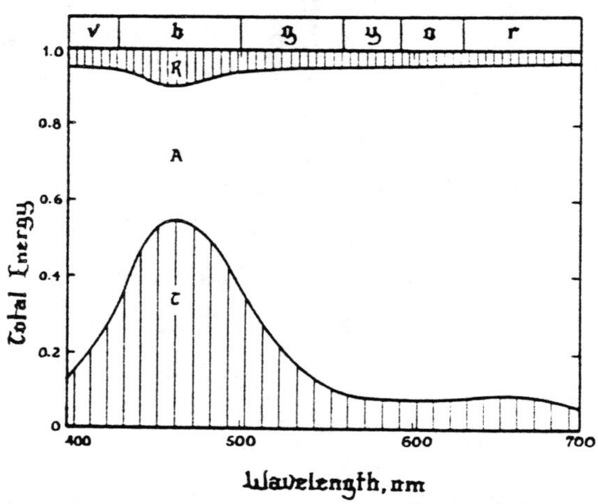

Figure V-3. Curves showing the reflectance (R), absorbance (A), and transmittance (T) of a blue material with high transparency.

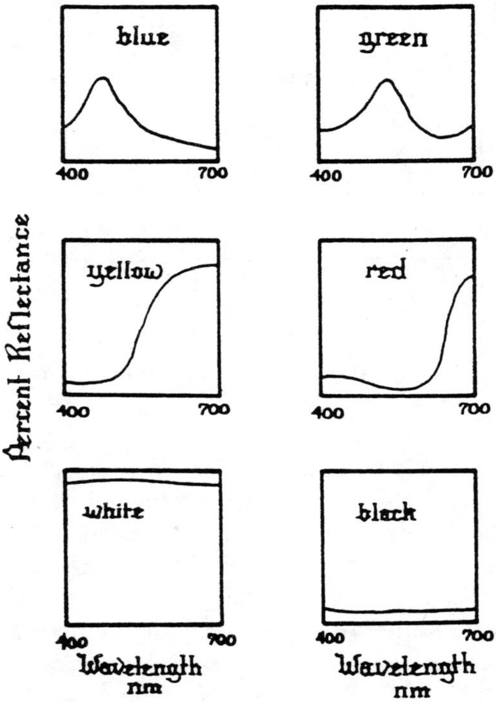

Figure V-4. Spectral reflectance patterns of four colors and white and black.

(white and black). The colors can be seen to be in accordance with the wavelength regions given in Table V-1. White is the result of equal reflectance of all colors at high efficiency. Black results from equal reflectance of all colors but at low efficiency. Gray is intermediate between these two. The relative amount of absorption of light roughly correlates inversely with the reflectance of light. A black surface is strongly absorbing and weakly reflecting, while a white surface is weakly absorbing and strongly reflecting. If the plots are simply inverted and the vertical scale is relabeled as absorbance, then they become *spectral adsorption* curves. For each color there is a small but definite amount of achromatic reflectance across the visible range. The color-producing reflectance is superimposed on this achromatic reflectance.

PERCEIVED COLOR

The colors actually perceived by the human eye are not always those indicated by the experimentally determined wavelength reflectance pattern of an object. In addition to the spectral reflectance by the object,

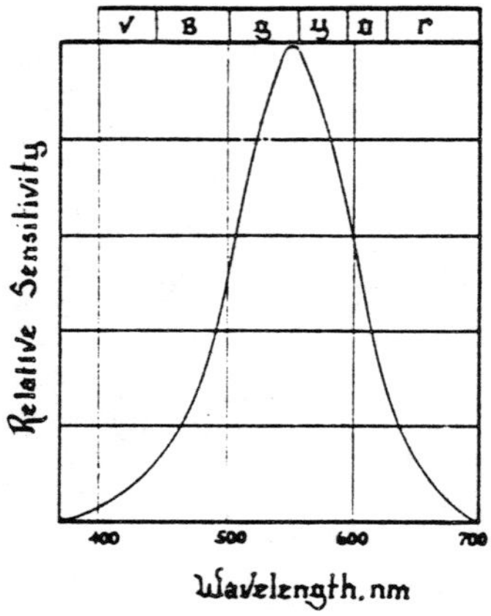

Figure V-5. Spectral sensitivity of the typical human eye in bright light.

the color perceived also depends upon the spectral sensitivity of the light detector and the spectral energy distribution of the illuminating source. The wavelength distributions of primary light sources used in illumination were discussed in Chapter II and reference should be made to these sources for their particular details. The detector of light could be a camera film, a phototube, or the eye. Of most interest here is the eye as a detector because understanding visual perception of appearance is the goal. The nature of human vision is a complex subject which involves advanced psychology and physiology. Fortunately, Figure V-5 gives sufficient information to handle the subject at the level needed for topics in this book. Figure V-5 shows the sensitivity of the eye in the visible region at each wavelength in bright illumination. This curve represents the eye of a "typical observer." The retina actually has red-, green-, and blue-sensitive cones, and when the sensitivity curve of each is overlapped with the others and the result summed, one obtains the curve shown.

The sensitivity of the human eye to different wavelengths of light is nonlinear over the visible range. The eye is most sensitive to the intermediate wavelengths in the visible region (the yellow-green region) and reaches a maximum sensitivity in bright light at 553 nm. It is least sensitive to the extreme wavelengths of the visible region, which represent violet and red. Hence much less light intensity is needed in the yellow-green energy range than in the red and violet ranges for the eye to perceive the same apparent brightness of these colors. It is interesting to note that the maximum sensitivity of the human eye to bright light is near the most prevalent wavelength of sunlight (Figure II-3) received at the surface of the earth. This fact has led to interesting deductions about the course of evolution of vision.

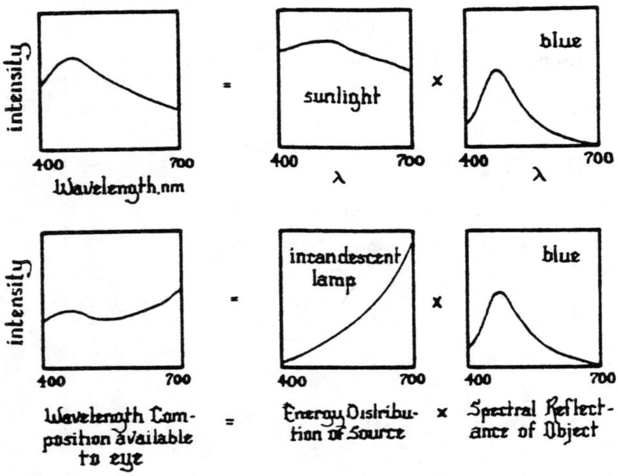

Figure V-6. The difference in the wavelength composition reflected from a blue object as a result of changing the illuminating sources. (The sensitivity of the eye is not included.)

The color actually perceived by the eye can be expressed as the numerical product of the spectral reflectance plot of the object (Figure V-4), the spectral sensitivity of the eye (Figure V-5), and the spectral energy distribution of the illuminating source (Figure II-4, II-6, II-7, and II-9 to II-13). One can now analyze why a color may appear to be different when the source of illumination is changed. If we assume the spectral response of the eye is constant in all cases, some graph displays of the differences in the perceived colors for an object can be made as shown in Figure V-6. It is difficult to predict from the spectral energy distribution in the figure the actual color which will be perceived. But clearly the blue perceived in each case is somewhat different because the incandescent light and sunlight used for illumination emit a different wavelength composition. Table V-2 more broadly summarizes the distortions of the natural color of an object due to different colored light sources.

The light emitted by an illuminating source can affect the color of the object or picture being viewed. One of the most vivid examples of this fact is the "alexandrite effect" (Chapter X). The mineral alexandrite changes color from red in incandescent lighting to green in daylight. Paintings show similar effects, but in a less dramatic way. Paintings can be divided into several classes according to the colors the artist used. The colors from bluish to yellowish could be classified according to the black-body standard as cold, neutral, warm, or hot. The warmer the picture is, the lower its color temperature would be. The silvery-toned landscape paintings that appeared late in Corot's career, particularly those having

Table V-2. The Effect of Colored Light on the Appearance of Colored Objects

Natural color of object	Color of light illuminating the object					
	Red	Orange	Yellow	Green	Blue	Violet
Black	Red-black	Orange-black	Yellow-black	Green-black	Blue-black	Violet-black
White	Red	Orange	Yellow	Green	Blue	Violet
Gray	Red shade	Orange shade	Yellow shade	Green shade	Blue shade	Violet shade
Red	Red	Scarlet	Orange	Brown	Purplish black	Reddish black
Orange	Red	Orange	Yellow-orange	Greenish yellow	Black	Black
Yellow	Orange-red	Yellow-orange	Yellow	Yellowish green	Greenish black	Black
Light green	Red shade	Yellow-green	Greenish yellow	Green	Blue-green	Bluish shade
Deep green	Black	Greenish black	Yellowish green	Green	Greenish blue	Blue-black
Light blue	Violet	Dark gray	Yellowish shade	Blue-green	Blue	Violet
Deep blue	Purple	Blue-gray	Gray	Blue-green	Blue	Blue-violet
Violet	Reddish black	Red-purple	Gray	Blue	Violet-blue	Violet
Purple	Red-shade	Red shade	Red shade	Black	Blue	Violet
Rose	Red tint	Red tint	Red tint	Greenish black	Blue shade	Violet shade

bluish hues, could be displayed under cold light such as that which certain fluorescent lighting provides. The warm colors characteristic of the landscape painting by the Barbizon School would be strengthened if displayed in the warm lighting provided by the incandescent lamp. However, the subtle blues of the Impressionists would appear inky blue by incandescent illumination. An exhibitor might experiment with matching the color temperature of the light source to that of the object. However, attention should always be paid first to the potential for photoinduced damage that might occur with certain light sources. The ultraviolet energies of unfiltered fluorescent bulbs or daylight can rapidly destroy the color of some pigments and dyes, in addition to damaging paper and cloth. With sensitive materials it may be necessary to sacrifice an elegant method of illumination for the safest method of display.

ADDITIVE PRIMARY COLORS

Many of the formalisms in use today to describe color had their origin in the nineteenth century. The impetus for work in this area came primarily from the desire to obtain color photographs (Chapter XII) and the attempts to understand color vision. The terminology of "additive" and "subtractive" sets of colors was thus introduced. Thomas Young's early nineteenth-century theory of color vision referred to the three primary color sensations, and this formalism is still in use. If the visible region is divided into three equal parts, 400–500 nm, 500–600 nm, and 600–700 nm, by absorbing the remaining wavelengths, then these three regions define the additive primary colors of blue, green, and red as shown in Figure V-7. The term "primary" comes from the fact that these are the colors to which the cones of the eye respond. More colors can be matched additively with blue, green, and red than with any other three colors.

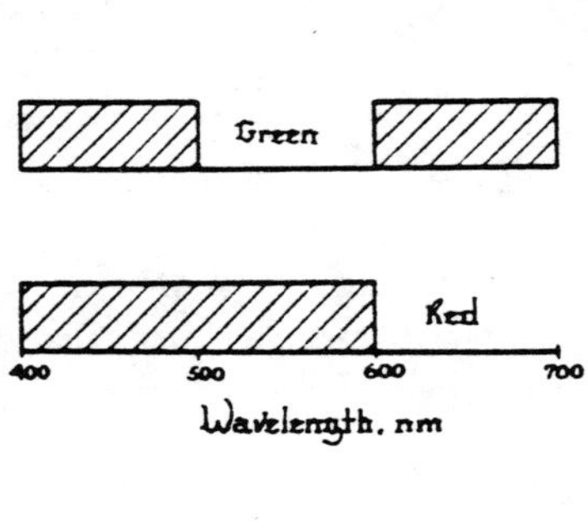

Figure V-7. Energy ranges defining the additive primary colors. Light in the cross-hatched regions is absorbed, while that in the blank region is passed.

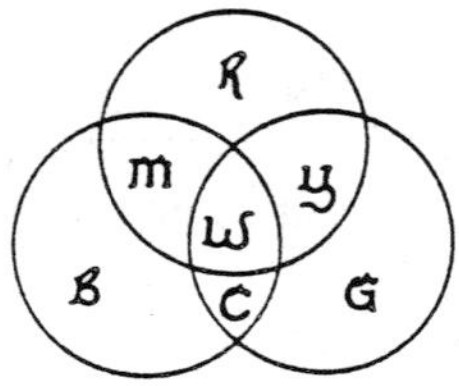

Figure V-8. Lights produce additive color mixtures. Red, green, and blue-purple beams superimposed on a white screen produce cyan, yellow, and magenta. Overlapping correct amounts of all three produces white.

If three projectors with red, green, and blue filters on them are arranged such that the light of these colors is projected with equal levels of intensity on a white screen, then the process of mixing additive colors can be demonstrated (Figure V-8). Where red and green overlap, wavelengths in the 500–700 nm range can pass, but those at 400–500 nm are blocked. Hence a yellow color results. Where red and blue overlap, the 400–500 and 600–700 nm ranges are transmitted, but the 500–600 nm range is blocked. This leads to a magenta color, and so forth. The resultant color is the sum of the two constituents used to produce it, and hence the term "additive" is used for the constituents.

If the intensity of the light in each projector is varied, virtually all other shades of color can be made to appear in the overlapping region. At first sight this would appear to be an excellent method of generating colors. However, the method suffers from several important disadvantages. One is that it can only be used in transparent media because it is based on the principle of transmitted light. The superimposition of colored lights, the mixing of dyes in a clear solution, or the mixing of colorants in transparent glass and glaze can employ additive color methods. Another disadvantage is that, because light filters are used, the total intensity of light is diminished.

It might be noted that additive color mixing leading to a so-called "optical mixture" is achieved when light of different colors is reflected from two or more closely spaced, different colored surfaces. This technique, as employed by Impressionist painters, is known as *pointillism*. Color television and the mixing of colored lights also are based on the additive primary mixing process. Another means of color mixing is usually employed in opaque paints and films. This other method is known as the subtractive process.

SUBTRACTIVE COLORS

In relation to photography, du Hauron in 1862 described the virtues of the subtractive color principle, which overcomes most of the disadvantages of the additive principle. Just as additive color mixing refers to the process of adding colored lights, the subtractive process refers to the removal of certain wavelengths of light as a result of absorption or scattering. In the transmitted light shown in Figure V-8, red, green, and blue are removed. However, in the regions where the colors overlap, three new colors appear. Yellow is produced where red and green overlap. Purple is produced where red and blue overlap, and blue-green appears where blue and green overlap. These three new colors are known as the subtractive colors. They have come to be known also by the names of dyes of the same color. Blue-green is often called cyan, purple is referred to as magenta, and yellow remains yellow. The wavelength regions which define the subtractive primaries are related to the additive primaries but involve transmission of two-thirds of the spectrum while the remaining one-third is absorbed, as shown in Figure V-9.

Overlapping any of the subtractive primaries (Figure V-10) gives the additive primaries and black in reflected light. It can be seen that cyan and yellow give green, cyan and magenta give blue, and yellow and magenta give red. Cyan, yellow, and magenta mixed together equally give black. Consequently, the subtractive process can be used in mixing of colors in opaque painting media and photographic film. Mixing of different amounts of each subtractive primary produces different colors so that about 95% of all colors can be obtained.

It is interesting to note the relationship of the additive and subtractive colors to one another on the color rosette in Figure V-2. The color rosette could be viewed as a triangle instead of a circle. The three additive colors are at the corners and the three subtractive colors are on the edges, as shown in Figure V-11. The colors opposite one another are complementary. The sum of any two additive primaries at the corners in transmitted light gives the subtractive color located between them. The sum of any two subtractive colors on the edges in reflected light gives the additive primary

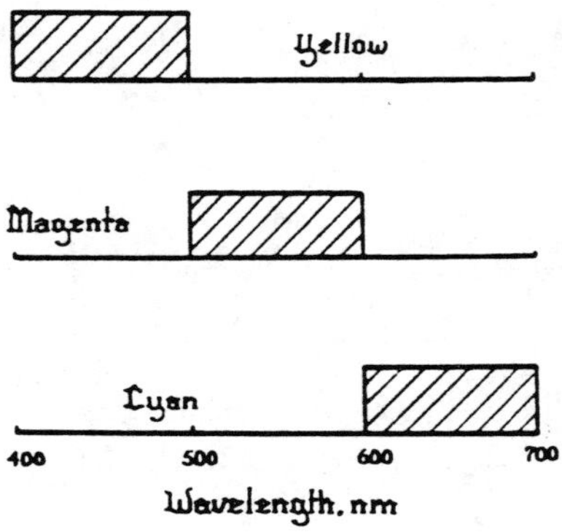

Figure V-9. The subtractive colors and their generation by removal of the additive primaries. Light in the cross-hatched regions is absorbed.

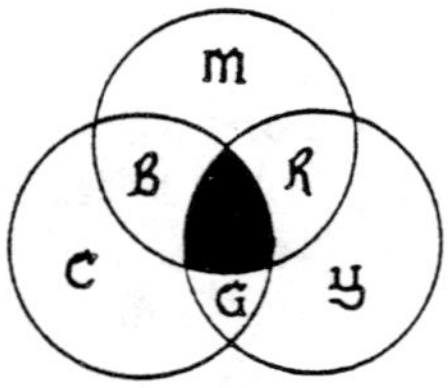

Figure V-10. Paints produce subtractive color mixtures. Superimposing cyan, yellow, and magenta on a white field produces red, green, and blue. Overlapping correct amounts of all three produces black.

lying between them. Since reflected light is more common in application to art materials, the subtractive primaries are more broadly useful.

An important deduction from the color rosette as it relates to the mixing of colors is the fact that there is a considerable difference of wavelength among the colors from green to blue which is not accompanied by a difference in visual color sensation corresponding to that occurring in their complementary colors, yellow to violet (Table V-1). For this reason, two bluish-green shades which are hardly distinguishable by the eye call for a difference between their complementary colors which seems disproportionately large. Some practical problems encountered in the mixing of actual pigments are taken up in Chapter XI.

Consideration of the simple mixture of primary colors used in producing the net color of an object is not the final story. How does one speak of gray, brown, and olive green, for example, in this framework? In order to make the discussion of color more complete, a series of additional subjective definitions have been introduced over the years. These are generally applicable to all methods of color matching, but they are often brought into play in different ways. The *hue* is the actual wavelength classification of color (red, yellow, green, blue, etc.). The *brightness* or *value* is the lightness or darkness relative to a series of grays ranging from white to black. Brightness can be thought of as a hue ranging from dim to dazzling. The *saturation* is a measure or the difference between the color (other than white and black) and a gray having the same brightness.

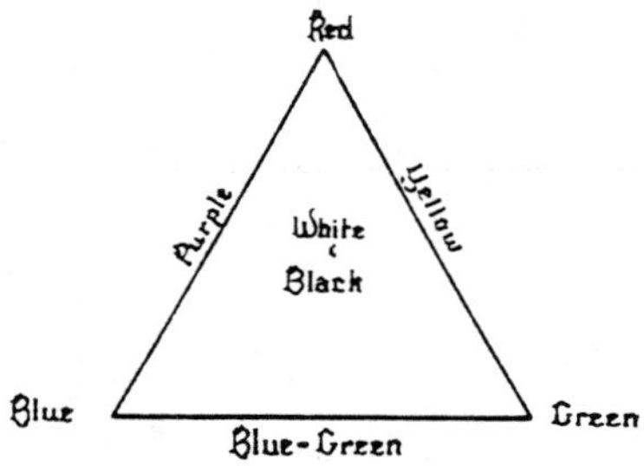

Figure V-11. The color triangle showing the relationship of the additive and subtractive colors.

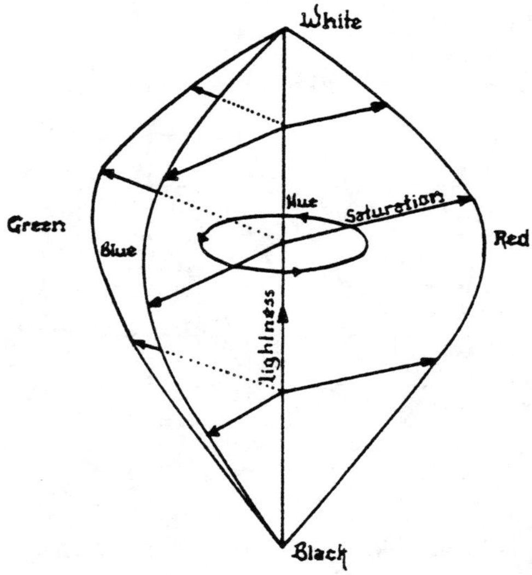

Figure V-12. The three-dimensional basis of the Munsell color system.

The description of color in terms of hue, brightness, and saturation is more complete than that in terms of hue alone. The reflected light from a film of brown paint and that from a film of yellow paint may have exactly the same wavelength composition, but they look very different to the eye. The difference between yellow and brown goes beyond hue alone. It results from a difference between lightness and/or saturation. The reflected light from brown has a lower degree of lightness than does yellow. Similarly, pink and red can have the same hue but differ in saturation, the pink being desaturated red. The wavelength composition of reflected light from the achromatics (white, gray, and black) have the same wavelength composition but they differ in their degree of lightness. The lightness progressively increases from black through a series of grays to white.

Color appearances should be specified in terms of hue, brightness, and saturation. Two films with the same hue, brightness, and saturation will appear identical. The surfaces may appear different, however, as a result of texture and composition differences. These latter factors will be taken up in Chapter VI, where materials are discussed.

A method of color description in common use is the Munsell color system. The Munsell system involves human judgment in matching colors and in this context it has been labeled as subjective. It is accurate enough for mixing of paints on an artist's palette and has a high degree of objectivity. The system is based on the three-dimensional "tree" or double cone shown in Figure V-12. The hue is represented as a circle around the

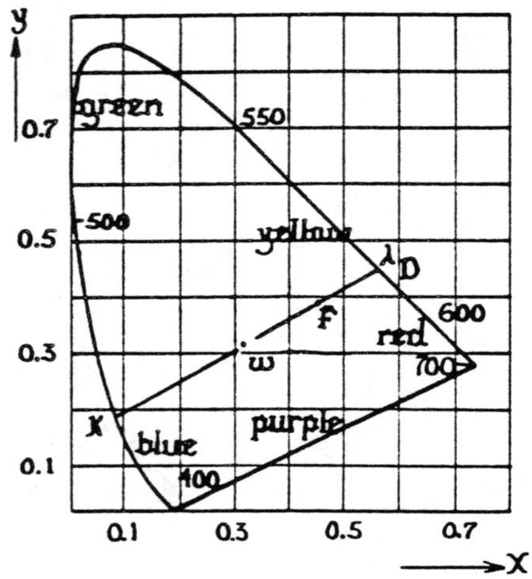

Figure V-13. CIE chromaticity diagram. W, neutral point (white); F, color point of the dominant wavelength λ_D; K, complementary color.

axis much like the color rosette shown in Figure V-2. The brightness or lightness is established by positions perpendicular to the planes of hue. The saturation of a color is determined by the distance from the center of the hue circle along lines of radius.

The most objective method for specifying colors is the trichroma or tristimulus system developed by the Commission International de l'Eclairage (CIE) in 1931. The CIE system serves as the international standard for color theory and measurement. It employs photometric instruments. The subjective terms of hue, brightness, and saturation are redefined in terms of objective ones. In the CIE system *wavelength* is the objective term that replaces hue, *luminance factor* replaces brightness, and *purity* replaces saturation. The system is based on additive color mixing where colors are matched by mixing colored lights. The standard response curves for the normal eye in the red, green, and blue are multiplied by the reflectance of a particular color at each of these colors and the illuminating source intensity at each of the colors. The result is a set of three tristimulus values X (red), Y (green), and Z (blue). A chromaticity diagram can then be made with the chromaticity coordinates x, y, and z defined respectively as the fractional amounts of X, Y, and Z in the color of interest, using the equations

$$\frac{X}{X+Y+Z} = x, \qquad \frac{Y}{X+Y+Z} = y, \qquad \frac{Z}{X+Y+Z} = z \tag{V-1}$$

The sum of x, y, and z equals unity, so x and y alone are sufficient to describe the color. When the values of the chromaticity coordinates x and y

are plotted, the diagram shown in Figure V-13 results. The colors lie on an almost parabolic curve, the ends of which are connected by a straight line representing purple. The achromatics—white, gray and black—lie on a line perpendicular to the plot passing through the center *W*. This line is the luminosity coordinate. A color can be defined in terms of its chromaticity coordinates x and y as a *chromaticity point*. Alternatively, when two chromaticity coordinate values x and y are plotted, their intersection point corresponds to the chromaticity of the given color, say *F*. The dominant wavelength λ_D of that color is obtained from the intersection point on the parabolic curve by a line drawn from the center of the diagram *W* through the chromaticity point *F*. The complementary color *K* and its dominant wavelength are obtained by extrapolation of that line to the opposite side of the spectral curve. The purity is found from the shortest distance between *F* and the spectral curve. The nearer the chromaticity point lies to the curve, the greater the purity value is.

The CIE system is the most widely adopted method of matching colors, but it works best for homogeneous colors such as those produced by colored lights. It has little direct application to heterogeneous systems, such as pigment particles dispersed in a binder or vehicle, for reasons which will be discussed in Chapters VI and XI.

BIBLIOGRAPHY

Billmeyer, F. W., and Saltzman, M., *Principles of Color Technology*, Interscience, New York (1966).

Bouma, P. J., *Physical Aspects of Color*, Macmillan, New York (1971).

Clulow, F. W., *Color: Its Principles and Application*, Fountain Press, London (1972).

Crollay, E. K., Berim, R. E., and Gersonskaja, V. E., Investigation and Classification of Works of Painting on Color and Color Rendering, in *ICOM 4th Triennial Meeting*, Venice, 1975 (75/19/5).

Evans, R. M., *An Introduction to Color*, Wiley, New York (1948).

Judd, D. B., and Wyszecki, S. G., *Color in Business, Science and Industry*, Wiley, New York (1975).

LeGrand, Y., *Light, Color and Vision*, Wiley, New York (1957).

Munsell Book of Color, Munsell Color Co., Baltimore, Maryland (1966).

Murray, H. D., *Color in Theory and Practice*, Chapman and Hall, London (1952).

Ruhemann, H., Experiences with artificial lighting and paintings, *Stud. Conserv.* 6(2/3), 83 (1961).

Wright, W. D., *The Measurement of Color*, Macmillan, New York (1969).

Wyszecki, S. G., and Stiles, W. S., *Color Science*, Wiley, New York (1967).

VI

The Optical Appearance of Materials

Only within the last 150 years or so has the world of science, not to mention that of artists and craftsmen, been provided with a sound physical basis for the understanding of light generation, light measurement, the origin of color, optical effects, the existence of ultraviolet and infrared energies, and even the meaning of light itself. Electromagnetic energy and the phenomena it produces have, of course, been present since the origin of the universe and have imposed themselves on every aspect of nature. Lack of knowledge about the nature of light did not deter scholars, artists, and craftsmen from making marvelous use of light. But now in the present state of technology and science, one can often systematically analyze in a fundamental manner how light causes materials to appear and behave as they do. In the past the need for a "wait-and-see" period plagued the creative artist and craftsman whenever a new method was tried. For example, one of Leonardo da Vinci's experiments in his "Last Supper" developed major defects during his own lifetime. His creations were his research and were subject to the same technical complications experienced by any researcher. Fundamental principles of science and a store of technical information on many materials can be put to use by the artist of the present day. One can sometimes *predict* what is likely to happen when new materials are combined or aged in a valued work.

Practically all of the physical principles covered in this chapter were introduced in Chapters I-V. Since in most cases it is the *interplay* of these various principles which determines the fate of light illuminating an object and eventually produces the object's appearance, this chapter rests on the base provided by the earlier chapters. It deals more with specific materials than new principles. The practical meaning and uses of the index of refraction of a material are discussed in more detail than was done earlier. The important role of particle size in the appearance of composite materials containing particulate matter is brought into consideration. The concepts of reflection, refraction, and transmission of light are revisited, this time with an eye toward understanding why materials yield a wide variety of appearances. Paints, glasses, ceramics, metals, fibers, and various transparently coated materials are examined within this physical framework. The molecular origin of color and the changes that light induces in organic and inorganic materials are covered in greater detail in Chapters VII-XII. An evaluation of the aesthetic quality of an object is, of course, subjective and is not a subject of this book.

INDEX OF REFRACTION

A singularly useful parameter for discussing how a material affects light is the refractive index η of the material. In Chapter IV, η was defined as the velocity of light in air divided by the velocity of light in the medium of interest. The index of refraction for a material was shown to play a role in reflection, refraction, transmission, and polarization of light by the material. We now wish to draw upon these physical phenomena and apply them more specifically to understanding the appearance of art and historical materials.

Practical Use of the Index of Refraction

The refractive index of a given material is a constant, but occasionally values differ somewhat, depending upon the source of the data and differences in the method of the measurement. Differences in measurement usually originate in three ways. (1) η depends upon the wavelength of light used in the measurement (dispersion). Figure IV-2 showed this for a series of transparent glasses. Obviously η values measured with a light wavelength of 450 nm will differ from those measured at 600 nm. Most values are measured using the sodium *D* line at 589.3 nm. (Because of strong absorption at that wavelength by some compounds, the lithium line at 670.8 nm is occasionally used.) (2) Refractive indices vary slightly in value with temperature. Most measurements are made at 25°C, which approximates room temperature, but measurements at 15° and 20°C are also quite common. (3) Particle size, particle composition, and degree of opaqueness can affect the measurement of η in solids. Strictly speaking, a completely opaque material has no measurable optical index of refraction because light does not penetrate the surface. Metals are the most opaque

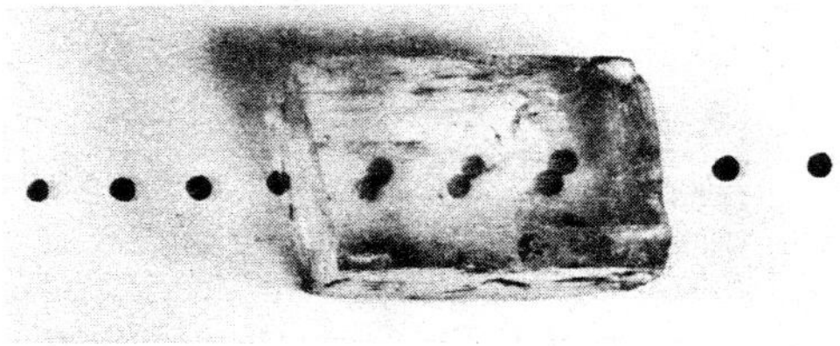

Figure VI-1. Double refraction produced by a crystal of calcite $CaCO_3$.

materials toward visible light, and while no metal is completely opaque, their optical indices of refraction are not measurable. Values can be calculated from electrical measurements, but they are not generally useful for optical purposes.

Another possible reason for differences in reported refractive index values has to do with a physical property of many crystals and not with difficulty in the measurement. Many crystals have more than one refractive index. The effect, known as *anomalous refraction*, was recorded in the Middle Ages when sailors visiting the eastern fjords of Iceland brought Iceland spar ($CaCO_3$) back to Europe. Iceland spar has the property of making anything seen through it appear double. Figure VI-1 shows this phenomenon. In order to understand how this occurs in crystalline materials, crystal optics needs to be considered briefly.

In most crystalline substances the velocity of a portion of the light transmitted through them is found to differ for different orientations of the crystal, while the velocity of the remainder of the light is the same in all orientations. Since the refractive index of the crystal depends upon the velocity of light traveling in the crystal, it follows that the index of refraction will be different in different directions. Hence most crystalline materials have either two or three characteristic η values. Those crystals that have two indices are called *uniaxial*. The tetragonal, trigonal, and hexagonal crystal systems are uniaxial. The two indices of refraction for these crystals are conventionally labeled ε and ω. The majority of crystals in nature have three indices of refraction and are called *biaxial* crystals. They belong to the orthorhombic, monoclinic, and triclinic crystal systems. The three η values characteristic of biaxial crystals are labeled α, β, and γ. Cubic

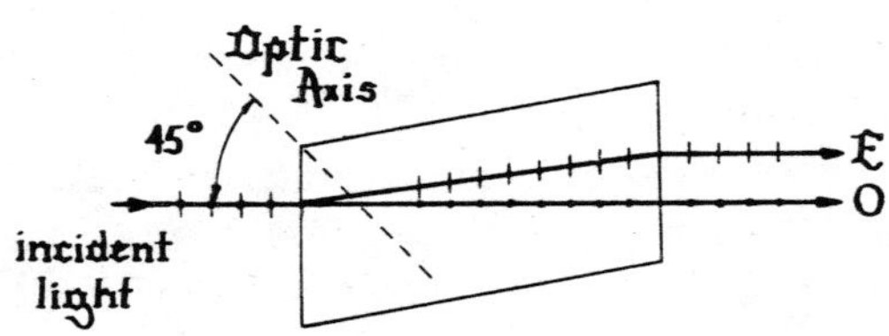

Figure VI-2. Generation of ordinary (O) and extraordinary (E) rays of light by a uniaxial calcite crystal. A double image results.

crystals are the only ones which have a single η value for all orientations of the crystal. They are called *isotropic* crystals.

To describe more specifically how the multiple indices of refraction originate, we can consider a uniaxial crystal of Iceland spar (calcite). Whenever a beam of light is incident at a right angle to the face of a uniaxial crystal, it is split into two beams, one of which has the same direction as the incident beam and passes straight through the crystal, while the other has a velocity that varies with the particular direction of vibration of the beam. This second beam is deflected upon entering and emerges parallel to the direction of the original beam. The paths of the light rays are shown in Figure VI-2. The invariant ray is called the *ordinary ray*, while the one that changes direction is the *extraordinary ray*. These two rays are polarized in planes perpendicular to one another, as indicated by the lines and arrows in the figure. The ordinary ray follows the ordinary laws of refraction (Chapter IV), while the extraordinary ray does not. Since the velocity of light for the two rays differs, the refractive indices measured for the two rays must also differ. For the ordinary ray, the ordinary index ω or η_O is defined. It is sometimes also referred to as the *major principal index of refraction*. The index of refraction for the extraordinary ray ε or η_E varies between the value of ω and some lower or higher value. The extreme value of ε relative to ω is called the *minor principal refractive index*. The maximum difference between the values of ω and ε is called the *birefringence* of the crystal. If $\varepsilon > \omega$, then the crystal is *positive* uniaxial, and if $\varepsilon < \omega$, then it is *negative* uniaxial.

The origin of the term "uniaxial" comes from the fact that there is only one direction (the optic axis) for which ordinary and extraordinary rays do not appear. Biaxial crystals have two directions for which ordinary and extraordinary rays do not appear. This is equivalent to saying that light vibrating in any plane will have the same velocity in two of the three mutually perpendicular directions in the crystal. Biaxial crystals have two axes which correspond approximately to the optic axis in uniaxial crystals. Massicot is an example of a weakly birefringent pigment, while vermilion is more strongly birefringent. Both are uniaxial. Chrysocolla is a weakly birefringent, biaxial pigment. Table VI-1 contains values of the refractive index for selected natural pigments, adhesives, oils, resins, and other materials found in works of art.

In using tables of refractive indices for crystals, confusion might result if some consistency is not followed. For example, comparing the ε index of one crystal to the ω value of another would have little practical meaning. However, most pigment materials are polycrystalline, which means that they are used in a powdered form. Since a powder contains a random orientation of crystallites, it is reasonable to use the weighted mean

Table VI-1. Index of Refraction η for Some Materials of Interest in Art[a]

Material	η	Material	η
Titanium dioxide (rutile)	2.61, 2.90	Sienna, raw	1.87–2.17
Titanium dioxide (anastase)	2.56, 2.49	Indian yellow	1.67
White lead (lead carbonate)	2.09, 1.94	Vermilion	2.854, 3.201
Zinc white (zinc oxide)	2.00, 2.02	Realgar	2.54, 2.68, 2.70
Chalk	1.66, 1.44	Red lead	2.40, 2.44 (Li)
Gypsum	1.21, 1.52, 1.53	Cadmium red	2.64–2.77
Aluminum hydrate	1.50–1.56	Umber, raw	1.87–2.17
Cerulean blue	1.84	Van Dyke brown (Cassel brown)	1.62–1.69
Azurite	1.730, 1.758, 1.838	Water	1.33
Cobalt blue	1.74	Gum arabic solution, 10%	1.334
Prussian blue	1.56	Gum arabic solid	1.476
Ultramarine, natural (lazurite)	1.50	Egg tempera	1.346
Chrysocolla	1.575, 1.597, 1.598	Glue solution, 10%	1.348
Smalts	1.45–1.52	Spirits of turpentine	1.47
Green earth	2.5–2.7	Poppyseed oil	1.469
Chromium oxide green	2.5	Walnut oil	1.477
Cobalt green	1.94–2.0	Linseed oil	1.478
Malachite	1.65, 1.87, 1.91	Tung oil	1.517
Orpiment	2.4, 2.81, 3.02 (Li)	Venice turpentine	1.53
Cadmium yellow	2.35–2.48	Colophony	1.525
Massicot (litharge)	2.67, 2.54 (Li)	Copal (Congo)	1.545
Naples yellow	2.01–2.28	Dammar (Djakarta)	1.515
Ochre, yellow (goethite)	2.27, 2.40, 2.45	Mastic	1.536
		Shellac	1.516
		Beeswax (molten, 74°C)	1.442
		Diamond	2.418

[a] Uniaxial crystals are given in the order ω, ε. Biaxial crystals are given as α, β, γ. From Gettens and Stout, updated from the *Handbook of Chemistry and Physics*, CRC Publishing Co., Cleveland (1976).

refractive index when comparing pigments to one another. The mean index can be calculated by the following equations for uniaxial and biaxial crystals, respectively:

$$\eta_{\text{uni}}^{\text{mean}} = \frac{2\omega + \varepsilon}{3}; \qquad \eta_{\text{bi}}^{\text{mean}} = \frac{\alpha + \beta + \gamma}{3} \tag{VI-1}$$

For example, titanium dioxide in the rutile form is uniaxial and has as refractive indices $\omega = 2.61$ and $\varepsilon = 2.90$. Hence the mean index for practical usage is $[2(2.61) + 2.90]/3$ or 2.71. The index of refraction of a liquid is easily measured and used because there can be only a single value of η characteristic of the randomly oriented molecules themselves.

It is useful to note that predictions of values or, in many cases, relative trends in the indices of refraction given in Table VI-1 cannot be made from the formula for the pigment alone. It is helpful to also know something about the structure. A few rough generalities can be made about η values, however. Organic materials have indices of refraction in the 1.3–1.6 range, which is typical of compounds composed of the lighter elements —carbon, hydrogen, oxygen, nitrogen, etc. Organic crystals (none shown) containing molecules that are structurally very different in one dimension than in another (as in a large, flat molecule) have high birefringence. A symmetric ion or molecule does not produce strong birefringence. For inorganic materials, both the structure and the atomic numbers of the components contribute to the refractive index values. Lattices containing large, highly negative anions, particularly sulfides, tend to give relatively high indices of refraction because the ions are easily distorted by electromagnetic waves. Vermilion, HgS, has the highest refractive index of the substances in Table VI-1. The mineral galena, PbS, has an even higher value, $\eta = 3.912$. Large, relatively polarizable Pb^{2+} ions are added to a glass matrix to increase the overall index of refraction in the manufacture of crystalware. When cut, the crystalware exhibits rainbow-like colors because of its high dispersion. The large index of refraction also produces a great deal of reflection from the cut faces.

Lattices composed of a regular, tightly packed array of atoms or ions have high density and thus relatively high refractive indices regardless of the atomic number of the elements. Titanium dioxide and diamond are examples of this. The high index of diamond causes strong dispersion and reflection of light and, when coupled with its transparency, gives brilliance and color to its facets. In general it can be said that the ability of a substance to deflect electromagnetic radiation is a function of the density of that substance. The more space per given volume that is filled by its atoms, the more it affects the light and in turn the higher the refractive index becomes.

Hiding Power

An important concept for understanding the appearance of objects comes from the fact that the amount of light scattered or reflected from the interface of a composite of two materials depends strongly upon the difference between the indices of refraction of the two media. Air–surface and pigment–vehicle interfaces are typical light scatterers in art objects. If the difference in the index of refraction is large at an air–surface interface, then the surface will be a good scatterer of light and its transparency will be reduced. If there is no difference, then no light will be scattered at the interface and all the light will pass into the material where it can be absorbed by or transmitted through the object. A popular trick in science magic shows is to make a glass object "disappear" when it is immersed in a clear, colorless "secret potion." The secret is only that the index of refraction of the potion is matched to that of the glass so that no light is deflected at the glass–solution interface. Hence the glass cannot be seen. Practical use of the same principle can be made to distinguish different types of glass from one another. A set of solutions of organic solvents is made which have the same refractive indices as various types of glass. The type of glass is quickly identified by finding the solution in which the glass cannot be seen when immersed.

Consider the pigment–vehicle interface in a paint. If the difference in the refractive indices $\Delta\eta$ of the pigment and vehicle is large, then the pigment will be a good scatterer of light and its opacity or hiding power will be large. *Hiding power* is the measure of a paint's ability to opaquely cover a surface. The light-absorbing power of the pigment particles will also affect the opacity of the paint. $\Delta\eta$ is defined by

$$\Delta\eta = \eta_{\text{larger}} - \eta_{\text{smaller}} \qquad \text{(VI-2)}$$

Note that the use of $\Delta\eta$ in this way is just a superficial application of equation (IV-8). A large value of $\Delta\eta$ leads to an opaque object or a paint with good hiding power, while a small $\Delta\eta$ gives a transparent object or a paint with poor hiding power. One must also consider the amount of pigment in the vehicle. Some encaustic paint films consist primarily of wax and a small amount of pigment. Even though the $\Delta\eta$ value at the wax–pigment interface may be large, the medium can retain a high degree of transparency due to the small number of light-deflecting pigment particles. It is assumed in this section that there is a sufficient number of pigment particles to encounter most of the light rays entering the paint.

In order to have a useful white paint a large $\Delta\eta$ between the pigment

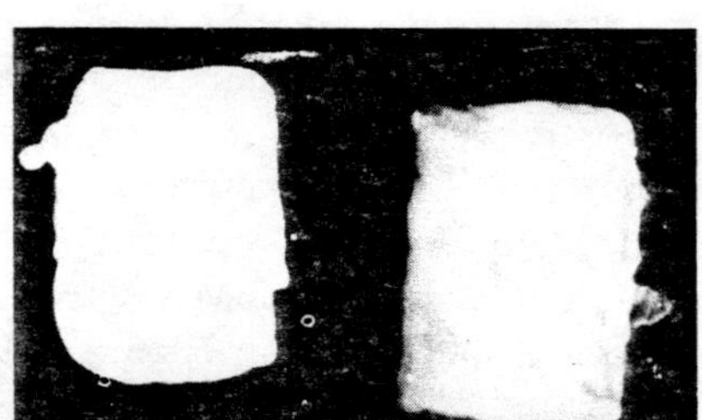

Figure VI-3. Chalk in glue (left) and linseed oil (right). The difference in transparency against a black background can be traced to a difference in $\Delta\eta$.

and vehicle (causing good hiding power) and a small absorption factor uniformly distributed across the visible range (creating white) are needed. For example, chalk ($CaCO_3$) in a glue solution can be brushed onto wood or canvas and has reasonably good hiding power, but chalk in linseed oil gives a grayish-white appearance that is rather transparent. Figure VI-3 shows the comparative transparency of $CaCO_3$ in glue and linseed oil against a dark background. The reason for this difference can be understood by determining $\Delta\eta$ in each case as defined by equation (VI-2):

$$\begin{aligned}\Delta\eta &= \eta_{\text{chalk}} - \eta_{\text{glue}} \\ &= 1.60 - 1.35 = 0.25\end{aligned}$$

$$\begin{aligned}\Delta\eta &= \eta_{\text{chalk}} - \eta_{\text{linseed oil}} \\ &= 1.60 - 1.48 = 0.12\end{aligned}$$

More light penetrates the chalk–linseed oil mixture and eventually reaches the ground than occurs with the chalk–glue mixture. The ground shows through in the chalk–oil mixture and gives an unsatisfactory white paint. $\Delta\eta$ is larger in the chalk–glue mixture and so its covering power is greater. The "Tingelstad I" altar frontal dated 1260–1300 A.D. from Norway has a ground composed of chalk in glue, the $\Delta\eta$ for which renders it moderately opaque. A semitransparent, bluish painted glaze on the frontal also contains some chalk. The binder for the chalk is an egg–oil emulsion and the $\Delta\eta$ value is sufficiently small to render the glaze semitransparent. Another example of this effect involves "whitewash" commonly used on exterior walls. The whitewash material is calcium carbonate, $CaCO_3$, mixed with water and a little glue. [Whitewash used on henhouses for antiseptic purposes is usually hydrated lime, $Ca(OH)_2$, which eventually turns into $CaCO_3$ by reaction with CO_2 from the atmosphere.] When first painted onto a surface, whitewash has an off-white cast and is not entirely opaque. However, as it dries, its opacity increases and so does its whiteness. The $\Delta\eta$ relationships explain these changes. When wet, $\Delta\eta = \eta_{\text{chalk}} - \eta_{H_2O} = 1.60 - 1.33 = 0.27$. When dry, $\Delta\eta = \eta_{\text{chalk}} - \eta_{\text{air}} = 1.60 - 1.00 = 0.60$. The increase in $\Delta\eta$ upon drying gives whitewash increased opaqueness. Following the same line of reasoning, we can see why watercolors tend to be more transparent when first applied than when dry. This is because η_{H_2O} is greater than η_{air}, so that $\Delta\eta$ involving the watercolor pigment becomes larger upon drying. That fact, however, makes it difficult to use watercolors over a dark or colored background. The transparency of the watercolor when first applied allows the colored background to show, but, of course, filtered by the watercolor layer. As the painting dries, the background is seen to a lesser degree and this changes the color match and harmony of the painting.

The relative hiding power of various whites in a common binder, say linseed oil with $\eta = 1.48$, can be evaluated using their indices of refraction:

$$\text{titanium dioxide:} \quad \Delta\eta = 2.71 - 1.48 = 1.23$$

$$\text{lead white:} \quad \Delta\eta = 2.01 - 1.48 = 0.53$$

$$\text{zinc white:} \quad \Delta\eta = 2.00 - 1.48 = 0.52$$

Titanium dioxide produces the most opaque paint of the three, with the opacity of lead white and that of zinc white being quite similar to one another.

Black paints and some colored paint films tend to have greater hiding power than do whites because light, in addition to being scattered from the surface, penetrates the pigment particles and is absorbed. To obtain a high degree of saturation of color in a paint, one might select a pigment with a low rather than a high index of refraction because more light will penetrate the surface and be selectively absorbed. Yet the transparency of the paint will be high if its index of refraction is small. In general, an increasing amount of electromagnetic radiation is absorbed as (1) the saturation of the color increases, (2) the darkness of the color increases, and (3) the glossiness decreases. The relative amount of light absorption is given in Table VI-2 for surfaces with various light-absorbing capabilities. Black, of course, absorbs broadly over the visible region and white does not, so they are found at the extremes of the list. Darker colors absorb more intensely than do lighter colors and blues absorb higher energy than do reds, so they order themselves accordingly in the register of total light energy absorbed. Watercolor paintings sometimes show this effect on their reverse side. The light absorption capability of certain pigments is less than that of others. A pattern often like the painting design itself appears

Table VI-2. Visible Light Absorption by Various Colors

Color	Relative absorption of visible energies
Black	100
Dark blue, brown-green	85–90
Gray	75–85
Red, light browns, light blues	70–75
Cream	50–55
White	40–50

Table VI-3. Temperatures of Light-Irradiated Panels

Surface	Temperature of panel, °C
Black	100
Unpainted iron	69
Aluminum paint	62
Yellow	41
White	28

on the *verso* as a result of discoloration and deterioration of the paper by the transmitted light.

The light energy absorbed by the coloring agents must be accounted for eventually, and most materials convert it into heat energy. The amount of heat generated and retained depends upon the visible-light-absorbing power of the surface, as indicated in Table VI-3, and the heat capacity of the material. It is apparent that the darker the surface is, the better its hiding power, but the more energy it absorbs and thus the hotter it becomes.

In the repair of objects such as glasses and ceramics, a conservator may be confronted with the problem of matching the index of refraction of an adhesive to that of the original surface in the hope of avoiding an unsightly mend. Of course the transparency and color match are also necessary to consider. If color is sought, then pigments can be added. If a decrease in transparency is sought, the surface can be scuffed. However, only when the index of refraction of the adhesive and that of the object are closely matched will the repair appear to be satisfactory in terms of reflection and refraction. The indices of refraction for several synthetic adhesives are given in Table VI-4. The "disappearing act" of the magic show mentioned earlier is the objective in such repairs.

Table VI-4. Index of Refraction of Synthetic Adhesives

Adhesive	η
Epoxy	1.58
Opticon UV-57[a]	1.532
Elmer's[b]	~1.48

[a] Opticon Chemical, Palos Verdes Peninsula, CA.
[b] Borden Co.

Change in Hiding Power with Age

As oil paintings age, the hiding power of the paint layers often decreases. This change can affect the appearance of the painting in two ways, although commonly only one or the other is found in a given painting. In one case, as the paint ages there may be a change in the tonality among the colors originally used by the artist. Since changes in tonality manifest themselves in subtractive color relationships between films, the discussion of this point will be deferred until later in this chapter when colored films are considered. In other cases, the underlying features, such as drawings and underpaint, which have been painted over by the artist may become apparent because of an increase in the transparency of the top paint layer. These underlying features are called *pentimenti* or "second thoughts." There are many examples of revealed pentimenti available and Figure VI-4 shows one of them. Changes can occur, however, in the surface character of a painting which permit less light to reach the paint layer and thus nullify the effects of the increased paint transparency.

The change in hiding power of a paint layer can be traced to two sources. First, the index of refraction of the natural oils, particularly linseed oil, used as a vehicle for the pigments tends to increase with age. Pigments have larger indices of refraction than linseed oil to begin with,

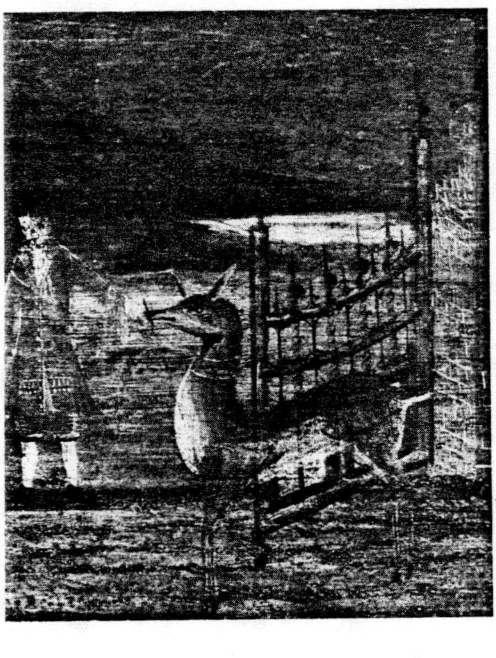

Figure VI-4. Pentimenti. The gate of the fence can be seen through the doe. (Courtesy, Henry Francis du Pont Winterthur Museum.)

and an increase in the index of refraction of the oil causes the value of $\Delta\eta$ between the pigment and oil to decrease. As $\Delta\eta$ decreases, less light is scattered at the pigment–oil interface and the opacity of the paint layer decreases. White pigments and colored pigments desaturated with white pigment may show pentimenti because these pigments absorb little light. The opacity of the paint layer may be largely dependent upon the value of $\Delta\eta$ and when $\Delta\eta$ becomes small, the underlying features may become observable. However, dark-colored pigments also show pentimenti. Dark pigments, such as carbon black, often require more oil than do white pigments in order to become mobile on the brush. A paint with a large oil/pigment ratio will be more affected by an increase in the index of refraction of the oil than one with a small oil/pigment ratio. This is because aging produces chemical changes which diminish $\Delta\eta$ between the pigment and the oil. When the oil/pigment ratio is large, little pigment is present to absorb the light, and hence a decrease in hiding power occurs. Underlying features can then be seen.

A second cause of increasing transparency of paint layers with age is the formation of a "soap" by an ion of the pigment and the carboxylate groups ($RCOO^-$) of species in the oil or egg tempera. The product has a lower index of refraction than the original pigment, so the opacity of the paint layer diminishes. The pigment most commonly found to form a soap in a paint is lead white. The Pb^{2+} ion in $2PbCO_3 \cdot Pb(OH)_2$ reacts with fatty-acid anions to form lead linoleate and lead stearate having the general formula $(RCOO)_2Pb$. These lead carboxylate compounds in paintings have a lower index of refraction than does the inorganic salt from which they were derived, and this leads to a reduction of hiding power as they form. The term "soap" comes from the fact that the same reaction of fatty acids from animal fat and the sodium ion from lye has been used for centuries to make common cleansing soap.

When grime and yellowed varnish are present on the surface of a painting, less light reaches the paint layer. This has the effect of nullifying the increase in transparency of the paint with age. If the surface remains transparent, then an intermediate situation is often noted, particularly in egg tempera paintings. A slight increase in transparency of the tempera-base paint layers produces not pentimenti, but rather an increase in the saturation of color of the layer because more light can penetrate the pigment particles and thus be absorbed. Hence, unusually vivid colors can at times be seen in old tempera paintings.

Anomalous Dispersion

Colored materials create an aberration with regard to the index of refraction which is known as *anomalous dispersion* in the visible range.

Although first noticed in 1840 by William Henry Fox Talbot, the inventor of modern photography, the description of the phenomenon was first published by Le Roux in 1862 in France. It was studied more extensively by Christiansen in Denmark after 1870. Normal dispersion of light is the term applied to the fact that the index of refraction of light in colorless matter increases slightly as the wavelength of light decreases through the visible range. This behavior was shown in Figure IV-2. Anomalous dispersion occurs each time an energy absorption band is encountered. In the energy range of an absorption band an *increase* in the index of refraction occurs with an *increase* in the wavelength of light. On the short-wavelength side of the absorption band, η decreases in the normal way with increasing wavelength. However, the decrease becomes more rapid as the absorption band is approached. Within the absorption band little light is transmitted (in a hypothetical case of total absorption, no light at all is transmitted, and no η value could be measured). On the long-wavelength side of the absorption band, η takes on a larger value than on the short-wavelength side, but decreases normally as the wavelength increases. On either side of the absorption band the dispersion is normal, but the discontinuity at an absorption band interrupts the curve. This behavior is shown in Figure VI-5 for a violet compound where yellow–green light is absorbed. The refractive index curve is shown as a dashed line through the absorption band. The strong absorption makes it very difficult to assign a value to the index of refraction in the band. The dispersion is said to be anomalous because η is larger for the longer wavelengths than for the shorter wavelengths. In Figure VI-5 the dispersion of orange light is seen to be greater than that of blue light. Thus the spectrum can be thought of as folding

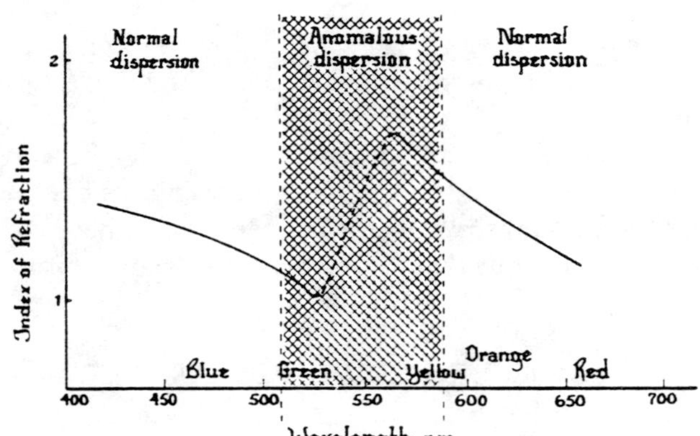

Figure VI-5. Anomalous dispersion for a violet compound. The absorption band occurs in the 510–580 nm region.

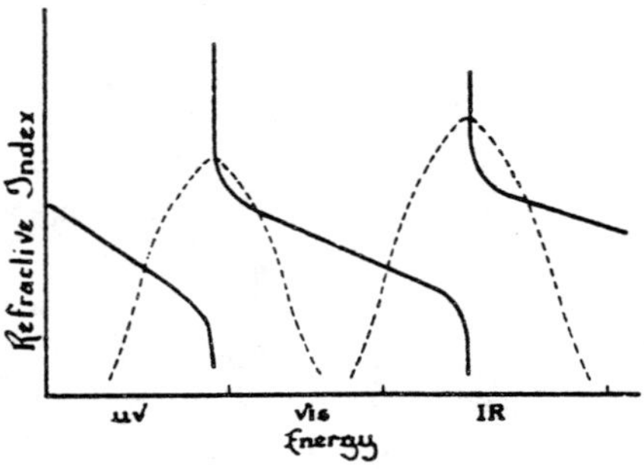

Figure VI-6. The index of refraction of light (solid lines) in the vicinity of energy absorption bands (dashed lines) in a hypothetical compound.

over on itself such that the dispersion of light produced by a purple prism might produce a color progression of blue–indigo–violet–red-orange–yellow, with green not appearing because it is absorbed. The index of refraction on one side of the absorption band can be very different from that on the other. For example, in solid cyanin the index changes from 1.2 to 2.2 in a 120-nm-broad range centered at the visible absorption band.

Anomalous dispersion is not really anomalous at all. It is a property of all materials somewhere in the spectrum of electromagnetic radiation. It can occur in the vicinity of any and all absorption bands. Colorless materials exhibit anomalous dispersion, but not in the visible region. It occurs instead at infrared and ultraviolet absorptions. Figure VI-6 shows a general plot of the index of refraction for a hypothetical compound having absorption bands in these regions. At the risk of seeming paradoxical, one might even say that anomalous dispersion is the cause of normal dispersion! The "normal," negative-sloping curvature of η as the wavelength increases in the colorless glasses shown in Figure IV-2 results from anomalous dispersion due to absorption bands in the ultraviolet region in these colorless glasses.

Anomalous dispersion is a partial source of some properties of pigments. One notes, for example, that the average hiding power of red and orange pigments is frequently better than that of blue and violet pigments. This is because the index of refraction is greater on the long-wavelength side of an absorption band than on the short-wavelength side. Hence the pigments that reflect the longer wavelengths often have higher η values and appear somewhat more opaque than those that reflect shorter wavelengths.

THE EFFECT OF PARTICLE SIZE ON APPEARANCE

In the foregoing discussion of opacity and surface appearance, one of the most important physical characteristics of particulate materials was not considered. This factor is the size of the particles. Particle size is extremely important in light scattering and needs careful attention for many applications. For centuries artists have known from experience that excessive grinding of certain pigments leads to loss of the color or a change in the color. They also have known that the opacity of a pigment could be varied by grinding for different lengths of time. Craftsmen in glassmaking have made use of different particle sizes to obtain beautiful effects of opalescence in glass. Indeed, particle size considerations, whether invoked knowingly or unknowingly, in some way touch on most areas of art and create many appearances in nature. A practical way of discussing the effect of particles on light waves is to separate the particles into two size classes: those of size less than the wavelength of the light being scattered, and those greater. Particles in the first size range produce Rayleigh scattering, while those in the second follow the conventional rules of reflection presented in Chapter IV.

Rayleigh Scattering

If a particle is smaller than 0.1λ, where λ is the wavelength of the light to be scattered, then the particle cannot disperse that light. The light will bend around the particle and be scattered in a manner different from that for larger particles. The theory of light scattering by small particles was developed by Lord Rayleigh in 1871 and is known as *Rayleigh scattering* or, equivalently, *Tyndall scattering*. Rayleigh found that for small particles the intensity of light scattered at right angles to the incident ray varies (1) directly with the intensity of the incident light; (2) directly as the square of the average volume of the scattering particles; (3) inversely as the fourth power of the wavelength of the incident light. The most dramatic and broadly applicable result of Rayleigh scattering theory is the concept involved in 3. That is, that the wavelengths of light encountering these very small particles are scattered according to $1/\lambda^4$. This means that shorter wavelengths are more strongly scattered than are longer wavelengths of light. For example, light at 400 nm is scattered about ten times more effectively than light at 700 nm.

The color of the sky has interested natural philosophers for centuries. The efforts of investigators no less than da Vinci, Newton, and Clausius were directed toward explaining the blueness of the sky. Rayleigh's work settled the question, however. The sky is not inherently blue. Water droplets and dust particles in the 0.1λ size range (λ in the visible light range) are plentiful in the earth's atmosphere. Sunlight containing all wavelengths of visible light encounters these particles, but the shorter wavelengths (violet and blue) are scattered more effectively toward the observer than are the longer wavelengths (red and yellow). Consequently, the eye receives a preponderance of shorter wavelengths when viewing the sky at right angles to the sun ray direction. The fact that blue color and not violet color is observed has mostly to do with the lower sensitivity of the eye to violet compared to blue (Figure V-5). When viewing sunsets and sunrises the eye is sensing light passing *through* the particulate matter in the atmosphere, and not light traveling at right angles to the sun rays. In the morning and evening the blue light is scattered to the side, leaving the less efficiently scattered yellow and red wavelengths to be seen in the direction of the sun. Hence the vivid hues of the morning and evening sun also result from the effects of Rayleigh scattering by particulate matter in the atmosphere, but differ in color from north sky light because of the angle of observation.

In a related way distant mountains often appear blue to the eye rather than the natural color of the trees and meadows. The Blue Ridge Mountains in Virginia are so named for their color when viewed at a distance. The fifteenth century painters Jan Van Eyck and Hans Memling tended to paint distant mountains blue rather than green. Artists doing this are bringing into their works the effects of Rayleigh scattering as well as some of the blue hue of the sky. The fine haze in the air causes a bluish tinge to the dark backdrop of the mountains because of the Rayleigh effect.

Apart from the blue of the sky, the blue of blue jays and other blue-feathered birds is the most elegant example of Rayleigh scattering in nature. The barbs of blue-jay feathers have transparent cells which contain fine particles with air interstices. These particles scatter blue light more effectively than other colors. Pigments in the cells absorb the longer wavelengths that are not reflected. Hence only blue color is seen.

The effects of Rayleigh scattering are seen in vivid ways in solids and liquids. For example, when milk is greatly diluted with water it takes on a bluish tinge by reflected light but has an orange-yellow tinge when light is viewed passing through the solution. This is because colloidal particles of butter fat in the milk having sizes in the range of 0.1λ scatter the short wavelengths most strongly, giving the blue color. The transmitted light has lost the blue and therefore has an orange-yellow color. A drop of colophony dissolved in alcohol and in turn dropped into water will produce a similar bluish tinge when viewed from the side. Colophony, alcohol, and water have no color themselves, but together they appear with color as a result of the Rayleigh effect. Italian varnishes used in the sixteenth through eighteenth centuries, particularly on musical instruments, show red and yellow hues and an occasional flash of purple, depending upon the angle of viewing. These color variations are believed to result from a colloidal suspension of solid oxidation products of linseed oil in the varnish. They scatter light in accordance with the Rayleigh law. Opalescence in stones, glass, and ceramics is often due to small, insoluble particles dispersed in the matrix which produce Rayleigh scattering.

Pigment Particles

One is not very often concerned with particles in the 0.1λ size range for use as artists' pigments. These small particles can produce color effects by light scattering, but they are transparent and not able to impart their natural color. Pigments involve somewhat larger particles. For transparent media containing particles whose dimensions are about equal to or exceed the wavelength of light to be scattered, the fractional loss in intensity of transmitted light (I/I_0) due to scattering is given by

$$\frac{I}{I_0} = \exp(-KN\pi r^2 d) \tag{VI-3}$$

N is the number of particles per unit volume, r is the radius of the particles, and d is the optical path length. K is known as the scattering factor and it ranges in value from 0 to 4.3. The value of K mainly depends upon the ratio of the particle diameter to the wavelength of light (D/λ) and the relative index of refraction of the particles and the medium. K follows a $1/\lambda^4$ dependence for very small particles having dimensions much smaller than λ. As the particles become larger, K increases and reaches a maximum when the particle size is about equal to the wavelength of light. K decreases as the particles become still larger. Therefore, particles having dimensions about equal to the wavelength of light intended to be scattered are the most efficient at scattering that light. The wavelength range of visible light is 400–700 nm, so 400-nm or larger particles should be reasonably opaque,

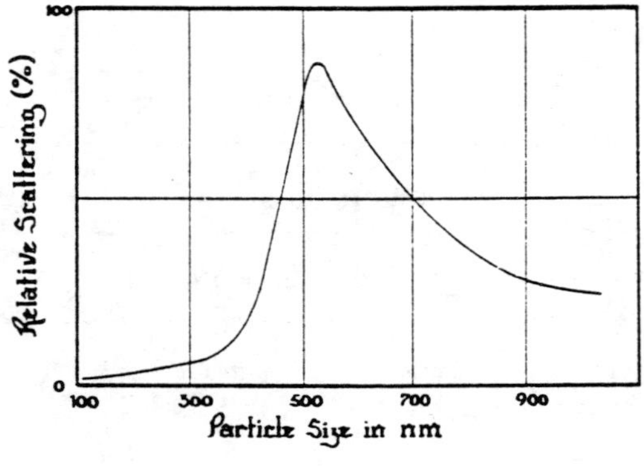

Figure VI-7. Light scattering from various particle sizes of zinc oxide for light at 530 nm.

provided their index of refraction exceeds that of the medium in which they are dispersed. Figure VI-7 shows the relative intensity of scattered light as a function of particle size for a common white pigment, zinc white, using light at 530 nm. Zinc oxide particles in the 100–1000 nm size range scatter the most light (λ = 530 nm) in the 530-nm size. The light scattering decreases somewhat for particle sizes greater than this, but there is a much more rapid dropoff in light scattering for smaller particle sizes.

Figure VI-8 summarizes the appearance and characteristics of particles of various sizes. The entry labeled "number of particles" in Figure VI-8 is obtained by reducing the amount of matter equal to 1 cm^3 to particles whose maximum dimension is given as the size.

Pigments are not all composed of the same sized particles. For example, indigo, lamp black, and Prussian blue are used in the size range below 1000 nm. These are called "fine"-grain pigments. White lead, vermilion, and cadmium yellow are used in the 1000–10,000 nm size range and are classed as "medium"-grain pigments. Verdigris and cobalt blue are used in sizes of greater than 10,000 nm and are classed as "coarse"-grained.

Particle sizes in old paintings usually average 30,000–40,000 nm, with some particles larger and some smaller than this range. Modern pigments are often sifted to obtain a uniform particle size and this fact can often be used to distinguish old paints from newer ones. Not all pigments can be ground to the same degree of fineness, however. For example, cobalt blue, malachite, and azurite are pigments which lose their

<table>
<tr><td></td><td colspan="5">nanometers</td><td colspan="4">centimeters</td></tr>
<tr><td>size</td><td>0.1</td><td>1</td><td>10</td><td>100</td><td>1000</td><td>0.001</td><td>0.01</td><td>0.1</td><td>1</td></tr>
<tr><td>number of particles</td><td>10^{24}</td><td>10^{21}</td><td>10^{18}</td><td>10^{15}</td><td>10^{12}</td><td>10^{9}</td><td>10^{6}</td><td>10^{3}</td><td>1</td></tr>
<tr><td>appearance in solution</td><td colspan="4">transparent</td><td>cloudy</td><td>turbid</td><td>translucent</td><td>semi-opaque</td><td>opaque</td></tr>
<tr><td>optical observability</td><td colspan="3">sub-microscopic</td><td colspan="6">microscopic (real form can be seen)</td></tr>
<tr><td>form</td><td colspan="2">molecular</td><td colspan="2">colloidal suspension</td><td colspan="5">mechanical suspension</td></tr>
<tr><td>recoverability</td><td colspan="4">Pass through filter paper</td><td colspan="5">Retained on filter paper</td></tr>
<tr><td>pigment size range</td><td colspan="4">none</td><td colspan="3">(hatched)</td><td colspan="2">none</td></tr>
</table>

Figure VI-8. A summary of the characteristics of various particle sizes. 1 nm = 10^{-3} microns.

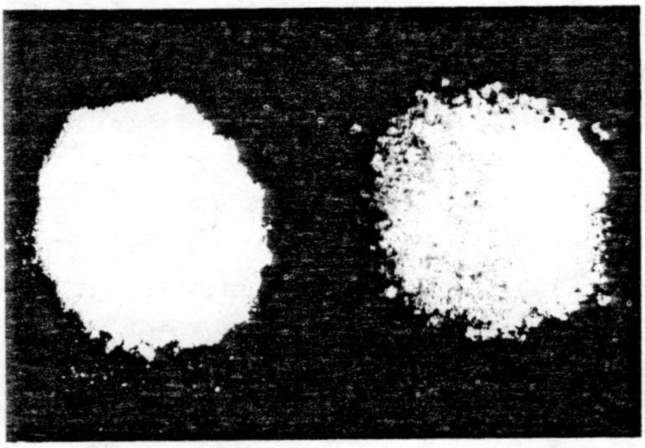

Figure VI-9. Two crystal sizes of potassium phosphate. The small crystal size is opaque because of the large amount of surface reflection. The larger crystal size produces less surface reflection and hence more transparency.

color (are *desaturated*) when ground too finely. This is because the amount of reflection of white light from their surface is greatly increased relative to the amount of light that is absorbed by the body. Since only the absorbed light produces color, the amount of color is reduced relative to the amount of white light reflected. The effect of crystal size on the amount of light reflected from a mass of crystals is shown for colorless crystals of potassium phosphate in Figure VI-9. Very small crystallites produce an opaque mass, whereas the larger particles exhibit a greater degree of transparency.

Desaturation also occurs when a white pigment is mixed with a colored one. Figure VI-10 shows what happens to the spectral reflectance of a magenta-colored pigment as one adds white pigment to it. A decrease in intensity of the color from magenta to light pink is observed as white pigment is added.

Two important considerations in the preparation of a paint are the particle size of the pigment and the amount of vehicle required. Particle size affects smoothness, brightness, and uniformity of a paint. For some pigments, coarse particles tend to enhance color, but may have poor

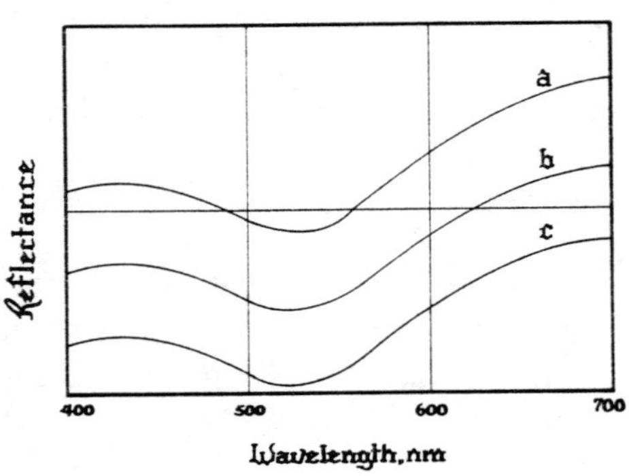

Figure VI-10. The decrease in the intensity of reflectance from (a) to (c) of magenta as white pigment is added.

covering power. Fine particles may give good covering power, but poor tinting character. In other pigments, the saturation of color is not decreased by grinding and good covering power as well as good tinting power can be obtained with small particle sizes.

It is possible to vary the reflecting power of the paint coating by varying the ratio of the amount of pigment to the amount of vehicle. The more pigment used, the more the surface will be matte. Flat or satin paints contain large amounts of pigment and fillers. Enamel contains varnish as well as less pigment than most paints and thus can produce a glossy glaze. The amount of oil required to thoroughly wet an artist's pigment and convert the pigment to a paste is given by the *oil absorption value* of the pigment. Some pigments require little oil to become usable, while others require much more. Lead white requires only about 9–12% oil by weight, while raw sienna and lamp black require almost 50% oil. The color of the paint is less affected by yellowing of the oil if the oil absorption value is low. The oil absorption values depend upon the surface characteristics of the pigment particles, which are controlled by the size and shape of the particles, the density of the pigment, and the molecular or ionic features of the pigment surface. High surface area, low density, and an electrostatically incompatible surface of a pigment will usually cause it to have a high oil absorption value.

The concept of light scattering by particles is not limited to the visible region alone. The same type of scattering can occur with all wavelengths of electromagnetic radiation. This fact is put to use in infrared photography, where the underlying features of certain paintings can be revealed. Since the scattering power of pigment particles is at a maximum at the wavelength of light to be scattered, it follows that the larger the particle size is, the longer the wavelength of radiation must be to undergo maximum scattering. Hence particles that are opaque to visible light are often penetrated by infrared or longer wavelength radiation. The underdrawings of paintings may become observable using infrared photography. Pigments are not all equally transparent to infrared radiation, however. Malachite and vermilion continue to be as opaque in the infrared as they are in the visible. However, verdigris is quite transparent in the infrared.

A Summary of Opacity

We have encountered several factors in this and earlier chapters which render a nonmetallic surface opaque. It is worthwhile to consolidate and summarize these factors now before they become lost amid the other phenomena that can occur. Physical opacity can arise in varying degrees from any or all of the following factors. Usually more than one factor is present in a given opaque material and not all factors create the same degree of opacity.

1. A high degree of absorption of light by a colored object reduces the amount of light available to be transmitted. In metallic objects an extensive amount of light is absorbed and usually also reflected and this is responsible their high degree of opacity.

2. The presence of particulate bodies dispersed in a medium can cause light to be scattered randomly and this reduces the ability of the medium to transmit light. White paints often achieve opacity by this mechanism. The size of the particles is an important factor in determining their ability to scatter light.

3. A large difference between the index of refraction of a material and that of its surroundings creates a high degree of reflection at the interface. Hence less light is available to be transmitted through the media involved. White pigments may rely on this mechanism for opacity.

4. A large index of refraction for a material creates a large critical angle of incidence such that light energy, once inside the material, has less chance of escaping before it is absorbed or attenuated.

5. A high degree of diffuse reflection from the surface will diminish the amount of light available to be transmitted by the material. The increase in the amount of surface area available for reflection as well as the randomness of the surface will increase the degree of reflection and diffusion. Since a high percentage of light is reflected, less light can be absorbed or transmitted (the RAT law). "Frosted" glass is an example of this factor.

Physical transparency occurs in the opposite situations. However, one can *perceive* transparency only by being able to see both the transparent material and objects behind it. Otherwise, the so-called "invisible shield" popularized in a toothpaste advertisement is the result.

THE REFLECTION, TRANSMISSION, AND ABSORPTION OF LIGHT

In Chapter IV the physical laws governing reflection, transmission, and absorption of light by materials were introduced. We stopped short of considering properties of surfaces, such as texture and composition, so that the final picture could be formulated here with particular application to individual materials. The ideas developed in Chapter IV will be heavily relied upon and amalgamated with what is presented in this section.

Gloss and Matte Surfaces

Color can be defined completely in terms of hue, value, and saturation. However, the same color may not look the same on surfaces of metal, paper,

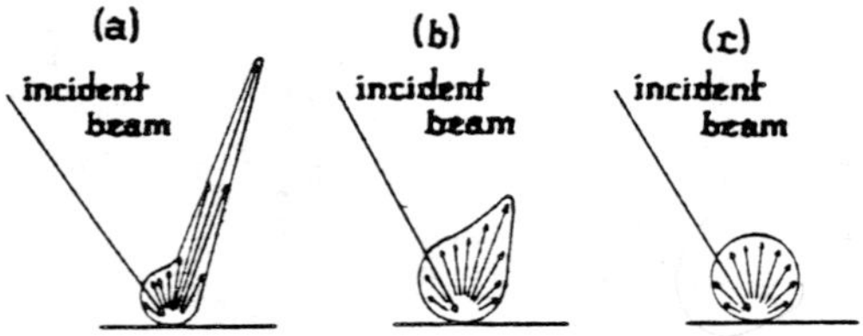

Figure VI-11. (a) Specular reflection from a glossy surface, (b) a combination of specular and diffuse reflection, (c) diffuse reflection from a matte surface.

textiles, wood, plastic, and glass. New attributes of the color emerge. The appearance of all surfaces is closely related to the complex ways they reflect light and to the amount of light reflected. The degree of smoothness is generally the controlling feature in reflection. A perfectly smooth surface, such as a pane of glass, is said to *specularly reflect* light (Figure VI-11a). Most of the light incident onto the surface is reflected at the same angle as the incident light. The perception of *glossiness* results from a high degree of specular reflection. There are other kinds of glossiness corresponding to different patterns of specular and diffuse reflection. Luster and sheen are two of these.

As the surface becomes roughened, the amount of specularly reflected light decreases, but the amount of diffusely reflected light increases. *Diffuse reflection* of light occurs when light is scattered in all directions as a result of a randomly irregular surface. A *semimatte* surface has significant specular reflection, but with sufficient diffuse reflection so that glare does not result. Figure VI-11b shows the reflection profile of a semimatte surface characteristic of many paper and painted surfaces. If the amount of specularly reflected light is further decreased so as to be at the level of the diffusely reflected light, then a *matte* surface results (Figure VI-11c). The appearance of a matte surface has little dependence upon the angle of viewing, because light is uniformly scattered in all directions.

Pastels, watercolors, and some fresco paintings are examples of art objects having a matte surface. Watercolors and fresco a secco paintings involve pigments in a water medium containing a small amount of glue or egg as a binder. Buon frescoes do not contain a binder. Pastels are composed of concentrated pigment with gum acacia, tragacanth, honey, etc., as a binder. Although the artist may be able to create superb work with these media, the colors have somewhat less depth than those achievable in oil because pastels and watercolors rely mostly on surface reflection of light to produce color. The colors may be saturated, but lack the depth that can be created by glazing and scumbling in oil and tempera painting. The appearance is matte because light striking the surface is either diffusely scattered or passes between the surface particles and strikes particles underneath. The diffuse scattering from interior particles causes the light to be lost for the purpose of viewing. Frescoes may have a moderate degree of gloss when the pigment is drawn into the plaster ground. The reaction of the plaster with carbon dioxide from the air causes a glassy skin of calcium carbonate to form. A fine sheen is characteristic of fresco paintings in which the surface has been finely trowelled.

The angle of projection of light beams is an important factor in the appearance of a surface. Normal light illumination exists when light beams of equal intensity are incident onto the surface at more or less a right angle. No textural shadows appear in this light. However, if light is projected across a surface by locating the lamp source to one side of the object, surface textures cast shadows and show sharp relief. This situation is known as *raking illumination*.

Surfaces with Transparent Coatings and Films

Among the most common surfaces created by artists and craftsmen are those having transparent coatings. Whether done for the purpose of protection or for artistic effects, the coated surface is found on most paintings, furniture, and ceramics and on some metals. Varnish, ceramic glaze, wax, and lacquer are typical transparent films that might be applied to an object. The coated surface regularly occurs in nature as well.

Light beams can penetrate the transparent surface, where they may be diffusely reflected by pigment particles or other coloring agents as illustrated in Figure VI-12. Selective absorption of light may occur, causing the reflected light to be colored. The characteristic surface appearance is produced by differences between the light reflected from the outermost surface and the light reflected from the underlying features. The specular reflection of white light from the surface is mixed with diffuse reflection from the underlying features to give a perception of texture and color in the object. The human visual system is able to separate the signals of the specularly reflected white light from those of the diffusely reflected colored light so as to perceive depth and richness. Ruskin in *Modern Painters* noted this vividly:

> Every color, wet, is twice as brilliant as it is when dry; and when distances are obscured by mist, and bright colors vanish from the sky, and gleams of sunshine from the earth, the foreground assumes all its loveliest hues, the grass and foliage revive into their perfect green, and every sunburnt rock glows into an agate.

Although Ruskin's observation was probably inspired by the beauty of the landscape after a rain, the same thoughts also apply to more mundane things, such as sidewalks and snow. A wet and a dry sidewalk appear quite different. The wet sidewalk is a matte surface (the concrete) covered by a

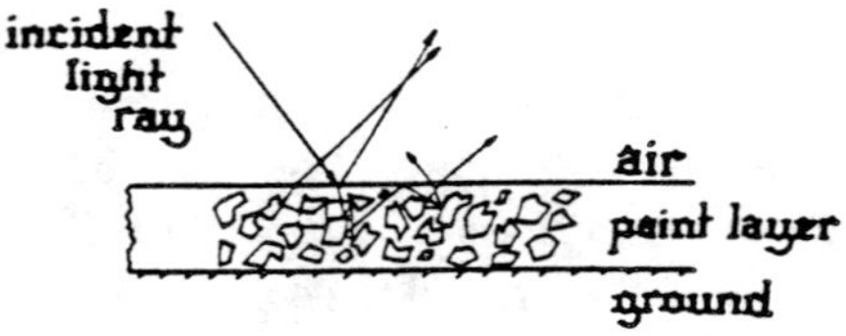

Figure VI-12. Reflection of light from a glossy, transparent surface having underlying colored or textural features.

smooth, glossy coating of water. The wet sidewalk is therefore darker than a dry one because less light reaches the cement, as a result of reflection from the water–air interface. Multiple reflections in the water layer bring out contrast in surface features of the concrete that are not seen when it is dry. The optical difference between fresh snow and partly melted snow is obvious to those who have seen it. Fresh snow is very white and bright, but as it begins to melt it takes on a grayish cast. The reason for this difference is that liquid water fills the interstices between the ice crystals during the melting process. Much less light is reflected from the water–ice interface ($\Delta\eta \simeq 0$) than from the ice–air interface ($\Delta\eta \simeq 0.33$). Hence, wet snow has less brilliance than does dry snow.

All transparently coated surfaces can be conceptually discussed in the same way. A darker appearance, deeper colors, and more contrast to the colors are often characteristic of a surface covered with a transparent, glossy film compared to the bare, uncoated surface. Varnish will be used here as the example of a coating over a paint layer in order to discuss the reasons for these differences. Spirit varnish is resin dissolved in turpentine or other solvents and has little or no visible-light-absorbing potential when freshly applied. (Mastic and dammar do have a pale yellow cast, however.) The refractive index relationships for an unvarnished and a varnished painted surface might be as shown in Figure VI-13. The paint layer is mostly pigment with some oil. The index of refraction of the pigment is used in this example. The value of $\Delta\eta$ between the air and the pigment is 1.5, but when the varnish layer is applied, two interfaces exist. One is between the air and varnish, with $\Delta\eta = 0.5$, and the other is between the varnish and paint, with $\Delta\eta = 1.0$. The unvarnished paint surface is matte in appearance, but the varnished paint surface is glossy. Therefore, more

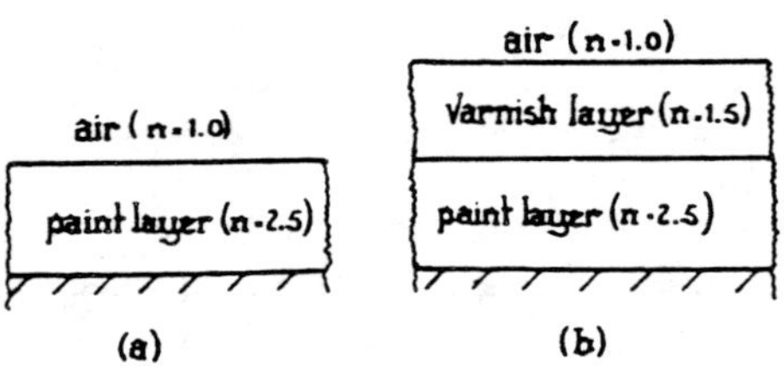

Figure VI-13. The index of refraction relationships for (a) an unvarnished paint layer and (b) a varnished paint layer.

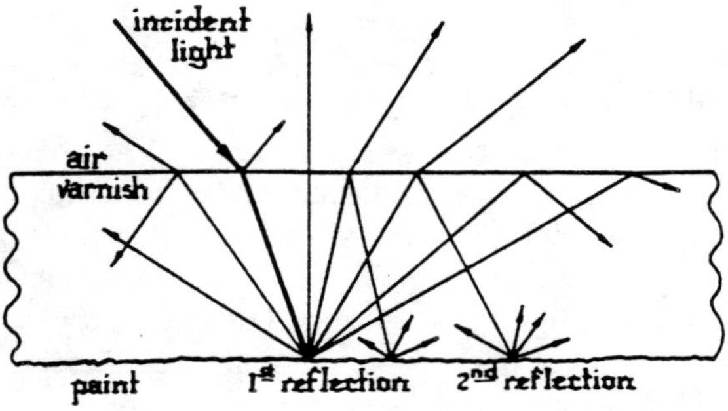

Figure VI-14. Internal reflections in a transparent varnish layer.

light is reflected from the varnish surface, leaving less light to be transmitted to and absorbed by the paint layer. This is the source of the *darkening effect* of a freshly varnished surface. Darkening results from the lower light intensity at the paint layer under the varnish. The reflection of light at the paint layer under a coat of varnish is less than that for the unvarnished paint layer because $\Delta\eta$ is smaller at the paint–varnish interface than at the paint–air interface. Once inside the varnish layer, a greater percentage of the light is transmitted into the paint. As a result, more light can be absorbed by the pigment particles and this will lead to a *deeper* and *richer* color. The increase in *contrast* of the colors comes from consideration of internal reflections occurring in the varnish layer. Figure VI-14 shows what is meant by internal reflections.

The incident white light strikes the varnish layer and penetrates it. The direction of propagation of the light beam is slightly altered by refraction at the interface. The beam of light is then scattered more or less diffusely from the paint layer in the first reflection. These light beams will

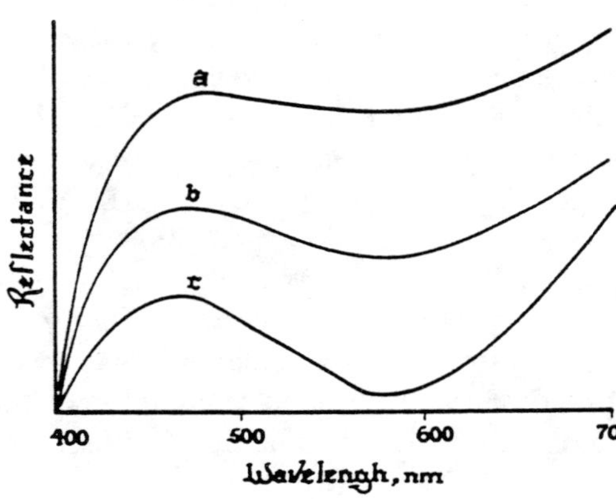

Figure VI-15. The selective intensification of color contrast by internal reflections. (a) First reflection, (b) second reflection, (c) third reflection.

either escape the varnish layer, carrying with them the color produced by selective absorption of light in the paint layer, or they will remain in the varnish layer as a result of internal reflections from the varnish–air interface. Note that this internally reflected light is now colored light and does not have the same spectral composition as the incident light. The discussion in Chapter IV on the critical angle of incidence applies here. That is, light in the varnish layer does not cross the varnish–air boundary and is not available for viewing if the angle of incidence is greater than the critical angle for the varnish. The light beams remaining in the varnish layer now will strike the paint layer again, be selectively absorbed according to the pigment characteristics, and then be scattered again. This process is repeated over and over. The net effect of these repeated events is to selectively intensify the observed color of the pigment. This intensification is depicted in terms of the percent reflectance for each reflection in Figure VI-15.

The incident white light becomes colored in the first reflection. However, increasing selection of the color-producing wavelengths occurs in the second and third reflections. Light leaving the material after many reflections thus has much greater color contrast than if a single reflection were all that occurred.

The most commonly encountered natural resins used for varnishing paintings are dammar and mastic. Mastic is more yellow than dammar because of the tail of several strong ultraviolet absorption bands, which extends into the blue of the visible region. Both dammar and mastic are fairly durable and transparent, but, as described in Chapter IX, they tend to become brittle and yellow with age. They also develop an obscuring fogginess, called *bloom*, with age. Shellac and copal are other resins, which have been used mainly on wood.

Synthetic resins are often employed now in place of the natural resins. Their physical properties, such as viscosity, hardness, solubility, permeability, drying rate, and wetting power, could be expected to and do differ appreciably from the natural products. As a group, though, they are more chemically stable toward the effects of heat and light and do not tend to yellow and become brittle with age. The best known of these resins with respect to artists' materials are the acrylics (Acryloid B-67, B-72, F-10), poly(vinyl acetates) (the AYAA, AYAB, AYAF, AYAT series), poly(vinyl chloride) (Vinylite), epoxy, and the polyesters. All have acquired various uses as replacements for natural oils, gums, glue size, egg, and varnish, but the acrylic resins are the most widely accepted. Acrylic plastics are marketed with the product names Plexiglas and Lucite. Shavings of these plastics can be dissolved in a suitable organic solvent and used to great advantage as a binder. The acrylic-based resins are fast-drying, highly transparent, colorless, and light-stable. However, Lucite has a serious disadvantage in that it tends to cross-link and become very insoluble as it ages. The indices of refraction of synthetic resins are similar to those of the natural resins. For example, the acrylics have indices in the 1.48–1.52 range, while those of the poly(vinyl acetate)s are 1.46–1.47. These values compare well with dammar and mastic, which have indices in the 1.51–1.54 range.

The higher viscosity and thus higher pigment holding power of synthetic resins permits their use in much thinner films than is possible with the natural resins. When used as a varnish, the synthetic resins dry quickly and tend to follow the surface irregularities in the paint layer. In contrast, the natural resin varnishes have lower molecular weight and lower viscosity and are slower drying. They fill the porous spots in impasto coatings and dry with a flat surface. The difference between the natural and synthetic resins is apparent in the optics of the surface. The synthetic resins give less glossiness to the surface and less darkening and contrast in the paint. Some synthetic varnishes dry so quickly that they must be sprayed onto the surface. This causes beads to form instead of a smooth, continuous film, with the result that the color contrast, darkening effect, and glossiness created by the natural transparent film are lost altogether. Although these problems are correctable, some painters and conservators prefer to continue to use dammar rather than the synthetic varnishes.

At times metals are lacquered to prevent corrosion and reduce the necessity of frequent polishing. Polishing removes layers of metal and eventually will lead to loss of detail in ornamental or engraved areas. Unfortunately, lacquering often creates an artificial "plastic" look, particularly if the lacquer is not colorless or if the surface does not have the same degree of gloss as the object it covers.

Very thin transparent films can create color effects in the form of interference patterns as a result of selective amplification and suppression of the intensities of various wavelengths of light. Interference of light was discussed at length in Chapter IV. Antique glass exposed to the effects of rain water or ground water often has a thin surface layer of silica flakes caused by dissolution of part of the silica glass. These flakes produce a range of bright, pearly colors. *Lusterware* is a kind of glass which has been deliberately given a thin layer of $SnCl_2$ in varnish deposited into the surface in order to produce color by interference effects.

The iridescent colors characteristic of many ornamental feathers, such as those of the peacock, result from many thin layers of essentially colorless keratin built up on the barbs of the feather. The colors of the peacock feather depend upon the angle of viewing, because the effective thickness of the horny keratin varies with the angle of viewing. The orientation of the barbs and barbules relative to the viewer is important in the interference patterns observed, but it also plays a role in the aerodynamics of the feathers. Peacock feathers are constructed so that the barbs and barbules are optimized for color production, but are ineffective for lift toward flight. Of course, the peacock is one of a number of flightless birds.

Colored, Transparent Films

The variable transparency of some paints led many of the great oil painters of the seventeenth and eighteenth centuries in particular to the indirect painting techniques of *glazing*. Such artists would use a colored, transparent layer of paint over an opaque, colored or neutral ground. The light reflected from the opaque layer would be filtered by the transparent layer and acquire a deep, lustrous color effect not achievable with a single paint. In many cases layer after layer of transparent glaze would be built up, giving spatial effects and atmosphere to the work. Masters such as Titian, Rembrandt, and Vermeer used as many as twenty glazes to capture the appearance of fabrics and human skin.

The colors produced by glazing techniques follow the subtractive color relationships. When a yellow glaze is applied over an opaque, blue layer, the yellow paint particles absorb the short, blue-producing wavelengths, while the blue pigment absorbs the long, yellow-and red-producing wavelengths. What remain to be reflected are the green-producing middle wavelengths.

The cross section of a glazed painting is shown in Figure VI-16. Traditionally used transparent glazes include ultramarine, viridian, burnt sienna, madder, ivory black, and ochres. Cadmium red, cobalt blue, and Naples yellow are useful in dilute preparations because they are rather opaque paint films. The pigment particles are suspended in the medium of wax–oil or resin–oil to give maximum transparency. The indices of refraction of the pigment and vehicle should be close to one another to facilitate the light transmission. A deep, full red can be generated by glazing madder lake in a vehicle over an opaque vermilion paint. A transparent yellow lake used over an opaque blue underpaint gives a deep,

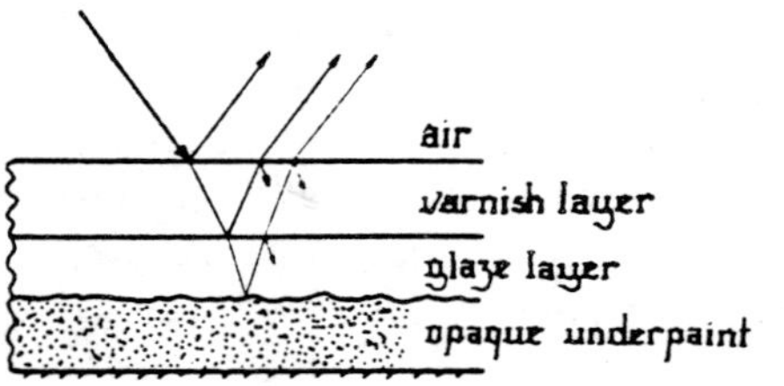

Figure VI-16. The cross section of a glazed painting where color filtering occurs.

lustrous green by subtractive color mixing. Striking full greens are also achieved by glazing copper resinate formed by mixing verdigris in a resin, such as Venice turpentine or balsam. However, in time the resin can yellow, causing the green to take on a brownish cast, as seen in many landscape paintings by Flemish, French, and Italian masters of the sixteenth and seventeenth centuries.

It is well known that drying oils have a tendency to yellow with age if they dry in the absence of strong sunlight. The resultant yellowing will be particularly apparent if the pigment with which the oil was mixed has only weak tinting strength. The result is a change in value, and the balance of the painting becomes altered. For example, the blue skies in some paintings from the Renaissance period in Italy and the North now have a grayish-green cast. The color change is sometimes traceable to the use of smalt as the blue pigment. Smalt is a stable pigment made from glass and has a low index of refraction and weak tinting strength. It the smalt particles are trapped in an aged medium having considerable yellow color, the weak blue color combines with the yellow of the resin to give a dull gray-green. Smalt has also been used in a form containing alkaline impurities. These impurities can react with the varnish and oil and alter the color. The more common blues, such as cerulean, azurite, and ultramarine, have stronger tinting power than does smalt, but can show this color change as well. In general, the red and yellow pigments are least affected by yellowing of the medium. This is because reds and yellows reflect about the same wavelengths as does the yellowed medium. Green pigments may be slightly affected by yellowing, but suffer less than blues. Blues suffer the most because they reflect the opposite end of the visible spectrum from yellow. This leaves only green light to be transmitted. The low level of reflection gives low brightness to the green color.

Changes in tonality also result when a painter has used a mixture of two subtractive colors to produce an additive primary color. If one of the subtractive colors fades, then the hue changes dramatically. Green foliage in eighteenth century Dutch paintings often has changed to a dull blue for this reason. A comparatively stable blue pigment, such as indigo, was combined with a fugitive yellow, such as gamboge precipitated on alumina, to achieve the green. The yellow has now faded, leaving the blue pigment to color the foliage.

The intermediate procedure between glazing and the use of an opaque film is *scumbling*. A thin, translucent film of a color or white is used to partially obscure and thus modify the color of a paint layer previously laid down. Scumbling can give a feeling of spatiality to a painting. By combinations of glazing and scumbling an ingenious artist can reproduce the surface texture of materials like marble, fur, and feathers.

Colored, transparent coatings have also been used on metals and glass for artistic purposes for many centuries. The most brilliant effects on metals are achieved when a transparent paint is applied to a highly reflecting metal. The strong reflectance of the metal can then be put to use for producing bright, lustrous colors. This technique was used in Burma and Kashmir, where gold objects were covered with a red varnish, which produces a ruddy color because of the yellow reflectance of the gold. Old iron and brass were often protected from corrosion by applying an East Indian palm resin known as Dragon's Blood. Silver and tin become yellow, and iron becomes a bronze color when covered with Dragon's Blood. In their writings on painting, Theophilus discussed shellacked and varnished tin leaf and Cennini recommended shellacked and varnished silver leaf as a replacement for gold leaf. Tempera paint was first used on gold leaf, but oil-based glazes were used later. In fact, painting with transparent colors on gold and glass predated easel work and may have provided the impetus for painters such as Van Eyck to try the same techniques in easel painting.

Metals

The electronic structure of metals is not like that of most other materials. In fact, the difference is what leads to the classification of materials as metallic and nonmetallic. Metals can be thought of in the simplest sense as an ordered array of positive ions immersed in an electron "gas." That is, most of the outer electron density of atoms becomes a mobile pool of electrons in the lattice. The atoms possess a relative positive charge as a result. This pool of electron density behaves as a free-floating, delocalized "gas" of relative negative charge. Such a physical form appears to be quite satisfactory in terms of energy because fully 80% of all elements known in nature adopt it as their most stable arrangement.

Because of this electronic structure, metals are good reflectors of light and good conductors of electricity. Most bulk metals are "white," in that they tend to reflect all wavelengths of visible light with about equal efficiency. A few metals, most notably copper and gold, are colored. The electronic structure of the metals described above is responsible for the high reflectivity and the appearance of color. It is the free-electron gas that is mainly responsible for the high degree of reflection of incident light, yet it absorbs a relatively small amount of light. The core electrons, which remain on the atoms, are much more involved with absorption and have less to do with reflection. The absorption of electromagnetic radiation by metals in most cases occurs in the infrared and the ultraviolet regions rather than the visible. This renders most metals "white" to the human eye. The total absorption and reflection of light by metals is very great, however. Metals do not transmit visible light in thicknesses greater than a few wavelengths of the light. Because of this, metals tend to have indices of refraction which are poorly defined and very susceptible to change with changes in the wavelength of the incident light. The refractive index values known are usually calculated from electrical measurements and have little relation to the bulk metal of interest in art objects.

Figure VI-17 shows the spectral reflectance curves of some common "white" metals and graphite. Silver and aluminum have the best visible light reflectance characteristics and this accounts for their frequent use in the manufacture of mirrors. Since the extent of reflectance of light and the degree of electrical conduction by a metal are determined by the ease of mobility of electrons in the electron "gas," it is not coincidental that silver and aluminum are also among the best conductors of electricity. In general, the better the electrical conducting properties of a metal, the better it is able to reflect light. Chromium is used as a plating for steel in the automotive industry because of its light-reflecting qualities when polished and its ability to protect against corrosion. Graphite is an allotrope of carbon and is best described as a semimetal. Because of its laminar (i.e., layered) structure, it has a gray-black matte appearance when viewed in one direction but good reflective properties when viewed in another. Graphite is highly lustrous when viewed with light reflecting from the laminar planes, but is matte when the light reflects from the edges of the planes.

Copper and gold are both colored metals because, unlike the "white" metals, they tend to reflect longer wavelengths of visible light more effectively than they reflect shorter wavelengths. The spectral reflectance curves of copper and gold are shown for the visible region in Figure VI-18. The reflectance of longer wavelengths gives them a yellowish cast in the bulk metal.

It is interesting to note that the coinage metals—copper, silver, and gold—are members of Group Ib of the periodic table. Copper, in row 4, is yellow; silver, in row 5, is "white"; and gold, in row 6, is yellow. This

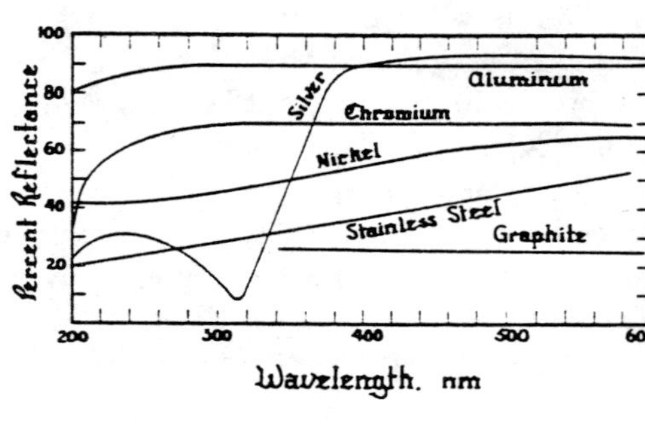

Figure VI-17. Reflection of light from several "white" metals and graphite.

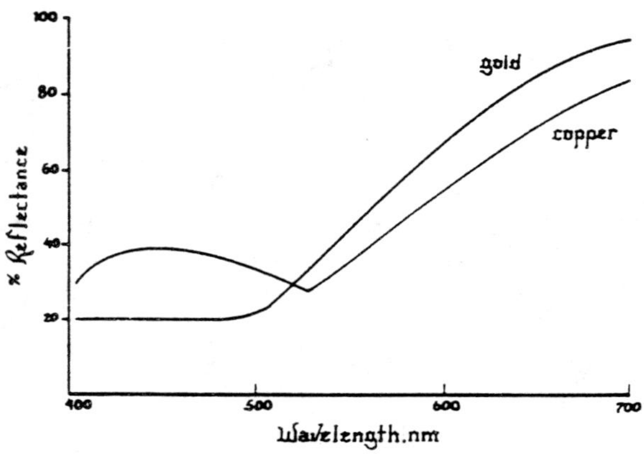

Figure VI-18. The spectral reflectance of the yellow metals, copper and gold.

apparent nonperiodicity in color has to do with the difference in energy of the inner-shell electrons in the three elements and is not easy to discuss on a rudimentary level. However, it can be noted from Figures VI-17 and VI-18 that the *shape* of the reflectance curve for each of the three elements resembles the others. All three elements exhibit a rather sharp increase in reflectance in the 300–500 nm range. This strong increase in reflection begins at about 320 nm in the ultraviolet for silver and hence it is "white" in the visible region. Copper and gold, on the other hand, do not begin to strongly reflect until the mid-visible range of wavelengths. The silver reflectance curve is merely shifted toward shorter wavelengths from the copper and gold curves.

The colors of metals are sensitive to grinding to submicroscopic particle size and to alloying. If gold, for example, is diluted with 25–80% silver, an ancient silvery–white alloy usually referred to as *electrum* is produced. The electronic structure of the metal in the alloy is altered to the extent that one almost entirely loses the yellow color of gold rather than merely desaturating it. Finely divided metals show striking colors characteristic of the electron gas. Finely divided gold is purple, while copper is red and silver is yellow. Other metals, such as mercury, become black when in the form of colloidal sized particles. The use of finely divided metals in producing color is important in glass manufacture and will be discussed in Chapter XI.

Shiny metal objects give a complex set of light signals to the eye. The color quality of local areas of the object is not always characteristic of the entire surface. For example, a silver teapot has differing shades of gray on the surface, and gold objects have hues ranging from orange to yellow in grooves and channels and highlights. The appearance of glossiness depends in part upon the eye and the brain perceiving three-dimensionality in the object. These perceptions are a subject of psychology and will not be considered further here.

Glass, Glaze, and Enamel

Glass and glazed objects are among the oldest forms of art. They date prior to the third millenium B.C. in Egypt. Glass, glaze, and enamel have a number of similar characteristics which permit them to be grouped together. They are fundamentally different from metals as a class of materials. Glasses are electrical insulators because of their electronic structure. When pure, they are transparent, hard but brittle, and reflect light efficiently, though not as efficiently as certain metals do.

Glasses are continuous solutions of silicate mixtures. The structure of glass is extremely complex and contains molecules of different sizes and shapes as well as many minor components. Glasses are not homogeneous chemical compounds, because the molecules lack a definite stoichiometry. They are not crystalline, because the molecules are arranged in a disordered fashion. Atomic ordering does exist because the SiO_4 polyhedra that form the basic building blocks retain essentially tetrahedral geometry. These SiO_4 units are linked together in a three-dimensional network by bridging the oxygen atoms. Pockets are present which can encapsulate cations in order to maintain a neutral electrostatic charge in the material. Figure VI-19 shows a two-dimensional representation of soda silicate glass.

Glasses can be thought of as vitreous silica and in this terminology they can be considered as extremely viscous solutions. The high viscosity of glass at room temperature is due to internal friction of atoms in the silica aggregates. The more cross-linking that occurs between Si–O–Si units in the network, the more viscous is the glass. High viscosity confers hardness, mechanical strength, and chemical inertness to glass.

Glazes are also glasses by their chemical and physical nature, but they are used differently. A glaze is a fused silicate mixture employed as an inert protective coating for ceramic clayware. Ordinary glaze is glossy and transparent. Glass and glaze owe their transparency to the lack of dispersed particulate matter in the matrix and the relatively unpolarizable nature of the atoms and bonds which make up the network. However, glaze can be made to take on a full range of appearances from transparent to opaque and from glossy to matte. The means by which these gradations in appearance are obtained are essentially the same as those for paints. Varying degrees of opacity can be obtained by suspending insoluble particulate matter in the glaze matrix. The greater the index of refraction of this opacifier, the more opaque the glaze becomes. For example, most

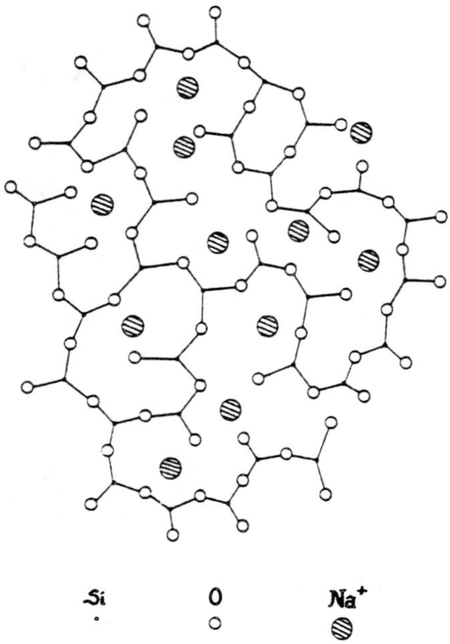

Figure VI-19. The structure of soda-silicate glass in two dimensions.

opacifiers are metal oxides. Probably the oldest and most commonly encountered opacifier in pottery is SnO_2. Its index of refraction ($\eta = 2.04$) is considerably greater than that of silicate glaze ($\eta = 1.55$), and this results in a great deal of light scattering at the surface of the tin oxide crystallites. The particle size range needed to achieve optimum light scattering and therefore the greatest opaqueness is equal to λ, where λ is the wavelength of light incident onto the particle. The cloudy opalescence that SnO_2 imparts gives a bluish tint to reflected light but a yellow-orange tint to transmitted light. This is because the smaller particles of SnO_2 are subject to Rayleigh scattering. As with paints, increasing the number of particles will give glazes a more matte appearance. Degrees of opacity in glaze also can be obtained by varying the cooling temperature. In a molten form the molecules and ions present move about freely. As the temperature drops, seed nuclei form. If the glaze cools rapidly, the crystallites are small and transparency results. Slow cooling causes larger crystallites and thus more opacity.

In glass of ancient Egypt, Phoenicia, Greece, and Italy, unusual effects were achieved by infusion and blending together of glass rods having different colors and opacities. Millefiori glass of Venice is an example of this method. The appearance of onyx, jasper, and agate were mimicked in Roman times by layering glass of different colors and then carving designs in the outer layer.

The term *enamel* is applied to glazes that are used in a relatively thin layer to hide underlying features from the observer. Usually an enamel is a densely opaque, white glaze applied to metals. Enameling has not always been done with the intention of hiding the underlying features of the metal, however. Enameling metals with a translucent glaze can enrich the quality of the metal surface. Translucent enamel on silver keeps its inherent color, but when used on gold, the yellow color of the gold is filtered through the enamel coating.

In general, ions and metals can be placed in an enamel matrix to give striking colors and effects. Blue enamels containing cobalt, purple enamels containing manganese, and green enamels containing chromium are known. A more detailed discussion of this facet of glass and glazes is deferred to Chapter XI, where the origin of color in inorganic materials is taken up.

Fibers and Woven Materials

Fiber surfaces found in textiles and other woven materials can have widely differing appearances. The gloss, texture, and transparency are the most notable features other than color. The angle of viewing or illumination plays a role in the appearance. The differences in visual appearance can be traced to variations in the size, shape, regularity, and roughness of the individual fibers and to differences in the orientation of the fibers caused by the weave. Both the nature of the fiber and the method used in weaving are

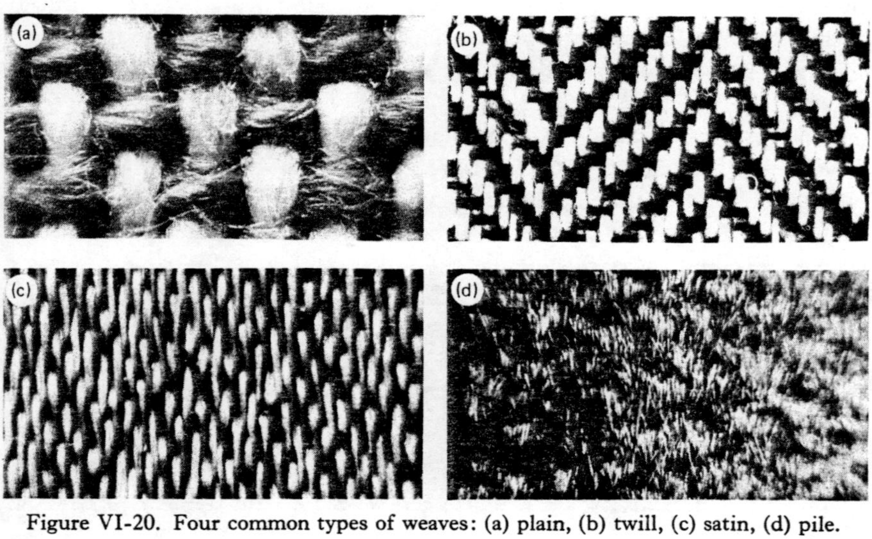

Figure VI-20. Four common types of weaves: (a) plain, (b) twill, (c) satin, (d) pile.

necessary to consider in attempting to understand the surface appearance of fabrics.

The most commonly encountered waves are the *plain* weave, the *twill* weave, the *satin* weave, and the *pile* weave. Figure VI-20 shows these four types of weaves. The plain weave is the most widely used because of its strength and simplicity. The pattern is a checkerboard one. In the twill weave the arrangement of the yarn gives a diagonal checkered pattern. The satin and sateen weaves have relatively long yarn exposure and give a smooth, lustrous surface. The pile weave orients the long axis of the fiber perpendicular to the surface of the fabric.

There are four major natural fibers which have been woven into textiles over the centuries. These are cotton, flax, wool, and silk. Cotton and flax originate from plants, while wool and silk are animal products. *Cotton* in textiles has been used for over 5000 years in India. The cotton fiber is a single, long, cylindrical cell composed mainly of cellulose. *Flax* is used to make linen and has been known for this application since the Stone Age. The flax fiber is a cellulose–lignin composite and is more irregular than most other natural fibers. However, the fabric obtained can be smooth and flat as well as strong. *Silk* is a long, slender protein fiber spun by the silkworm, and has been used as a fabric in China for more than 4000 years. The importance of the orientation of a fiber in achieving different appearances is nowhere more clearly evident than with silk. Plain weaves give broadcloth and chiffon, satin weaves give lustrous satin, twill weaves give foulard and peau de soie, and pile weaves give velvet and plush. Brocade and damask are other important silk fabrics which vary in appearance depending upon the weave. *Wool* is the protein fiber produced by sheep. It has been employed in the manufacture of textiles for nearly 6000 years. Wool consists of an inner shaft of bundled, elongated, spindle-shaped cells which are surrounded by a scale of shingle-like cells. The scale may be lost by repeated washings, however.

Gloss in fabrics occurs when the fabric is smooth and laid flat with the direction of illumination opposite the viewer. Smoothness is often mechanically achieved by pressing the fabric between rollers or plates. The process is generally known as *calendering* the fabric and must be repeated after laundering or cleaning. Figure VI-21 illustrates the appearance of the fabric surface before and after calendering. The principle behind calendering has been known for centuries and was used by certain Indian cultures. Damask linen is a good example of luster in cloth achieved by calendering.

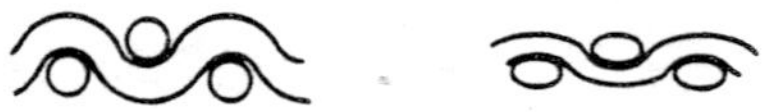

(a) (b)

Figure VI-21. A cross section of a fabric (a) before and (b) after calendering.

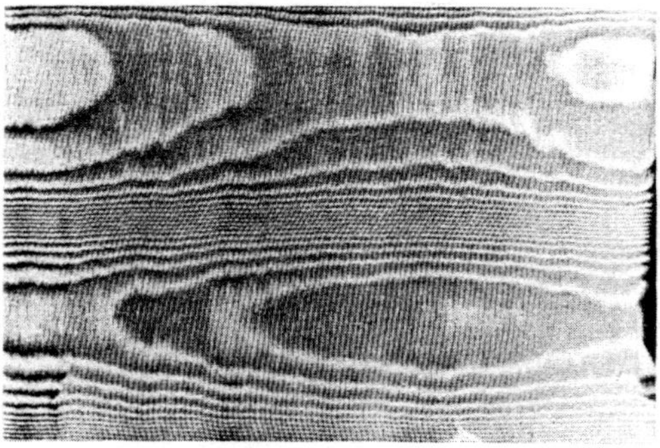

Figure VI-22. A Moiré fabric.

Different spots of lightness and darkness can be observed in damask, depending upon the direction of viewing. Glossiness, in general, comes from viewing the fibers from the side rather than from the end. Pile fabrics such as velvet and plush are matte when viewed into the pile, but glossy when viewed at an angle or on the folds. Moiré fabrics, also called "watered" fabrics or "watered silk," have a soft luster in the form of wavy, irregular-shaped ribs as seen in Figure VI-22. To create this effect, which has been practiced since the fifteenth century in France, the fabric is folded so that the ribs of yarn do not lie parallel. The fabric is calendered to make impressions of the yarn ribs on each other. When unfolded, the irregular-shaped ribs form a wavy, lustrous pattern. Interference and reflection of light from the misaligned parallel lines in the fabric lead to the optical effect. Moiré patterns can be seen sometimes when canvas and a finer fabric are overlayed during the relining of paintings.

The light reflected from silk fabrics is generally more intensely colored than that from wool fabrics, primarily because the silk fibers are smoother. The amount of specular reflection of light and the amount of light able to penetrate the surface of the fiber, where it can be selectively absorbed by the dye molecules, are thus greater in silks than in wools. The smoothness of the silk fibers causes their appearance to depend partly upon their orientation. Wool, on the other hand, diffusely reflects much of the light from the scales and bunched, irregular, crumpled fibers. Its appearance is much less dependent upon the angle of viewing. However, wool often has little ends of fiber sticking up from the surface that have a selected orientation. Therefore some degree of textural and color variation still can be noted as a function of the angle of viewing.

When repairing fabrics one must pay careful attention not only to color matching but also texture matching. It is possible to match a texture

Table VI-5. Ultraviolet Reflectance of Dry White Pigments, Bases, and Extenders

Compound	Ultraviolet reflectance factor at 300 nm, %
Lead-free zinc oxide	3
Leaded zinc oxide (65% ZnO–35% $PbSO_4$)	4
Zinc sulfide	6
Titanox B (TiO_2–$BaSO_4$)	6
Lead titanate	6
Titanium dioxide	7
Lithopone (70% $PbSO_4$–30% ZnO)	8
Zirconium oxide	41
Diatomaceous silica (Celite 110)[a]	45
Basic sulfate white lead (70–75% $PbSO_4$, 20% PbO, 5% ZnO)	48
China clay[a]	54
Aluminum oxide[a]	55
Basic carbonate white lead (Dutch process)[b]	62
Aluminum hydroxide[a]	67
Zirconium oxide	78
Magnesium carbonate[a]	81
Calcium carbonate	83

[a] An extender or base.
[b] Also called "flake white." A corroded form of $PbCO_3$ in $2PbCO_3 \cdot Pb(OH)_2$.

at one angle of viewing and have it not match at another. The orientation of the fibers must be carefully controlled.

Virtually all natural fabrics have an imitating synthetic counterpart. Acetates, nylon, and rayon can be made to imitate the appearance of silk, while acrylics can be used to imitate wool. Rayon can be made as an imitation for cotton. The trained eye can usually distinguish synthetic fabrics from natural ones because the synthetic yarns have a controlled fiber structure and size. Their physical characteristics can be adjusted to a particular method of weaving or desired appearance.

Reflection of Ultraviolet Radiation

We have been concerned up to this point with reflection of visible light as one of the dominant features in surface appearance. One may also need to consider the reflection of invisible ultraviolet radiation. Ultraviolet radiation can cause a great deal of damage to many types of organic materials, including paper, textiles, some dyes, and wood. To a lesser extent, inorganic materials can be altered by the effects of ultraviolet radiation. Direct exposure of light-sensitive materials to strong ultraviolet radiation is very inadvisable. However, a fractional percentage of the radiation emitted from common visible light sources used for illumination may occur in the ultraviolet region. Some direct exposure may be unavoidable. More often art objects are illuminated by indirect lighting. That is, the light encountering the object has undergone one or more reflections from a neutral wall or ceiling. In this situation it is the *ultraviolet* absorption and reflection characteristics of wall and ceiling paints that are of considerable interest. The absorption and reflection characteristics of paints in the visible energy range are no guide to their performance at ultraviolet energies. For example, titanium dioxide reflects visible light very well, but is a poor reflector in the ultraviolet. White plaster reflects both ranges very well. In general, the amount of ultraviolet radiation reflected from materials used in white walls is 3–80% of that incident onto it. Table VI-5 gives the percentage of ultraviolet radiation reflected from various white pigments and bases. Titanium dioxide and zinc oxide are the most effective white pigments at absorbing ultraviolet radiation and probably should be the white pigments of choice on museum walls. Other common whites include lead white and calcium carbonate, and they should be used with caution if ultraviolet reflectance is of concern.

BIBLIOGRAPHY

Beck, J., *Surface Color Perception*, Cornell University Press, Ithaca, New York (1972).

Berger, G., Weave interference in vacuum lining of pictures, *Stud. Conserv.* **11**, 170 (1966).

Brill, T. B., The appearance of materials, *Chemistry* **50**(10), 3 (1977).

Brommelle, N., Bloom in varnished paintings, *Mus. J.* **55**, 263 (1956).

Conservation of Paintings and the Graphic Arts (IIC Lisbon Conference) (1972), pp. 133, 139.

Cooke, H. L., *Painting Techniques of the Masters*, Watson-Guptill, New York (1972).

Feller, R. L., Stolow, N., and Jones, E. H., *On Picture Varnishes and Their Solvents*, The Press of Case Western Reserve University, Cleveland, Ohio (1971).

Fry, G., *The Varnishes of the Italian Makers of the 16th, 17th, and 18th Centuries and Their Influence on Tone*, Stevens and Sons, London (1904).

Gettens, R. J., and Stout, G. L., *Painting Materials*, Dover, New York (1966).

Harman, C. G., *Ceramic Glazes*, Cahners, Boston, Massachusetts (1973).

Hartshorne, N. H., and Stuart, A., *Crystals and the Polarizing Microscope*, 4th ed., Edward Arnold, London (1970).

Jensen, L. N., *Synthetic Painting Media*, Prentice-Hall, Englewood Cliffs, New Jersey (1964).

Jones, G. O., *Glass*, Chapman and Hall, London (1971).

Joseph, M. L., *Introductory Textile Science*, 3rd ed., Holt, Rinehart and Winston, New York (1977).

Kingery, W. D., *Introduction to Ceramics*, Wiley, New York (1960).

Koller, L. R., *Ultraviolet Radiation*, 2nd ed., Wiley, New York (1965).

Kühn, H., Aspects of panel painting, *Ciba Rev.* **3** (1963).

Larney, J., *Restoring Ceramics*, Bornie and Jenkins, London (1975).

Madsen, H. B., A new product for mending glass, *Stud. Conserv.* **17**, 131 (1972).

Mattiello, J. J., *Decorative and Protective Coatings*, Vols. I–IV, Wiley, New York (1941–46).

Mauersberger, H. R., *Matthew's Textile Fibers*, 6th ed., Wiley, New York (19[illegible]).

Mayers, R., *The Artist's Handbook of Materials and Techniques*, Viking Press, New York (1957).

Metelli, F., The perception of transparency, *Sci. Am.* **230**(4), 91 (1974).

Minnaert, M., *The Nature of Light and Color in the Open Air*, Dover, New York (1954).

Oster, G., and Nishijima, Y., Moiré patterns, *Sci. Am.* **208**(5), 54 (1963).

Patterson, D., *Pigments: An Introduction to the Physical Chemistry*, Elsevier, London (1967).

Pettingill, O. S., Jr., *Ornithology in the Laboratory and Field*, Burgess, Minneapolis, Minnesota (1967).

Plesters, J., A preliminary note on the incidence of discoloration of smalt in oil media, *Stud. Conserv.* **14**, 62 (1969).

Pomerantz, L., *Is Your Contemporary Painting More Temporary Then You Think?*, International Book Co., Chicago, Illinois (1962).

Rood, O. N., *Color: A Textbook of Modern Chromatics*, Trukner, London (1904).

Ruhemann, H., *The Cleaning of Paintings*, Praeger, New York (1968).

Ruskin, J., *Modern Painters, I*, Wiley, New York (1885).

Thomson, G., Topics in the Conservation of Surfaces, in *Applications of Science in the Examination of Art Works*, Museum of Fine Arts, Boston, Massachusetts (1971), p. 73.

Van Asperen de Boer, R. J., *Infrared Reflectography*, Central Research Laboratory, Amsterdam (1970).

Welty, J. C., *The Life of Birds*, Saunders, Philadelphia, Pennsylvania (1975).

Weyl, W. A., *Colored Glasses*, Dawson's of Pall Mall, London (1959).

VII

The Origin of Color in Organic Compounds

Up until now in this book the concepts of reflection, transmission, and absorption of light by materials have been discussed more from a macroscopic physical point of view than from a microscopic chemical point of view. Now we delve further into the "hows" and "whys" of the absorption process on a molecular level. The color and photoreactions of materials result from absorbed light. Hence, the absorption of light is a topic of theoretical and practical interest to the serious artist as well as to most museum personnel.

The term "organic" is not intended to imply that the compounds necessarily are living or were once part of a living species, but rather it refers to the presence of a carbon atom backbone in the molecule. Organic compounds have this skeleton, but do not all originate in living species. The majority of organic coloring agents are dyes. Dyes are almost always homogeneously dispersed on a molecular level in the substrate they color.

In order to deal with the origin of color on a molecular level, it is worthwhile to follow the advice of the King of Hearts and "begin at the beginning, go on till you come to the end, then stop." We will begin with the structure of the atoms which bind together to form the molecules, and build from this. The end will come when the methods have been delineated by which molecules have the capacity to absorb electromagnetic energy.

ATOMIC STRUCTURE

At the 1904 World's Fair in St. Louis, Missouri, the foremost physical chemists met. They were divided for and against on the issue of the existence of atoms. It was not until 1906 that modern atomic theory got an important foothold with the experiments of Ernest Rutherford, which established the nature of atoms. They consist of a positively charged nucleus surrounded by a system of negatively charged electrons. The atom is held together by electrostatic forces. The effective volume of the nucleus is extremely small compared to the effective volume of the atom, yet almost all of the mass is concentrated in the nucleus. Immediately after Rutherford's model of the atom was presented, scientists turned to the question of the possible organization of the satellite electrons. The first major advance is credited to Niels Bohr in 1913, although an Oxford astronomer, J. W. Nicholson, arrived at some of the same conclusions a year earlier. Bohr pictured the electron in the hydrogen atom as moving in a circular orbit about the proton as shown in Figure VII-1. He was able to calculate the energy levels of the electron and its radius from classical mechanics and explain with remarkable accuracy the experimentally observed energy emission spectrum of the hydrogen atom. Bohr's model is relatively easy to grasp and is still taught today, but it was realized soon after its development that this classical picture was not a complete one. For instance, it could not account for the spectrum of hydrogen in a magnetic field, nor, more seriously, could it be extended beyond the hydrogen atom to correctly predict the properties of the multielectron atoms. A model which could be used only for a one-electron atom was not broadly useful.

A major advance came in 1924 when Louis de Broglie proposed that all matter possessed wave properties. De Broglie's matter waves, or wave particle duality, provided the foundation for a new theory which firmly linked the quantum properties of energy noted by Planck [equation (I-2)] to physical systems composed of matter. In 1926, Erwin Schrödinger advanced the well-known wave equation that relates the energy of an electron to its wave properties. This model serves as a basis for many modern quantitative attempts to "explain" atomic and molecular structure. Because of the rather complicated mathematical form of the Schrödinger equation and its solutions, we shall discuss the important results only qualitatively. Some of the crucial predictions of the model can be difficult to visualize at a low level. This is partly because the model applies to submicroscopic (quantum mechanical) bodies, which do not follow the laws

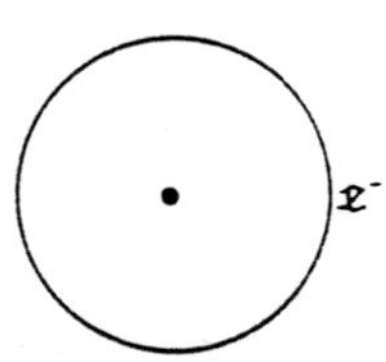

Figure VII-1. The Bohr atom.

and concepts of macroscopic Newtonian mechanics. Newtonian mechanics deals with the relatively large bodies of matter encountered in everyday experience. It is tempting, but incorrect, to try to transfer what we know to be true in Newtonian mechanics for large bodies of matter to the quantum mechanical world of atoms and molecules. There is a correspondence, but it is a subtle one.

One of the most important consequences of the wave–particle nature of matter is the uncertainty principle, which was proposed in 1927 by Werner Heisenberg. The *Heisenberg uncertainty principle* states that it is impossible to simultaneously know with certainty both the momentum and the position of an electron. This means that as a measurement of momentum is made more precisely, a measurement of the position of the electron becomes correspondingly less precise. Similarly, if the position of an electron is defined more precisely, its momentum becomes less accurately known. This is not a result of the poor state of the art with regard to the instruments that could be used for such measurements. It is a consequence of the quantum mechanical nature of very small particles. The practical lesson of the Heisenberg uncertainty principle is that we cannot simultaneously discuss electrons in terms of their exact energy and position. One must be satisfied with a "fuzzy" picture of electrons and think in terms of the probability of finding an electron in a given region of space rather than trying to determine its precise location.

The Schrödinger equation gives a picture of atoms in keeping with the Heisenberg uncertainty principle. It is a second-order differential equation, which can be solved exactly for an atom having one electron (e.g., the hydrogen atom or hydrogenlike ions such as He^{+}, Li^{2+}, Be^{3+}, etc.). A type of quantization is necessary to solve the Schrödinger equation. This quantization refers to a set of boundary conditions for the differential equation. The resulting solutions give specific wave functions (orbitals) and energies for electrons. An orbital can be likened to a region of space in which an electron resides. The orbitals are specified by three orbital quantum numbers which are in fact the boundary conditions that permitted a solution to the equation.

The first orbital quantum number is called the *principal quantum number* and is given the symbol n. In a general way, the effective volume of the electron orbital depends upon n, and the volume of the shell is related to energy. Thus the value of n is indicative of the energy of the electron. The larger n is for an electron, the greater is the energy of the electron. The allowed values for n for an atom are $n = 1, 2, 3, \ldots, \infty$. The second orbital quantum number is designated l. The l quantum number determines, in a general way, the shape of the region of space in which the electron moves. The quantum number l is called the *orbital angular momentum quantum number* or the *orbital shape quantum number*. Its allowed values are $l = 0, 1, 2, 3, \ldots, n - 1$ for each value of n. The third orbital quantum number m_l determines the orientation of the particular spatial configuration defined by l in relation to an arbitrary direction. The quantum number m_l is called the ***magnetic quantum number*** or sometimes the ***orbital orientation quantum number***. Its allowed values are $+l, +l - 1, \ldots, 0, \ldots, -l$ for a given value of l.

These three quantum numbers explicitly determine the size, shape, and orientation of the electron orbital in space. Each electron in an atom can be specified by a wave function corresponding to a solution of the Schrödinger equation for these three quantum numbers. There is a fourth quantum number, which does not evolve from the Schrödinger equation, but comes from the fact that an electron intrinsically has two different spin orientations with reference to an arbitrary axis. In a crude sense, the electron can be likened to a tiny bar magnet with a north pole and a south pole. The ***magnetic spin quantum number***, labeled m_s, specifies the direction of the electron spin in space with reference to the arbitrary axis. The two possible values of m_s for an electron are $+1/2$ and $-1/2$.

The Schrödinger equation can be written out for any atom, but, as already noted, it can be solved exactly only for one-electron atoms (H, He^{+}, Li^{2+}, etc.). Consequently, the three quantum numbers and the corresponding wave functions (orbitals) obtained by solving the Schrödinger equation of motion accurately apply only to hydrogen and these "hydrogen-like" ions. In spite of this, the concept of quantum numbers and orbitals remains broadly useful on a qualitative level in formulating the electronic structure of multielectron atoms. The orbitals of multielectron atoms are assumed to be analogous to those of the hydrogenlike orbitals, but are modified to account for the repulsive interaction between the negatively charged electrons. However, the "corrected" model is still only an approximate one.

The Pauli exclusion principle is the key to increasing the number of electrons to obtain a reasonable electronic description for multielectron atoms while retaining the simplicity of the one-electron description. Wolfgang Pauli stated that no two electrons in an atom can have all four quantum numbers the same. The process of building up the electronic description of multielectron atoms was called the "Aufbau" principle by Pauli. In Table VII-1 the electronic structure of atoms is built up through $n = 3$, using all of the allowed values for the four quantum numbers derived from the Schrödinger model and the restrictions of the Pauli principle. For example, with $n = 3$, the possible values of l are 2, 1, 0. Taking $l = 1$ to continue the example, the allowed values of m_l are 1, 0, -1. For each m_l value, m_s can be $+1/2$ or $-1/2$.

Symbols are attached to the sets of quantum numbers to distinguish

Table VII-1. The Possible Atomic Orbitals Through $n = 3$

n	l	Atomic orbital designation (n and l only)	m_l	Atomic orbital designation (n, l, m_l)	m_s	Maximum number of electrons in each orbital
1	0	1*s*	0	1*s*	$\pm 1/2$	2: $2e^-$ in *s* shells
2	0	2*s*	0	2*s*	$\pm 1/2$	2
	1	2*p*	1	$2p_x$	$\pm 1/2$	2
			0	$2p_z$	$\pm 1/2$	2 } $6e^-$ in *p* shells
			-1	$2p_y$	$\pm 1/2$	2
3	0	3*s*	0	3*s*	$\pm 1/2$	2
	1	3*p*	1	$3p_x$	$\pm 1/2$	2
			0	$3p_z$	$\pm 1/2$	2
			-1	$3p_y$	$\pm 1/2$	2
	2	3*d*	2	$3d_{xy}$	$\pm 1/2$	2
			1	$3d_{xz}$	$\pm 1/2$	2
			0	$3d_{z^2}$	$\pm 1/2$	2 } $10e^-$ in *d* shells
			-1	$3d_{yz}$	$\pm 1/2$	2
			-2	$3d_{x^2-y^2}$	$\pm 1/2$	2

each orbital in Table VII-1. The symbols *s*, *p*, *d*, and *f* carry over from the notation of early atomic spectroscopists. They used the terms sharp, principal, diffuse, and fundamental to describe lines in the energy emission spectrum of elements. The terms were abbreviated to *s*, *p*, *d*, *f*. This notation has been continued, so that an $l = 0$ orbital is always called an *s* orbital regardless of the value of n. An $l = 1$ orbital is a *p* orbital, an $l = 2$ orbital is a *d* orbital, and an $l = 3$ orbital is an *f* orbital. To distinguish *s*, *p*, *d*, and *f* orbitals arising from different n shells, the value of n is given preceding the letter, as in 2*p*, 3*p*, and 4*p* for $n = 2$, $n = 3$, and $n = 4$, respectively. The subscript listed after each orbital refers to the m_l quantum number. It gives the Cartesian direction(s) in space toward which the bulk of the electron density in the orbital points. For example, a 3*p* orbital can be oriented in the *x*, *y*, or *z* direction. The subscripts for *d* and *f* orbitals are more complicated than this because these orbitals direct electron density along two or more axes in space. The subscripts for the *f* orbitals are not given in Table VII-1.

Three-dimensional plots can be drawn of the wave functions for each orbital in Table VII-1. Those of the *s*, *p*, and *d* orbitals are shown in Figure VII-2. These plots are made for particular l and m_l values. Different n values will make the orbitals larger or smaller but will not change their shape or orientation. Smaller n values give smaller orbitals, while larger n values give larger orbitals. If signs appear on the lobes, they do not indicate electrostatic charges, but, rather, are the mathematical signs of the wave

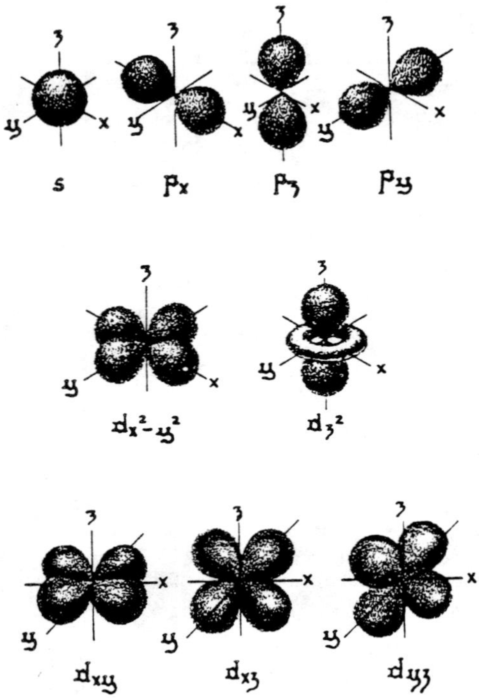

Figure VII-2. The three-dimensional shapes of the *s*, *p*, and *d* atomic orbitals derived from the solutions of the Schrödinger equation for the hydrogen atom.

function in that particular region of space. For example, half of the wave function of a *p* orbital has a negative sign, while the other half has a positive sign. The wave function changes sign at the origin. These signs have importance in chemical bonding. They must be considered when hybrid as well as molecular orbitals are constructed from the atomic orbitals.

Each of the atomic orbitals listed in Table VII-1 or shown in Figure VII-2 can hold a maximum of two electrons. The two electrons must have different m_s values ($+1/2$ and $-1/2$) according to the Pauli principle, and they are said to be paired electrons.

The next task is to assign electrons in proper sequence to the atomic orbitals to obtain a description of multielectron atoms. The orbitals fill with electrons in order of increasing energy. The lowest energy orbitals fill first and the highest energy orbitals fill last. The mnemonic device shown in Figure VII-3 is useful for remembering the experimentally observed sequence of the *n* and *l* quantum numbers in atoms. Figure VII-3 is constructed so that the *n* levels are ordered in rows with each successive row being offset one place to the right with respect to the row above it. Each row contains all of the allowed *l* values for the given value of *n*. The order of filling is obtained by moving vertically down the columns starting

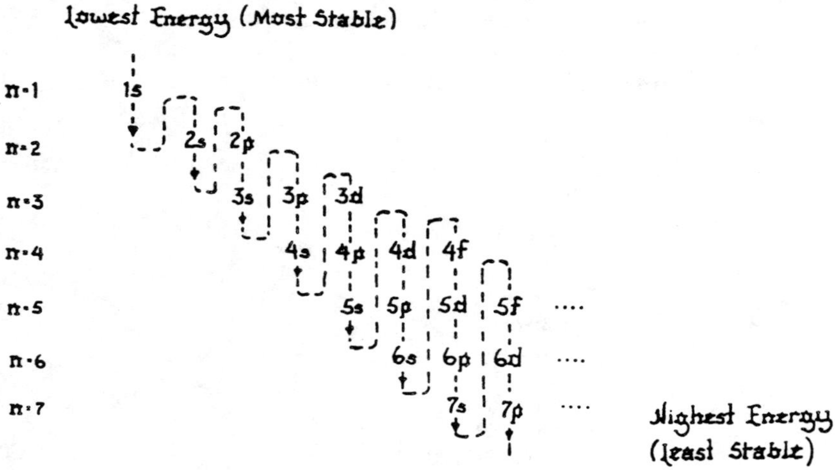

Figure VII-3. The experimentally observed order for assigning electrons to atomic orbitals in most atoms.

from the left-hand side and moving across to the right as shown by the dashed line. This sequence gives an energy ordering of shells as $1s2s2p3s3p4s3d4p5s4d5p\ldots$. The ordering follows a reasonable pattern through $3p$, but after this level the ordering changes and becomes increasingly irregular. However, the sequence shown is the experimentally observed one. As n becomes larger, the configuration of some atoms exhibits an exception even to this ordering because the orbitals become very close to one another in energy. These exceptions are not important for understanding color and photoreactions, so they will not be considered further.

The periodic table can now be built up in terms of electron shells, using the Aufbau principle. The electronic structure of some elements is shown in Table VII-2. A superscript of 1 or 2 is used to indicate the number of electrons in each level. Some details of energy ordering in the $2p$ level are given in this table for the elements boron through neon. Carbon, nitrogen, and oxygen are shown with unpaired electrons when it might be tempting to group these electrons together. However, *Hund's rule* states that within a given l shell, where the orbitals all have the same energy, one electron is placed in each m_l subshell before any of the electrons are paired. The filling of the d orbitals also follows Hund's rule. Hund's rule is another conceptual adjustment to the hydrogenlike model so as to be in accordance with experimentally observed behavior. Its validity can be proven mathematically, however.

Table VII-2. The Electronic Configuration of Several Elements Based on the Aufbau Principle

Element	Electron configuration	*p* Orbital filling
H	$1s$	
He	$1s^2$	
Li	$1s^22s^1$	
Be	$1s^22s^2$	
B	$1s^22s^22p^1$	$2p_x{}^1$
C	$1s^22s^22p^2$	$2p_x{}^12p_y{}^1$
N	$1s^22s^22p^3$	$2p_x{}^12p_y{}^12p_z{}^1$
O	$1s^22s^22p^4$	$2p_x{}^22p_y{}^12p_z{}^1$
F	$1s^22s^22p^5$	$2p_x{}^22p_y{}^22p_z{}^1$
Ne	$1s^22s^22p^6$	$2p_x{}^22p_y{}^22p_z{}^2$
Na	$1s^22s^22p^63s^1$	
Mg	$1s^22s^22p^63s^2$	
Sc	$1s^22s^22p^63s^23p^64s^23d^1$	

The modern periodic table has its origin in a time predating wave mechanics. In 1869 Dmitri Mendeleev in Russia and Lothar Meyer in Germany, working independently, published similar schemes for the classification of the elements. Their tables were the forerunners of the modern periodic table shown in Figure VII-4. The elements were arranged in order of increasing atomic number and it was observed that similar chemical and physical properties then recur periodically. It was later shown that elements and ions having the same outer-electron configuration have similar chemical properties. Thus the link between the outer-electron configuration of an element and the behavior of the element in bonding and chemical reactions was firmly made.

The various columns in the periodic table shown in Figure VII-4 are related to the electrons that are outermost. The Group Ia elements have ns^1 outer electron configurations and n increases upon descending each column of the table. The chemistry of all of these elements is similar. All Group IIa elements and helium have ns^2 outer electron configurations. Group IIIa–VIIIa elements have outermost electron configurations of np^1 to np^6. The elements in groups Ia–VIIIa (i.e., those with outer-electron configurations involving only s and p electrons) are commonly referred to as *main group* or *representative group* elements. The elements in Groups Ib–VIIIb are listed sequentially from IIIb to IIb and contain a block of nine elements in Group VIIIb for reasons which will not be extensively discussed here. The labeling is based on the observed highest formal oxidation that the element can achieve. Elements in the "b" block have outer-electron configurations involving nd^1 to nd^{10} from left to right. They

Representative elements ("a" block)

Transition elements ("b" block)

Noble gases

I	II	III	IV	V	VI	VII	VIII			I	II	III	IV	V	VI	VII	VIII
																1 H	2 He
3 Li	4 Be											5 B	6 C	7 N	8 O	9 F	10 Ne
11 Na	12 Mg											13 Al	14 Si	15 P	16 S	17 Cl	18 Ar
19 K	20 Ca	21 Sc	22 Ti	23 V	24 Cr	25 Mn	26 Fe	27 Co	28 Ni	29 Cu	30 Zn	31 Ga	32 Ge	33 As	34 Se	35 Br	36 Kr
37 Rb	38 Sr	39 Y	40 Zr	41 Nb	42 Mo	43 Tc	44 Ru	45 Rh	46 Pd	47 Ag	48 Cd	49 In	50 Sn	51 Sb	52 Te	53 I	54 Xe
55 Cs	56 Ba	57 La	72 Hf	73 Ta	74 W	75 Re	76 Os	77 Ir	78 Pt	79 Au	80 Hg	81 Tl	82 Pb	83 Bi	84 Po	85 At	86 Rn
87 Fr	88 Ra	89 Ac	104	105													

Inner transition elements

58 Ce	59 Pr	60 Nd	61 Pm	62 Sm	63 Eu	64 Gd	65 Tb	66 Dy	67 Ho	68 Er	69 Tm	70 Yb	71 Lu
90 Th	91 Pa	92 U	93 Np	94 Pu	95 Am	96 Cm	97 Bk	98 Cf	99 Es	100 Fm	101 Md	102 No	103 Lw

Figure VII-4. The periodic table of the elements.

are called *transition elements* or *transition metals* because all are metallic. The elements listed at the bottom of the table fit into the table after lanthanum (La) and actinum (Ac) and are called *inner transition elements* or more commonly the *lanthanides* and *actinides*. Their outer-electron configuration is nf^1 to nf^{14}. They are separated from the rest of the table merely for reasons of convenience. With them properly placed, the periodic table would have such an excessive width that it would not fit onto most book pages!

It is clear that the periodic table, which is based upon grouping elements together systematically according to their observed chemical properties, can be built up using the results of the Schrödinger model, electron spin, and the Pauli exclusion principle. Hence, a model recognizing the wave–particle duality of electrons must be basically the correct one. The Bohr model considered only the particle aspects. It remains now to use this electron formalism to understand how and why chemical bonds form between these atoms. Once this is done, the interaction of light with electrons in molecules can be developed and the origin of color and photoreactions discussed.

CHEMICAL BONDING AND MOLECULAR STRUCTURE

The first models using electrons to form chemical bonds were crude ones. In the most notable attempt, G. N. Lewis in years prior to 1920 introduced a shorthand notation referred to as "electron dots" to represent the outermost or valence electron description of atoms. The *valence electrons* are defined as those electrons in the highest n shell plus any unpaired d electrons. Only the valence electrons need to be considered in most discussions of chemical bonding and reactions. Lewis did not have the benefit of the atomic orbital theory just presented. He made his deductions from the stoichiometry and the observed chemical behavior of simple compounds. His deductions proved to be in basic accordance with the atomic orbital description and the Aufbau principle.

Examples of the Lewis notation for several elements and ions are shown below. The valence atomic orbitals corresponding to the dot formalism are shown in parentheses. The inner- or core-electron configuration of electrons does not need to be written because it is not involved in the chemistry of the species under normal conditions. It is chemically reasonable according to Hund's rule to leave the electrons unpaired insofar as is possible when using the electron dot notation.

lithium	$Li\cdot$	$(2s^1)$
lithium ion	Li^+	$(2s^0)$

nitrogen	$\cdot\underset{\cdot}{\ddot{\mathrm{N}}}\cdot$	$(2s^22p^3)$
phosphorus	$\cdot\underset{\cdot}{\ddot{\mathrm{P}}}\cdot$	$(3s^23p^3)$
chlorine	$:\underset{\cdot\cdot}{\ddot{\mathrm{Cl}}}\cdot$	$(3s^23p^5)$
chloride ion	$:\underset{\cdot\cdot}{\ddot{\mathrm{Cl}}}:^-$	$(3s^23p^6)$

Lewis was able to extend the electron dot model of atoms to include chemical bonding between atoms by noting that atoms usually seek a noble gas configuration of valence-shell electrons (i.e., they seek the configuration of the Group VIIIa elements) when forming ions or entering into chemical bonding. He referred to this tendency as the *octet rule* because many elements seek to have eight electrons in their valence shell. Hydrogen, of course, seeks to share only two electrons in a bond because the next higher noble gas element (helium) possesses only two electrons. The Lewis model proved to be fairly satisfactory for hydrogen and the second-row elements of the main group, but it often failed for the third and higher rows. Hence, the octet rule is hardly a general rule, but rather mostly applies to the second-row elements. For discussion of organic compounds, which are mainly composed of first- and second-row elements (H, C, O, N, etc.) the Lewis model is reasonably successful for rationalizing stoichiometries.

Problems are frequently encountered with inorganic compounds, which contain elements spanning the entire periodic table. Lewis attempted to incorporate the elements of rows 3, 4, and 5 into his model by suggesting that the third and higher row elements can "expand" their octet to include 10 or 12 electrons in the valence shell. A simplified picture of organic compounds is developed in the examples that follow. The expansion of the octet will not be needed in most cases because the elements in these examples come mainly from the second row. The Lewis model often leads to the correct prediction of the stoichiometry in simple compounds, and gives a valence-electron distribution consistent with other, more sophisticated models if one visualizes the electron pair arrangement in three dimensions. Naturally such an elementary model for complex molecular systems will be subject to many exceptions. A high degree of faith in the broad application of the Lewis model would be ill-advised.

In the Lewis model, chemical bonds are represented as pairs of electrons between atoms. The Lewis structure for two hydrogen atoms can be combined to form the hydrogen molecule, H:H, in which each hydrogen atom can be thought of as having the same number of electrons present in the noble gas element helium. The H_2 molecule is reasoned to be stable

Table VII-3. Lewis Structures for Some Simple Molecules

Compound name	Constituent atoms	Lewis structure
Hydrogen fluoride	$:\ddot{\underset{\cdot\cdot}{F}}\cdot + \cdot H$	$H—\overline{\underline{F}}|$
Methane	$\cdot\dot{\underset{\cdot}{C}}\cdot + 4H\cdot$	H | H—C—H | H
Ammonia	$:\dot{\underset{\cdot}{N}}\cdot + 3H\cdot$	H | |N—H | H
Water	$:\dot{\underset{\cdot\cdot}{O}}\cdot + 2H\cdot$	H | $|\underline{O}$—H
Carbon tetrachloride	$\cdot\dot{\underset{\cdot}{C}}\cdot + 4:\ddot{\underset{\cdot\cdot}{Cl}}\cdot$	$|\overline{Cl}|$ | $|\overline{\underline{Cl}}—C—\overline{\underline{Cl}}|$ | $|\underline{Cl}|$
Sodium chloride	$Na\cdot + \cdot\ddot{\underset{\cdot\cdot}{Cl}}:$	$Na^{+}\ |\overline{\underline{Cl}}|^{-}$

because of this fact. Fluorine atoms react with each other to form the fluorine molecule $:\ddot{\underset{\cdot\cdot}{F}}:\ddot{\underset{\cdot\cdot}{F}}:$ where each fluorine atom has the configuration of eight electrons present in the next higher noble gas, neon. Valence strokes are often used in place of the pairs of dots, so that fluorine also can be represented as $|\overline{\underline{F}}—\overline{\underline{F}}|$.

Continuing along the same line, we give in Table VII-3 examples of bonding in some small molecules according to the Lewis model. One or more lone pairs of electrons are present in nitrogen, oxygen, fluorine, and chlorine in these compounds. In the case of sodium chloride, a salt forms in order to best satisfy the octet requirement. The sodium ion loses an electron to achieve the noble gas configuration of neon, while chlorine gains an electron to achieve the configuration of argon. The six examples in Table VII-3 involve a single pair of electrons forming the bonds between atoms.

None of the compounds in Table VII-3 is colored. The eventual objective is the description of bonding in organic compounds which have

Table VII-4. The Lewis Structures for Simple Molecules Containing Multiple Bonds Between Atoms

Compound name	Lewis structure
Ethylene	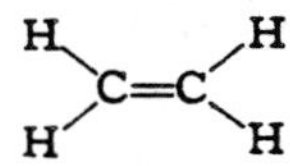
Acetylene	H—C≡C—H
Formaldehyde	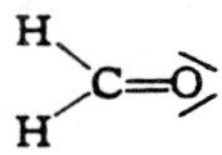
Oxygen[a]	$|\overline{\mathrm{O}}=\overline{\mathrm{O}}|$
Nitrogen	$|\mathrm{N}\equiv\mathrm{N}|$
Hydrogen cyanide	$\mathrm{H}—\mathrm{C}\equiv\mathrm{N}|$

[a] See text.

color. Colored compounds usually contain multiple bonds between atoms. In many cases of multiple bonding, the octet rule can still be applied. A double bond and a triple bond are represented by two and three pairs of electrons between atoms, respectively, as shown in the examples in Table VII-4.

Ethylene and formaldehyde contain double bonds, while acetylene, nitrogen, and hydrogen cyanide contain triple bonds. In each case the molecules are composed of atoms which have a noble gas electron configuration. Only one satisfactory Lewis structure exists in each case, including oxygen. However, a problem arises here, for the model is not in accordance with the known chemical properties of oxygen. The Lewis structure suggests that the molecule should be relatively unreactive because of the double bond and the absence of unpaired electrons. On the contrary, oxygen is chemically very reactive and the molecule is known to have two unpaired electrons. The reactivity arises from these unpaired electrons. A chemically consistent Lewis structure for oxygen is as follows, but note that it does not satisfy the octet rule:

$$|\dot{\mathrm{O}}\equiv\dot{\mathrm{O}}|$$

Thus the Lewis model fails to "explain" oxygen. Nitrogen, on the other hand, is extremely stable because it has a triple bond and no unpaired electrons. Hydrogen cyanide is said to be *isoelectronic* with N_2 because both molecules possess the same number of electrons. All of the molecules shown except for oxygen are well described by the Lewis model, but in general one should be cautious about drawing a Lewis structure for a molecule without also considering other known facts. Furthermore, there are many molecules and ions for which more than one Lewis structure is satisfactory. This latter situation actually can be very helpful in gaining an elementary understanding of the production of color in organic compounds.

The concept of *resonance* must be introduced to rationalize why several equivalent Lewis structures can exist for a molecule. Resonance occurs if two or more structures of the same total energy, which differ only in the position assigned to the electrons, can be drawn for a molecule. The actual electronic description of the molecule is not accounted for by any of these structures alone, but is intermediate between all possibilities. For example, if the position of the nitrite ion NO_2^- is fixed in space, then the distribution of the electrons can be drawn in two ways:

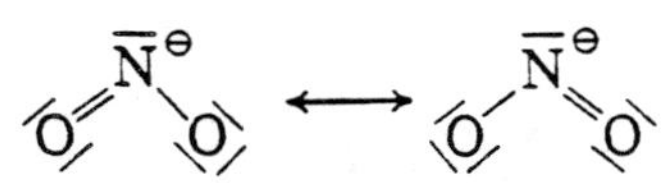

The molecule has not been flipped over, but the position of the electrons in the N—O bonds has been switched. The two structures have the same total energy and differ only in the position assigned to the electrons. Hence, they are called *resonance structures*, and NO_2^- is not described by either structure alone but is intermediate between the two. The N—O bond is effectively a "one and one-half" bond because half the time it is a single bond and half the time it is a double bond. Benzene, C_6H_6, is another molecule with resonance structures. The two structures shown below are equivalent in energy and differ only in the position assigned to the electrons:

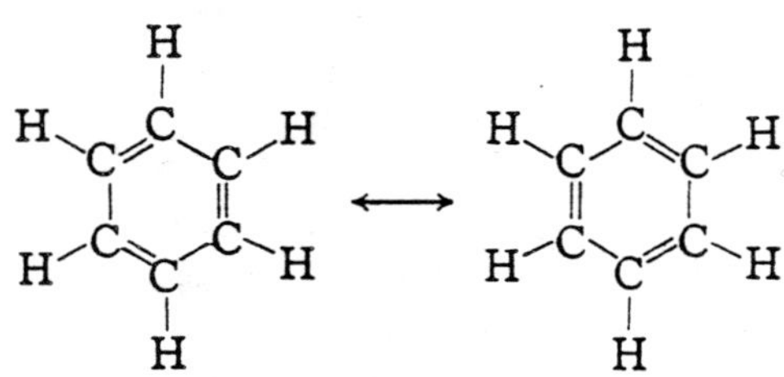

Para-nitroaniline, shown below, might be thought of as having two resonance structures, but these are energetically different and thus cannot be true ground-state resonance structures. The conversion of (A) into (B) requires an input of energy, but results from shifting around the existing electrons in a resonance-like manner.

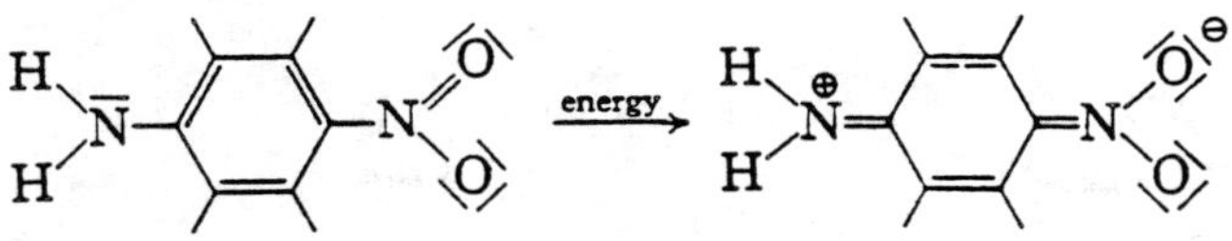

(A) (B)

The concept of such resonance-like structures can provide a simple means of understanding how some molecules absorb electromagnetic energy. *Para*-nitroaniline illustrates this point because absorption of energy causes a redistribution of charge in the molecule. On the other hand, the resonance structures of benzene cannot be used for energy absorption because the two structures shown have the same energy. Yet it is well known to chemists that benzene does absorb electromagnetic energy (but not visible light energy). After consideration of these facts and a large body of others, the conclusion must be reached that the Lewis concepts, even with resonance added, are a good start but they cannot give a complete description of bonding in molecules.

A more comprehensive and sophisticated model of the electronic structural properties of molecules is obtained by considering how the valence orbitals of atoms shown in Figure VII-2 might become involved in chemical bonds. The powerful method that evolves is known as *molecular orbital* (MO) theory, for which Robert S. Mulliken receives most credit. Mulliken and others in the 1930s and 1940s considered that the electron clouds of atoms lose some of their individual characteristics in a molecule and become "smeared out" over the molecular surface. Mathematically, they handled this problem by taking linear combinations of the atomic orbitals (LCAOs) on two or more bonded atoms to form a new set of orbitals called molecular orbitals. A linear combination of two equations (orbitals) is obtained by taking the additive and subtractive combinations to generate two new equations (orbitals). For example, if the wave function for an *s* orbital on one atom is combined linearly with an *s* orbital on another, then two new wave functions result. This process is illustrated pictorially in Figure VII-5a.

Two *s* orbitals, one on atom A and the other on atom B, are combined linearly along the x-axis direction. The signs of the wave function must be used to obtain the proper phasing. The positive combination is called a σ_s molecular orbital and results in *bonding* between atoms A and B because the electron density is maximized between the atoms. It is worthwhile to emphasize again that the signs shown correspond to the sign of the wave function in various regions of space and do not refer to electrostatic charge. The negative combination is called a $\sigma_s{}^*$ molecular orbital and does not result in bonding. It is, in fact, said to be *antibonding* because the electron

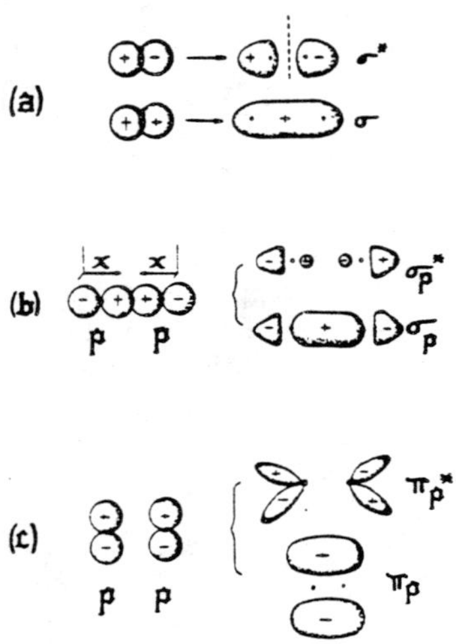

Figure VII-5. Contour diagrams for molecular orbitals formed from linear combinations of atomic orbitals. (a) Two *s* orbitals, (b) two *p* orbitals on the bonding axis, (c) two *p* orbitals perpendicular to the bonding axis.

density becomes zero (forms a node) in a region between atoms A and B. As the names "bonding" and "antibonding" imply, bonding results in fusing the atoms together, while antibonding means the atoms are forced apart. This point will be made more decisively later in this discussion when electrons are assigned to the orbitals.

The p_x orbitals on atoms A and B lie along the x axis (the internuclear axis) and point toward one another. As before, two molecular orbitals can be generated by taking linear combinations of the atomic orbitals. Figure VII-5b illustrates the results that are obtained. Because of the axis system employed, the positive combination gives a bonding interaction and is called a σ_{p_x} molecular orbital. The positive regions add to one another and concentrate electron density between the atoms. The negative combination is antibonding and is called a $\sigma^*_{p_x}$ molecular orbital. The $\sigma^*_{p_x}$ molecular orbital has the appearance of two distorted p orbitals but with a node between the atoms A and B. In general, a σ-bonding molecular orbital of any kind has no nodes containing the internuclear axis, while a σ-antibonding molecular orbital also has no nodes containing the internuclear axis, but has a node *perpendicular* to the internuclear axis.

The p_y and p_z orbitals of atoms A and B also can be combined to form molecular orbitals. The characteristics of these molecular orbitals are somewhat different than the σ–σ^* molecular orbital set formed from the p_x atomic orbitals. This is because the p_y and p_z orbitals on atoms A and B are perpendicular to the internuclear axis. Figure VII-5c illustrates the results of the linear combinations of such orbitals. The positive combination gives a bonding molecular orbital in the xy plane or xz plane, depending upon whether the p_y set or the p_z set of orbitals is chosen. These molecular orbitals are called π-bonding MOs. The negative combination of either set gives an antibonding molecular orbital having a node perpendicular to the internuclear axis. These molecular orbitals are called π^* MOs. The linear combination of p_y atomic orbitals or p_z atomic orbitals produces molecular orbitals of the same sign combination and shape. The p_y set is rotated 90° with respect to the p_z set in accordance with the axis system. In general, a π-bonding MO has one node containing the internuclear axis, and a π^*-antibonding MO has, in addition to this node, a node perpendicular to the internuclear axis. As before, the electron density goes to zero between atoms when the molecular orbital is antibonding.

In any particular set, the bonding molecular orbitals are lower in energy than the antibonding molecular orbitals. Even though the antibonding molecular orbitals may not contain electron density in the ground state and hence can be considered as only "virtual" orbitals, they will become important in discussions of color and photoreactions. Particular note of them should be made in each example.

The purpose of this exercise with molecular orbitals is to illustrate more realistically than was done with the Lewis model what happens to the electrons when atoms adhere together. The valence electrons of atoms lose some of their individual identity and are assigned to orbitals in the molecular unit rather than to individual atoms. The task now is to construct an energy level diagram for these new molecular orbitals just as was done for atomic orbitals earlier. Of course the structure of the molecule and the composing elements must be considered because the orbital ordering in each case is dependent upon the relative energy of the original atomic orbitals. If one considers only the $1s$, $2s$, and $2p$ orbitals on atoms A and B in the diatomic molecule AB, a total of ten molecular orbitals is generated. Five of these are bonding and five are antibonding.

Shown in Figure VII-6 is the experimentally determined energy level diagram for most diatomic molecules. The standard formalism for representing the MO energy levels is used. The ordering of the atomic orbitals on the original atoms A and B is shown on either side and the molecular orbitals generated by linear combinations of the atomic orbitals are shown in the middle. The energy of the orbitals increases from the bottom to the top of the figure. Electrons can be fed into this molecular orbital energy level scheme in the same way that they were fed into the atomic orbital energy level scheme depicted in Figure VII-3. As before, a maximum of two electrons can be placed in each orbital. Of course, only the number of electrons available in the original atomic orbitals on atoms A and B can be used in the MO scheme. The filling beings with the lowest

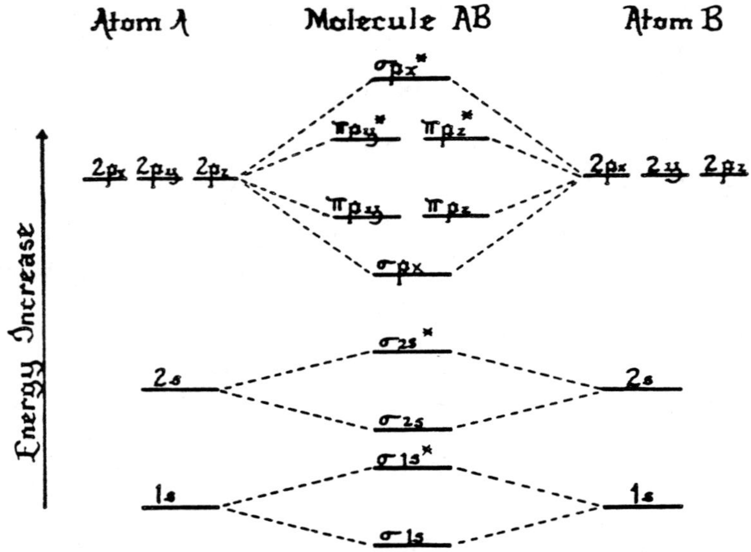

Figure VII-6. An energy level digram for the molecular orbitals of many diatomic molecules.

energy levels (those nearest the nucleus) and proceeds toward higher energy (upward in the figure) as necessary. For example, H_2 has two electrons. These both reside in the σ_{1s} level and are labeled $\sigma_{1s}{}^2$. There is one net σ bond as a result. The molecule He_2 does not exist, because it would have the configuration $\sigma_{1s}{}^2\sigma_{1s}{}^{*2}$. The two electrons in the bonding molecular orbital are canceled by two electrons in the antibonding molecular orbital, giving no net bond. The molecule Li_2 forms with a $\sigma_{1s}{}^2\sigma_{1s}{}^{*2}\sigma_{2s}{}^2$ configuration and has one net bond. The molecule Be_2 would have eight electrons and does not form for the same reason that He_2 does not. The molecules B_2 and C_2 both exist with one and two bonds, respectively, but they are more spectroscopic curiosities than chemically important molecules. The molecule N_2 has a configuration of $\sigma_{1s}{}^2\sigma_{1s}{}^{*2}\sigma_{2s}{}^2\sigma_{2s}{}^{*2}\sigma^2_{2p_x}\pi^2_{2p_y}$ $\pi^2_{2p_z}$, giving three net bonds. The Lewis structure and MO description for N_2 lead to the same result. Nitrogen, comprising 80% of the earth's atmosphere, is thermodynamically very stable because of the triple bond. The O_2 molecule has 16 electrons and thus two more than N_2. These two electrons become $\pi^{*1}_{2p_y}$ and $\pi^{*1}_{2p_z}$ according to Hund's rule. Hence, there is a net of only two bonds in oxygen because these two unpaired electrons in the antibonding levels cancel one of the three bonds. Oxygen is experimentally found to possess two unpaired electrons in accordance with this MO description. Recall that the Lewis structure of O_2 could not accommodate the unpaired electrons without violating the octet rule. The very high degree of reactivity of oxygen toward paper, textiles, wood, and dyes, particularly in the presence of light, results mostly from the reactivity of these two unpaired antibonding electrons. Completing the second row, F_2 and the peroxide ion, O_2^{2-}, both have 18 electrons and one net σ bond because the electrons in the two π-bonding molecular orbitals are canceled by the electrons in the two π^* molecular orbitals. The $\sigma_{p_x}{}^2$ pair of electrons forms the single bond. The molecule Ne_2 does not exist because the $\sigma_{p_x}{}^2$ pair of bonding electrons is canceled by the $\sigma_{p_x}{}^{*2}$ antibonding electrons.

The MO description elegantly and satisfyingly explains many details of the bonding and chemistry of diatomic molecules. It can also be extended to much larger molecules, such as dyes, fibers, and oils. Unfortunately, the MO description becomes much more involved for polyatomic molecules, but it continues to give a good account of electronic structure and properties. A variety of degrees of sophistication in a molecular orbital treatment can be used, but only the simplest one will be

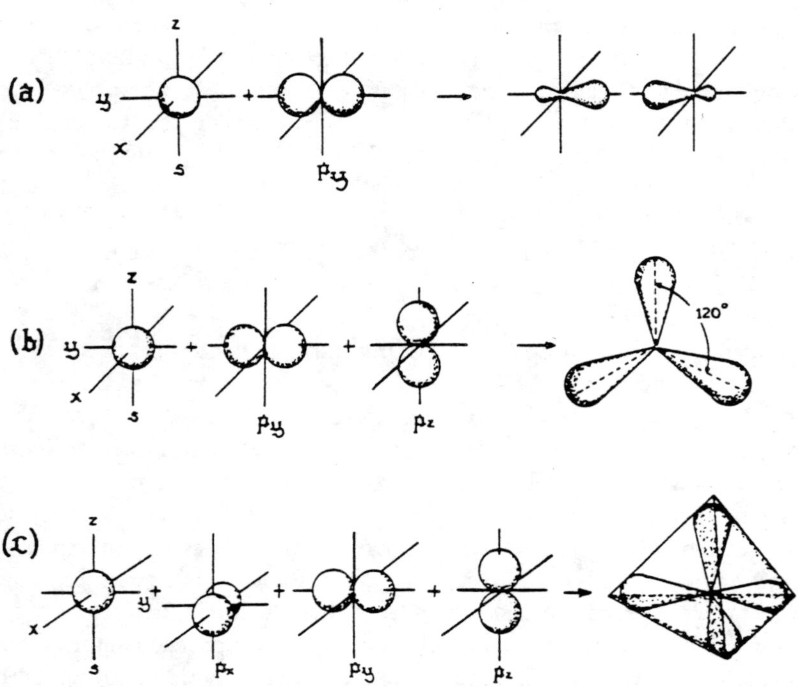

Figure VII-7. Contour patterns for the hybrid orbitals obtained by mixing s and p orbitals. (a) Linearly arranged sp hybrids, (b) trigonal planar sp^2 hybrids, (c) tetrahedrally oriented sp^3 hybrids.

applied here to rationalize the bonding in polyatomic molecules. First of all, it is usually convenient to mix or *hybridize* some or all of the atomic orbitals on an atom, particularly when forming the σ-bond framework, before forming molecular orbitals. This step is not necessary, but it makes the description less cumbersome in that it reduces the number of orbitals needed at any one time. Linus Pauling in the late 1920s developed the *valence bond theory*, which has as a central argument the concept of hybridization or mixing of atomic orbitals on an atom. The mixture of an s and a p orbital in linear combinations *on the same atom* gives two equivalent hybrid orbitals linearly disposed with respect to one another on that atom. These orbitals are called "*sp* hybrids" because one s orbital and one p orbital on the same atom are involved. The appearance of the *sp* hybrid orbitals is shown in Figure VII-7a. An s and two p orbitals combine to give three equivalent hybrid orbitals in trigonal planar disposition on the atom. These hybrids are shown in Figure VII-7b. They are called sp^2 hybrids, with the superscript indicating the number of p orbitals involved and not the number of electrons. An s and three p orbitals combine to give four equivalent hybrid orbitals in tetrahedral disposition on the atom as shown in Figure VII-7c. These orbitals are called sp^3 hybrids. The σ-bonding framework determines the basic structure of molecules and is usually responsible for most of the bond energy. Conceptually, the π-bonding framework can be derived separately from these σ-bond orbitals. It is formed from the atomic orbitals left unhybridized on the atoms. The π orbitals will be the most important ones in eventually describing the origin of color in organic compounds.

The bonding in ethylene, C_2H_4, illustrates the conceptual blending of hybridization and molecular orbital theory. The axis system shown in Figure VII-8 will aid in the description. Ethylene has a σ-bond system and a carbon–carbon π-bond system as noted earlier in the discussion of Lewis structures. The complete σ-bonding framework consists of a carbon–carbon bond and four carbon–hydrogen bonds all arranged in the xy plane. The ground-state electronic configuration of the carbon atoms is $1s^22s^22p_x{}^12p_y{}^12p_z{}^0$. The $1s^2$ electron pair is very stable and does not enter into the bonding picture. To achieve an orbital description of the carbon atoms which will accommodate the known bonding, each carbon atom can be energized by promoting one electron from the $2s$ orbital into the $2p$ level. This gives an excited-state electron configuration of $1s^22s^12p_x{}^12p_y{}^12p_z{}^1$. The $2s^12p_x{}^12p_y{}^1$ set of orbitals can be hybridized to a trigonal planar set of sp^2 hybrid orbitals in the xy plane. This hybridization procedure for carbon is an artificial exercise required in most valence bond treatments of bonding. One should not be overly concerned with how it occurs or even with its validity. The point is that now two of these half-filled hybrid

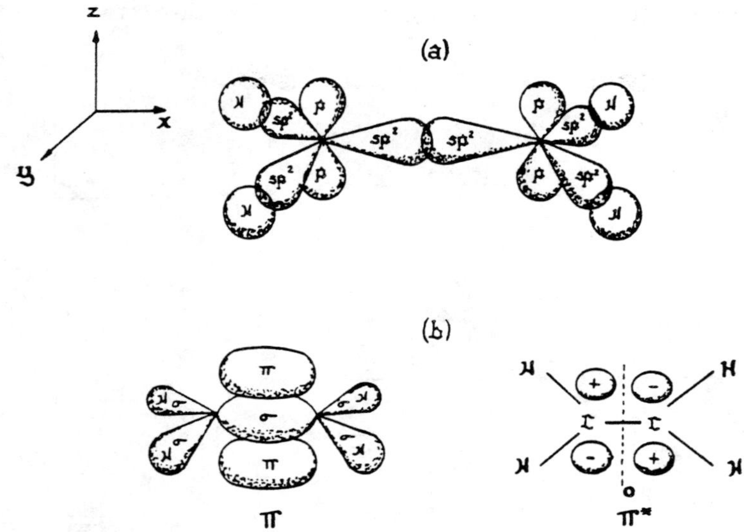

Figure VII-8. (a) Hybridization of the carbon orbitals in ethylene. (b) Formation of the π and π^* molecular orbitals by overlap of the carbon $2p_z$ orbitals.

orbitals on each carbon can bond to hydrogen atoms. Each hydrogen atom contributes one electron to the bond in order to form the C—H bonds. The electron in the remaining hybrid orbital on carbon combines with its counterpart on the other carbon atom to form the C—C σ bond. This completes the bonding in the σ-bond framework. The $2p_z^1$ orbital on each carbon atom remains unhybridized and is perpendicular to the plane of the σ-bonding framework. Pictorially, these ideas are summarized in Figure VII-8a. The $2p_z^1$ orbital on the carbon atom can be combined in linear combinations with the other carbon to form a carbon–carbon π molecular orbital. Along with the formation of the π molecular orbital, a π^* MO of higher energy also forms. The two π MOs in the carbon–carbon framework are shown in Figure VII-8b.

The energy level diagram for the orbitals of ethylene is based on the energy of the $2s$, $2p_x$, $2p_y$, and $2p_z$ orbitals of each carbon atom and the $1s$ orbital on four hydrogen atoms. Twelve electrons were originally available in these atomic orbitals (four from each carbon atom and one from each hydrogen atom). The total valence orbital energy level diagram for ethylene with the 12 valence electrons is shown in Figure VII-9. The π molecular orbital contains the two electrons originally present in the $2p_z$ orbitals of each of the carbon atoms, while the higher energy π^* MO contains no electrons. There are five σ-bonding orbitals of lower energy

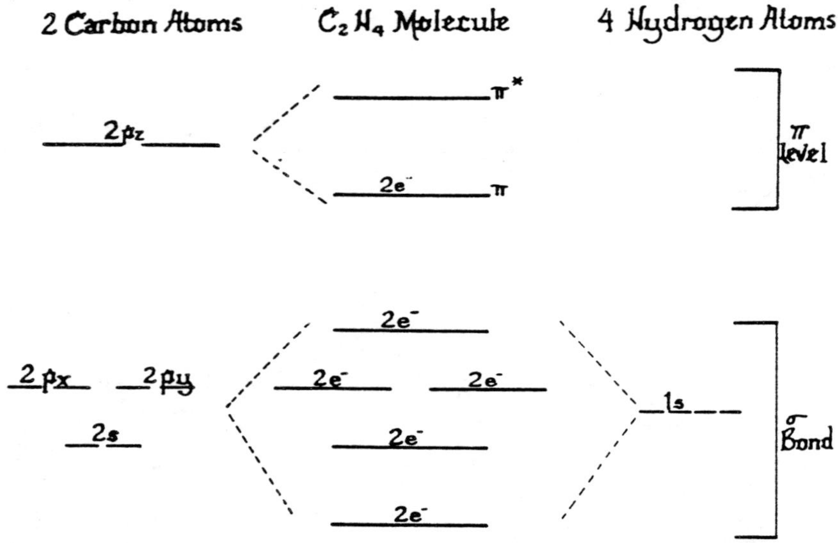

Figure VII-9. The energy level diagram for valence orbitals in the ethylene molecule.

in the C—H and C—C framework and each contains two electrons. This completes a simplified view of the atomic orbital contributions to bonding in ethylene.

A slight modification of the molecular orbital description of ethylene leads to a description of formaldehyde, H_2CO. The Lewis description of bonding in formaldehyde was shown previously. Formaldehyde is isoelectronic with ethylene; however, one of the CH_2 groups having a total of eight electrons is "contracted" to form an oxygen atom having the same number of electrons. Instead of forming five σ bonds as in ethylene, now only three σ bonds need to be formed along with two lone pairs on oxygen. In the same way as was done for ethylene, the carbon valence electrons can be energized to the $2s^1 2p_x^1 2p_y^1 2p_z^1$ configuration and then hybridized to the planar sp^2 hybrid set using the s, p_x, and p_y orbitals to

carbon		oxygen	
$s^1 p_x^1 p_y^1$	$+\ p_z^1$	$s^2 p_x^2 p_y^1$	$+\ p_z^1$
hybridize ↓	↓	hybridize ↓	↓
$(sp_a^2)^1 (sp_b^2)^1 (sp_c^2)^1$	p_z^1	$(sp_a^2)^1 (sp_b^2)^2 (sp_c^2)^2$	p_z^1
(forms three σ bonds)	(forms π bond)	(forms one σ bond and two lone pairs)	(forms π bond)

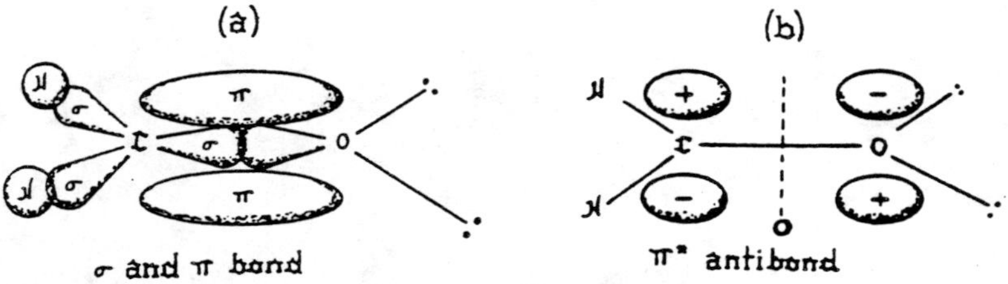

Figure VII-10. (a) Formation of the σ and π bonds in formaldehyde. (b) The π^* antibond orbital of formaldehyde.

form three σ bonds. The $2p_z$ orbital on carbon remains unhybridized to form a π bond. Oxygen can be treated in the same fashion, but now a pair of electrons occupies two of the three sp^2 hybrids. The $(sp_b{}^2)^1(sp_c{}^2)^1$ hybrid orbitals on carbon will bond to the $1s^1$ orbital on each hydrogen atom to form the CH_2 fragment, and $(sp_a{}^2)^1$ on carbon and $(sp_a{}^2)^1$ on oxygen will bond to form the C—O σ bond. The single electron in the p_z orbital on oxygen and carbon will combine in linear combinations to form the C—O π bond (and π^* antibond). There are two electrons in the π-bonding molecular orbital and no electrons in the antibonding molecular orbital. These orbitals are shown in Figure VII-10. The two pairs of electrons on the oxygen atom of formaldehyde are not involved in bonding and are sometimes cryptically indicated by the letter n. This n refers to a pair of nonbonding electrons, and should not be confused with the principal quantum number used earlier in the chapter.

The benzene ring will be discussed briefly as a final example of bonding in polyatomic molecules. The MO description of benzene, C_6H_6, is more complicated than that of C_2H_4 or H_2CO because the molecule has more atoms and thus more orbitals. It suffices merely to note the final results of the description. Valence bond theory can be used to account for the σ bonding. The σ framework of the carbon atoms is formed from the planar sp^2 hybrid orbitals on each carbon atom. Two of the three sp^2 hybrid orbitals on each carbon overlap with those on neighboring carbon atoms to form C—C σ bonds, while the remaining hybrid forms a C—H σ bond, as shown in Figure VII-11a. The molecule is entirely planar. The π framework is formed by overlapping the unhybridized p orbital (the p orbital perpendicular to the plane of the molecule) on each carbon atom with the equivalent p orbital on the adjacent carbon atoms. Since six p_π orbitals are being used, six π-molecular orbitals will be formed. Three of these will be bonding and three will be antibonding. These six MOs are shown in Figure VII-11b. The nature of the wave function for each is determined by considering the six unique linear combinations that can be made from the six p orbitals. Only the top view of each MO is shown.

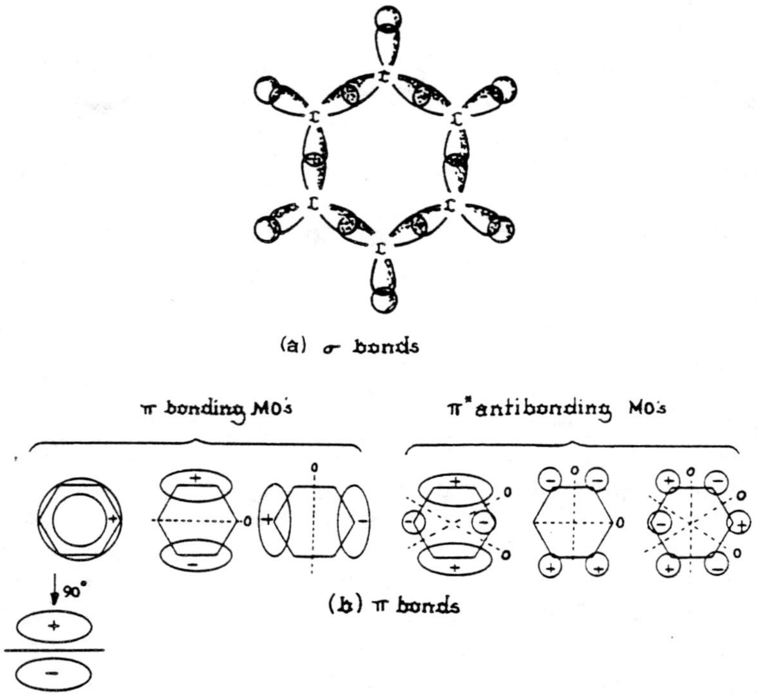

Figure VII-11. (a) The σ-bond framework of benzene formed from sp^2 hybrids on carbon. (b) The π molecular orbitals of benzene formed using the remaining unhybridized p orbital on each carbon atom. These p orbitals are perpendicular to the molecular plane.

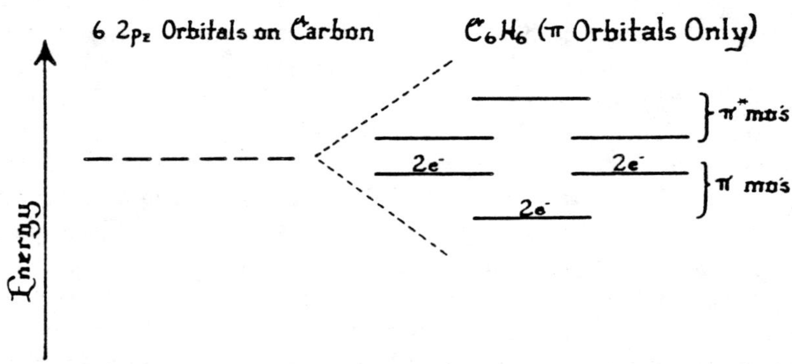

Figure VII-12. The energy level diagram for the π MOs of benzene.

If the molecule were flipped over, then the MOs on the opposite side would have the same shape as those shown but would have the opposite sign. As before, the next task involves filling an energy level diagram of the π and π^* MOs with the appropriate number of electrons. The energy level diagram for the π system of benzene including the six π electrons has the form shown in Figure VII-12. The six available π electrons group together in the three lowest energy orbitals, all of which are bonding molecular orbitals.

The structural formula of benzene is frequently drawn in the following Kekulé forms:

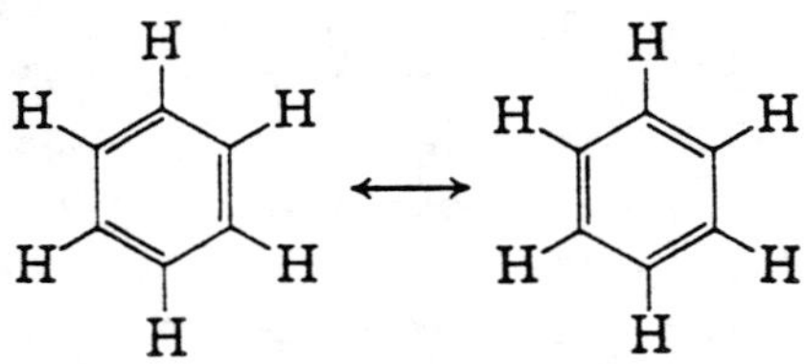

It would appear from these structures that there are three C—C double bonds in benzene. This is precisely what the MO description gives, although the MOs are "smeared out" over the molecular surface and have a more complicated appearance. More importantly, however, the MO description indicates that there are three higher energy, unfilled, π-antibonding MOs in addition to the π-bonding MOs. This new piece of information, not indicated in the resonance pictures of benzene discussed earlier, is of great importance in understanding how molecules produce color and engage in photoreactions. The antibonding levels give an idea of what the orbitals of higher energy (excited states) in molecules might look like.

All of these bonding models and theories give in varying degrees of sophistication a description of the molecular structure consistent with what is experimentally found. The MO model is most generally useful, but it is also most complicated. It should always be kept in mind that a theory is properly used when it is employed to interpret an experimental observation. It is unsound to reverse this sense. Sherlock Holmes said, "It is a capital mistake to theorize before one has data. Insensibly one begins to twist facts to suit theories, instead of theories to suit facts." In science, experimentally determined data should not be interpreted to fit a hypothetical bonding theory. If the experiments are done properly and the results do not fit a theoretical prediction, then the burden of proof lies on the theory. Electronic absorption spectroscopy is one of the most decisive experimental means for determining whether a bonding model of a molecule is basically correct. The origin of color and light-induced photodecomposition reactions in organic compounds also arises from energy absorption by electrons. Hence it is now possible to use the results of bonding models and electronic absorption spectroscopy to understand the origin of color and photoreactions in organic compounds.

ELECTRONIC TRANSITIONS IN ORGANIC MOLECULES: COLOR

The concept of the ground-state electronic energy levels in molecules has been introduced in several examples above. These energy levels are responsible in part for the geometric structure adopted by the molecules and the nature of the bonding. They contain the bulk of the valence-electron density. Electronic states of higher energy (usually obtained from the antibonding levels) are also present in every molecule. In addition, these higher energy states may contain some valence-electron density in the ground-state (lowest energy) description of a molecule. We have also arrived at the fact that these ground and excited states have well-defined energies, i.e., they are quantized. The energies involved in moving the electron density between the ground states and the excited states of a molecule are at the heart of the problem of understanding how dyes achieve their color.

Color in organic molecules is produced by the light-induced excitation of electrons from the ground state of the molecule to one of the many excited states. The excitation requires an external input of energy. The source of the excitation is usually the wavelengths (energies) of light that match the energy gaps in the molecule. From the information contained in Table V-1, we know that various wavelengths of light produce different colors. *Thus, absorption of a particular wavelength of light by excitation of electrons between two states whose energy difference corresponds to that particular wavelength removes light of that color from the spectrum.* Color is produced by the object because these wavelengths are missing from the reflected light. The color seen by the viewer is, of course, the complement of the color removed by this excitation process. For example, if an energy gap of 50 kcal exists between two electronic states in a molecule, then light of 575 nm (yellow) is removed from the spectrum. If the incident light was white, the object appears blue in color.

In order to continue to produce color, the electron in the excited state must quickly relax back to the ground state so that the absorption process can occur again. Otherwise the color would appear only for the time it takes for all of the electrons to reach the excited state. At that point the molecule could no longer continue to absorb those energies, and its color-producing potential would be gone. We know that dyes maintain their color on continuous exposure to light for a long time. This is because their electrons relax back to the ground state about as rapidly as they are excited to the higher energy states.

The example given above concerning the color observed in an object having an electron energy gap of 50 kcal contains an oversimplification. The impression might be that the wavelengths of electromagnetic radiation absorbed must exactly match the gap between the electron energy levels. If this were the case, then the wavelengths removed from the spectrum would occur only in very narrow ranges. Spikelike absorptions can often be found in isolated gas-phase atoms and molecules or in solids at very low temperature. However, at ambient conditions in solution or in the solid phase more typical of dyes and art materials, the absorption bands usually have wavelength breadths from 10 to 300 nm. These broad bands, rather than sharp spikes, result from a variety of sources. Of most importance in broadening the absorption band are the interactions between molecules in close proximity to one another, and the effects of vibrational motions in the molecules themselves. Both of these "smudge" the energy levels and increase the width of the associated absorption bands. A third cause of increased bandwidths comes from the fact that several absorptions may occur near one another. The bands may not be resolved, giving the impression that a single, broad absorption is present. The wavelength at which an absorption band has maximum intensity is labeled λ_{max}, and is usually listed when referring to bands having great breadth. A simplified diagram showing the "fuzziness" of the levels is shown in Figure VII-13. The resulting absorption spectrum gives rise to significant band width.

The intensities of absorption bands in compounds show considerable variation and are a topic which is difficult to treat adequately on a simplified level. Very intense bands usually result from "allowed" processes. An allowed process theoretically does not violate any rules of quantum

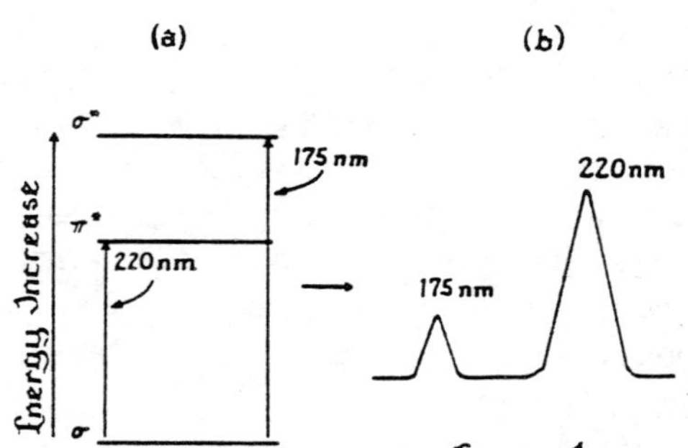

Figure VII-13. (a) A simplified energy level diagram for a molecule. (b) The resulting absorption spectrum. The intensities shown are not significant.

mechanics and thus represents a favorable molecular event. Weak bands are often assigned to "disallowed" processes, which violate one or more quantum mechanical rules. They gain their intensity from a breakdown in the rules of molecular absorption as a result of perturbation of the electronic states by other molecular events. Perturbations from the vibrational motion of atoms is a common reason for the appearance of theoretically disallowed absorption bands.

The intensity of an absorption band is often measured by the *extinction coefficient* of the absorption. The extinction coefficient is heuristically useful, but absorption spectroscopists use the theoretically sound term "oscillator strength" when discussing intensity. The concept of the extinction coefficient was introduced in Chapter IV. Values of the extinction coefficient in the 100–100,000 range usually arise from allowed processes, while extinction coefficients of 1 or less are usually from disallowed processes. Little can be said with certainty about the allowedness of transitions with extinction coefficients in the 1–100 range unless an accurate assignment of the bands can be made. If the assignment is known, the allowedness or disallowedness can be deduced theoretically.

Although it is not strictly correct to do so, electronic transitions might be considered in an elementary discussion to occur between the MO energy levels. Using the terminology developed in this chapter, a variety of electronic transitions are possible from the ground state to excited states. These are

$$\sigma \rightarrow \sigma^* \qquad \mathrm{n} \rightarrow \sigma^*$$

$$\sigma \rightarrow \pi^* \qquad \mathrm{n} \rightarrow \pi^*$$

$$\pi \rightarrow \pi^* \qquad \pi \rightarrow \mathrm{n}$$

As mentioned before, n represents a nonbonding pair of electrons. Figure VII-14 shows a highly simplified electronic energy level diagram summarizing the excitations in a hypothetical molecule. An excitation of a nonbonding electron n to a π-antibonding level (π^*) can exist. Excitations between σ and π bonding levels and antibonding levels commonly occur. To cause these excitations, light energy corresponding to the magnitude of the energy gap between the levels in the particular absorbing species must be incident onto the material. It is important to recognize that an electron is not driven out of the molecule during this excitation. The molecule retains the electron, but there is a momentary reshuffling of the arrangement of the valence-shell electrons. On the other hand, an input of a large amount of energy as can occur with far-ultraviolet radiation and X-rays can excite electrons beyond the ionization limit and thus strip them from the absorbing species. The atom or molecule becomes an ion in this case.

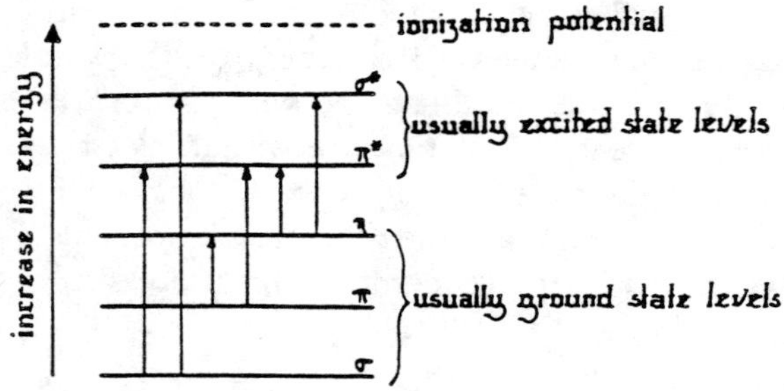

Figure VII-14. Some possible electronic transitions in a molecule. Initially, the ground-state levels usually contain electron density, while the excited states are devoid of electron density.

The most commonly encountered transitions in organic materials are $\sigma \rightarrow \sigma^*$, $\pi \rightarrow \pi^*$, and $n \rightarrow \pi^*$ with the energy ordering $E_{\sigma \rightarrow \sigma^*} > E_{\pi \rightarrow \pi^*} > E_{n \rightarrow \pi^*}$. The energy of the $\pi \rightarrow \pi^*$ and $n \rightarrow \pi^*$ transitions frequently falls in the visible region and hence these transitions are important for color production by dyes. Particular attention will be focused on these two types of electronic transitions. The $\sigma \rightarrow \sigma^*$ transition is not very important in dyes because it is usually caused by ultraviolet radiation of wavelengths less than 200 nm. Common light sources used in illumination do not emit significant amounts of radiation in this region.

A straightforward example of a $\pi \rightarrow \pi^*$ electronic transition is found in the ethylene molecule, whose bonding was described in the preceding section. An electromagnetic wave with an oscillating electric vector (Figure I-2) can interact with the carbon–carbon π bond and impart energy to that bond, provided the photon energy matches the energy gap between the π and π^* levels. This effectively results in a momentary disruption of the π bond. Figure VII-15 illustrates the process. The electric vector of the wave must be in the plane of the π bond as shown. The energy of the photon must match the π–π^* energy gap. The $\pi \rightarrow \pi^*$ transition in ethylene occurs in the ultraviolet at 180 nm. This wavelength of radiation is removed from the spectrum of light incident onto the molecule. The

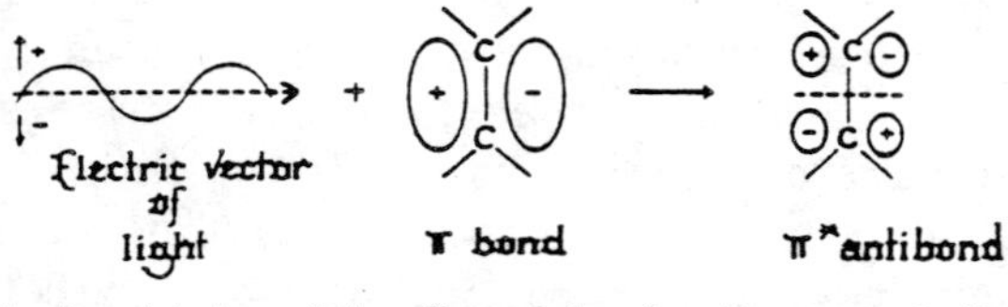

Figure VII-15. A pictorial view of the effect of the electric vector of a light wave on the π bond of ethylene.

$\pi \rightarrow \pi^*$ excitation in benzene occurs at 203 nm. Ethylene and benzene are colorless materials because neither absorbs energy in the visible range.

The $n \rightarrow \pi^*$ transition can be illustrated with the formaldehyde molecule. In this case, an electromagnetic wave with its electric vector oscillating perpendicular to the molecular plane excites one of the nonbonding electrons n on the oxygen atom into the π^* molecular orbital of the C—O bond. The Lewis structure representation of this process is

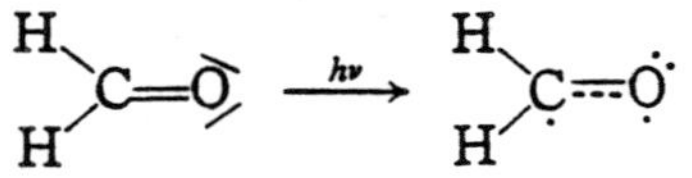

The π bond is weakened momentarily because of the presence of the electron in the π^* orbital. The energy of radiation for this excitation in formaldehyde corresponds to a wavelength of 190 nm, while in acetone, $(CH_3)_2C{=}O$, it occurs at 277 nm. These two compounds are colorless. The $n \rightarrow \pi^*$ transition can occur only if nonbonding electrons are present on one of the atoms and an adjacent π bond exists. Some fragments of organic molecules where these requirements are satisfied, and hence $n \rightarrow \pi^*$ transitions can be found, are given in Table VII-5.

In a way, the $n \rightarrow \pi^*$ transition can be thought of as being a *charge transfer transition* in which a net shift of electron density in the molecule appears to form a resonance structure of the molecule. It is not a true resonance structure because the two forms differ in energy. For example, the electron density shift can be produced in *para*-aminobenzaldehyde by the appropriate energy of light:

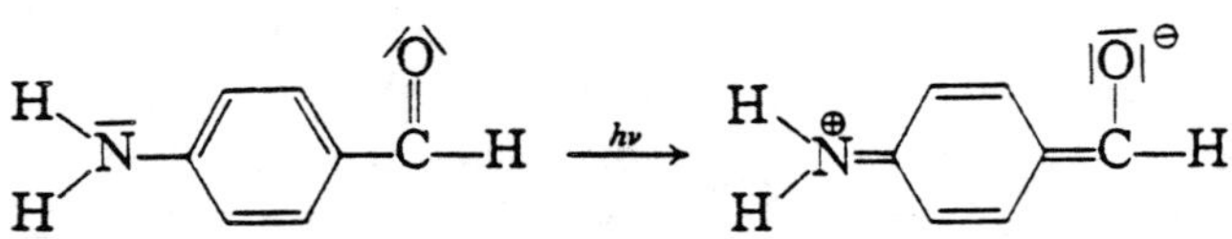

A number of n, π, and π^* orbitals become involved in the shift of electron density. The H_2N— group could be replaced by any electron-donating functional group and the —COH group could be replaced by any electron acceptor and the same type of transition could occur, albeit with a different energy. A system of alternating delocalizable double bonds must connect the groups.

We have just considered some of the different types of electron density changes which take place in organic molecules as a result of exposure to light. Attention can now be directed toward the relationship between the chemical structure of organic compounds and their color. Such an assessment must always be based on the absorption spectrum of the compound and not merely the visual appearance of color in the compound. It is possible to have a complicated absorption spectrum stimulate the same or

Table VII-5. Some Fragments of Organic Molecules Where $n \rightarrow \pi^*$ Transitions Occur

Fragment	Name	Fragment	Name
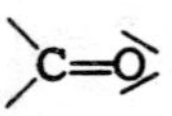	Carbonyl group	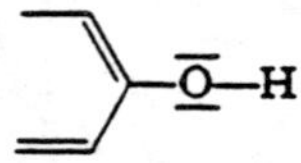	Phenolic group
$-\overline{N}=\underline{N}-$	Azo group	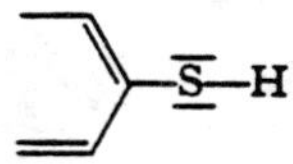	Thiophenolic group
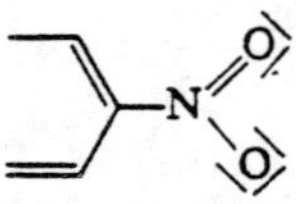	Nitro group	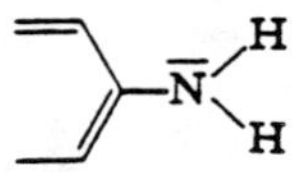	Amino group

nearly the same visual color as a simple spectrum produces. It is also important to recognize that one cannot accurately predict colors of compounds from elementary bonding models.

An early attempt by O. N. Witt in 1876 to systematize the observed colors in organic compounds produced the terms chromophore, auxochrome, and polyene. Organic dyes were said to have a chromophore and an auxochrome group separated by a polyene, although some dyes have the chromophore group, but no auxochrome group. The *chromophore* is an organic functional group that acts as an electron acceptor. It was said to produce the color. Some common chromophore groups are given in Table VII-6. The *auxochrome* is an electron-donor group, which was said not to produce color by itself, but to increase the intensity of the already existing color. Some examples of auxochrome groups are also given in Table VII-6. In using the terms auxochrome and chromophore to refer to color enhancement and color production, respectively, the distinction should be considered to be one of degree rather than kind. The two terms are old ones dating from a time when energy absorption processes in molecules were not understood. The terms have little absolute meaning.

The auxochrome and chromophore are linked together in the molecule through a polyene system. A *polyene* refers to a conjugated system of double or triple bonds. Conjugation exists in a system of alternating single and double or triple bonds connected in a chain or ring of carbon atoms:

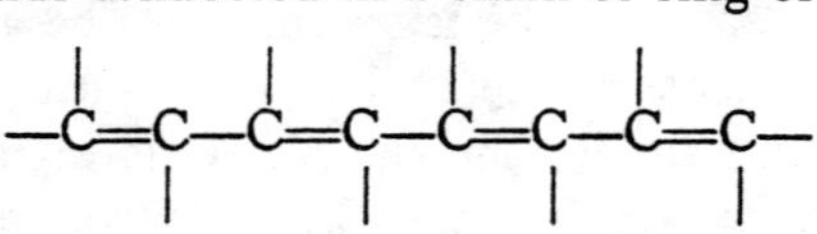

The bonds between carbon atoms in the benzene ring are conjugated, for example. Such networks are also termed *unsaturated* chains. The chromo-

Table VII-6. Some Chromophore and Auxochrome Groups

Chromophores		Auxochromes	
Functional group	Name of group	Functional group	Name of group
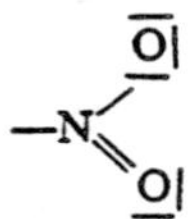	Nitro	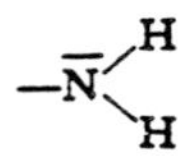	Primary amino
—N=N—	Azo	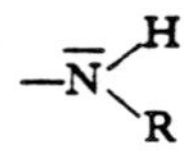	Secondary amino[a]
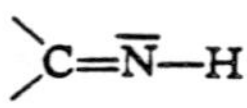	Imino	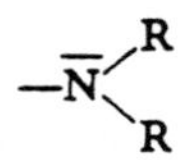	Tertiary amino[a]
—N=O	Nitroso	—O—H	Hydroxyl
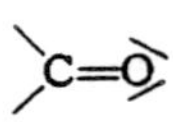	Carbonyl	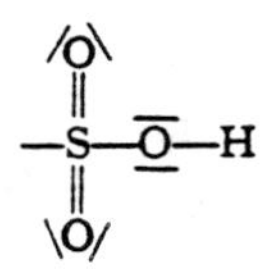	Sulfonic acid
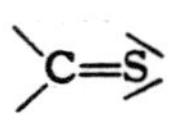	Thiocarbonyl	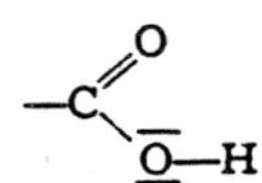	Carboxylic acid
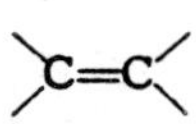	"-ene"	—Cl	Chloro
		—Br	Bromo

[a] R is an organic group.

phore and auxochrome group engage in conjugation with polyene. In this way, electron density is easily shifted about, or *delocalized*, when exposed to electromagnetic radiation.

Saturated organic hydrocarbons contain no double or triple bonds and do not absorb visible energy. Hence they cannot produce color. The electronic transitions in saturated hydrocarbons occur in the ultraviolet energy region and are of the $\sigma \rightarrow \sigma^*$ or $n \rightarrow \sigma^*$ type. *All colored organic hydrocarbons contain double bonds.* With the auxochrome–polyene–chromophore system, absorption of visible light can be accommodated in theory by shifting electrons in the molecule to various resonance-like forms. *Para*-aminobenzaldehyde, shown earlier, is an example of this. The colors of compounds are found to be sensitive to the number of conjugated double bonds. One of the simplest illustrations of this point is the diphenylpolyene series, C_6H_5—$(CH{=}CH)_x$—C_6H_5. The $\pi \rightarrow \pi^*$

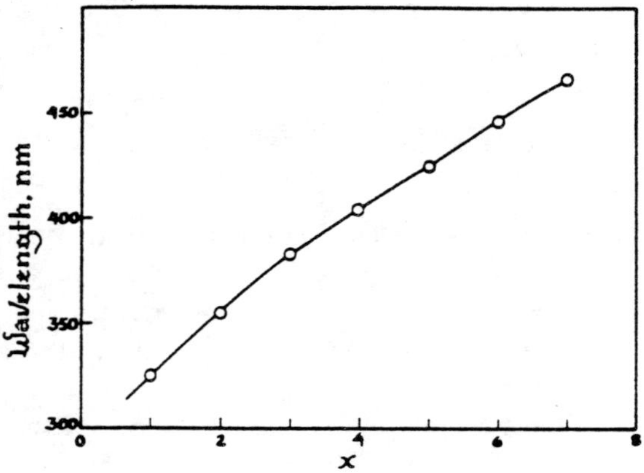

Figure VII-16. The variation in wavelength of the $\pi \rightarrow \pi^*$ electronic absorption band with the number of double bonds x in diphenylpolyenes, C_6H_5—$(CH{=}CH)_x$—C_6H_5.

transition shifts to lower energy (longer wavelength) as the magnitude of x increases. As x increases, it is energetically easier to excite the electrons because the energy levels become more numerous and closely spaced. Figure VII-16 shows the wavelength increase of the $\pi \rightarrow \pi^*$ transition as x increases in value.

The functions of the auxochrome and chromophore are best understood by looking at what happens to the electronic absorption spectrum of a conjugated polyene as an auxochrome and a chromophore are added to it. Azobenzene

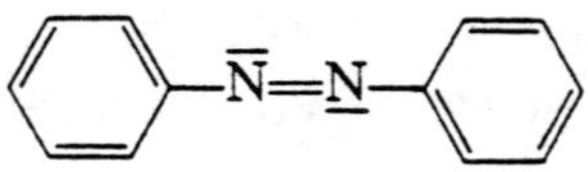

has a series of high-energy absorption bands in the 200–300 nm region which correspond to $\sigma \rightarrow \sigma^*$, $\sigma \rightarrow \pi^*$, and $\pi \rightarrow \sigma^*$ transitions. These will not be considered here because they are not involved in the production of color by the molecule. However, a $\pi \rightarrow \pi^*$ transition is present at about 330 nm, and a weak $n \rightarrow \pi^*$ absorption band occurs at the visible–ultraviolet edge at about 400 nm. The $n \rightarrow \pi^*$ band involves the nonbonding electrons on the nitrogen atoms, plus the π^* orbitals made up from the N=N unit and the two phenyl rings. This weak band imparts a pale yellow color to azobenzene. Now if an auxochrome group, such as dimethylamine, $(CH_3)_2N$—, is added to one of the phenyl rings in the *para* position, as shown below, the $\pi \rightarrow \pi^*$ transition at 330 nm moves to a lower wavelength at 410 nm:

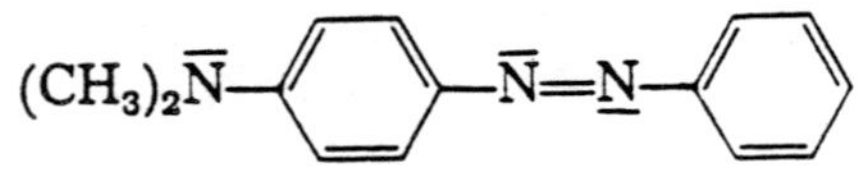

This wavelength is at the edge of the visible region, so the yellow color of this compound is considerably stronger than that of the unsubstituted azobenzene. Note that a new band has not been produced by the auxochrome. Instead the $\pi \rightarrow \pi^*$ transition has been merely shifted from the ultraviolet range to the ultraviolet–visible edge. If a chromophore such as an $—NO_2$ group is added to the other ring in the *para* position, a band at 480 nm appears:

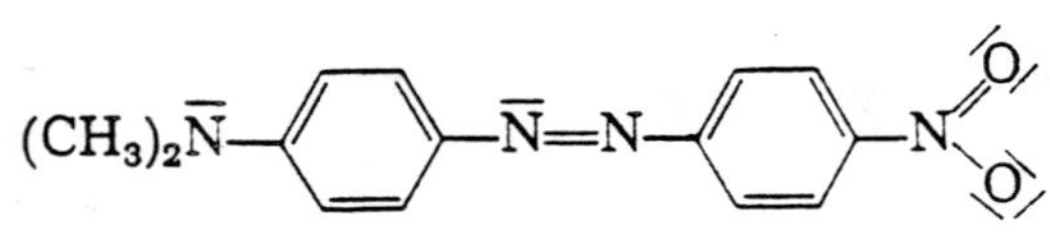

This is found to be the same $\pi \rightarrow \pi^*$ transition noted before but now moved into the visible region. It gives a very intense yellow color to the compound. From this example it can be seen that the *auxochrome and chromophore terminology has little to do with color enhancing or color production in a fundamental sense. It has more to do with the shifting of already existing electronic transitions in the compound from the ultraviolet region into the visible region.*

The direction of vibration of the electric vector of light relative to the molecular axes is important in light absorption. A molecule is frequently found to absorb radiant energy more strongly for certain angles of incident light than for others. This fact was noted earlier in regard to ethylene. In organic crystals, the absorption is greatest when the electric vector is vibrating along the direction of the bonds in the chromophoric groups. In *p*-azoxyanisole

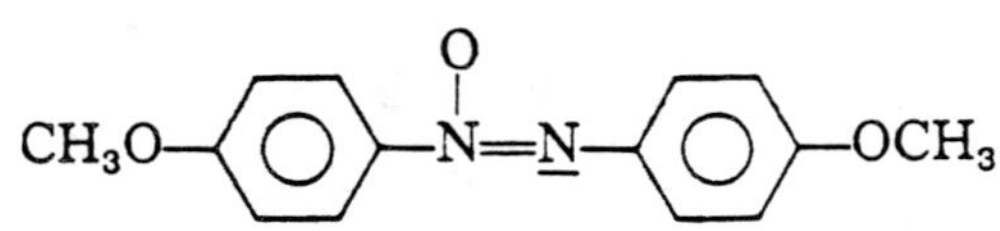

when the electric vector vibrations are parallel to the long dimension of the molecule, i.e., parallel to the N=N double bond, the absorption process is facile and the transmitted light is yellow. When the electric vector vibrations are transverse to the N=N bond, the absorption process does not occur and the molecule is colorless. This experiment can only be done with polarized light.

In summary, color is produced when molecules selectively absorb certain wavelengths of visible light because of electronic transitions between quantized electronic states. *The molecule is altered by absorption of light.* This extremely important fact needs to be considered further. Before taking up this aspect in Chapter IX, it is worthwhile to become familiar with some of the details of organic materials that are important because of their potential to absorb visible light. This is the topic of Chapter VIII.

BIBLIOGRAPHY

Ballhausen, C. J., and Gray, H. B., *Molecular Orbital Theory*, Benjamin, New York (1965).
Gray, H. B., *Chemical Bonds*, Benjamin, New York (1973).
Jaffe, H., and Orchin, M., *Theory and Application of Ultraviolet Spectroscopy*, Wiley, New York (1962).
Rao, C. N. R., *Ultraviolet and Visible Spectroscopy*, Butterworths, London (1961).
Wheland, G. W., *Resonance in Organic Chemistry*, Wiley, New York (1955).

VIII

Organic Dyes

Organic dyes are coloring agents which are homogeneously dispersed in the dyed medium. Frequently, the dye is molecularly dispersed in the substrate which is being colored, but this is not always the case. In contrast, a pigment is a coloring agent with particulate dimensions and is heterogeneously dispersed in the medium being colored. Hence the distinction between a dye and a pigment is not always a sharp one. Since the intent of both dyes and pigments is to color a substrate, the distinction between them is best thought of as being one of convenience, rather than of fundamental importance. Historically, organic compounds have been used as dyes and inorganic compounds used as pigments. Discussions of the two are separated along this line, with dyes being considered here and pigments considered in Chapter XI. However, it is noteworthy that in the last fifty years organic compounds have found increasing use as pigments.

One often finds the term "dyestuff" used in place of the word "dye." Both can refer to the coloring agent, but the term "dye" is replacing "dyestuff" in this context. "Dyestuff" is also used to refer to intermediate compounds in the dyeing process.

There are numerous ways to organize dyes into generic groups. Almost every method of classification has ambiguities or is incomplete. For instance, dyes could be organized into groups depending upon whether they are fugitive or stable, what their intended use is, their historical importance, their chemical structures or properties, their method of application, the fibers which they dye, or their color. No space will be devoted to justifying the distinctions between one listing or another.

Three methods of categorizing dyes will be presented in this chapter. One method of classification useful to the historian would involve a historical breakdown by age. A second method of classification, of value to the chemist, is separation of dyes according to the chromophore present. A third method, of value to the dyer and colorist, is separation of dyes along the lines of the method of application. The classifications here are by no means intended to be complete and exclusive, but rather to provide an introduction to this extensive field. The reader should consult the *Colour Index* for a comprehensive compilation of dyes. This is an internationally recognized reference work published by the Society of Dyers and Colourists in England. It first appeared in 1924 and is updated regularly. The *Colour Index* classifies colorants in two ways. One is on the basis of their application and properties. The other is by their chemical structure whenever this is known.

METHOD OF APPLICATION

The method of dyeing and the mechanism by which the dye is anchored in the fiber might not seem to be a particularly fundamental way to begin a discussion of dye classification. On the contrary, however, the color of the dye, the fibers which can be dyed, and the fastness toward light, washing, and pollutants depend wholly or in part upon how the dye is applied. Some of the terminology needed to classify dyes by other schemes comes from the method of application of the dye.

Both the dye and the substrate being dyed are chemical species having functional groups with "lumps" and "pockets" of electrostatic charge or acid–base characteristics. The dye and the fiber can interact with one another by ionic forces (positive and negative sites attract one another), by hydrogen bonding (hydrogen atoms bound directly to oxygen and nitrogen atoms have a secondary affinity for neighboring nitrogen and oxygen atoms), by van der Waals forces (weak secondary attractions generated by polarity of charge in the dye and fiber), and by actual covalent bonds.

Acid dyes are dyes with acidic functional groups, such as the sulfonic acid functional group, $—SO_3^-$. Usually the acid site is employed as a sodium salt. When placed in aqueous solution, these dyes will bind to fibers having exposed protonated basic sites as exist in proteins. The protein backbone structure of wool and silk is

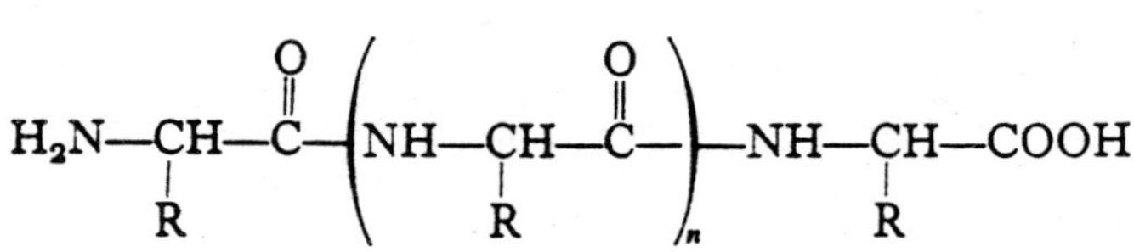

The R groups are organic side chains which characterize the amino acids. The chains are often cross-linked by disulfiide groups. The amine groups in the protein are basic and often are protonated. They can react with the acid functional groups of the dye to form saltlike linkages as shown:

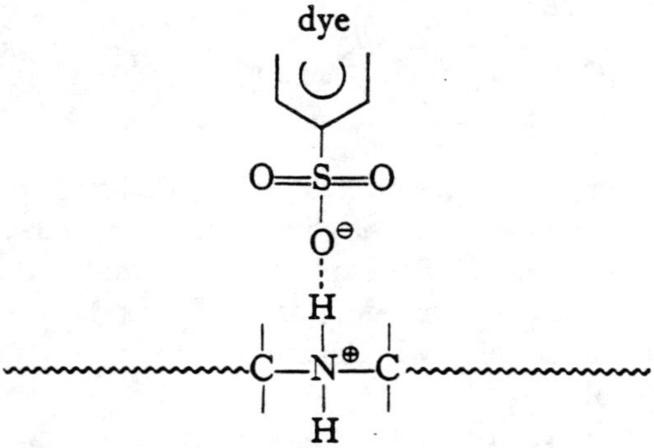

The dye becomes chemically attached to the fiber in this way. Nylon also contains exposed amine groups and can be dyed with acidic dyes. The azo, nitro, triphenylmethane, and carbonyl dyes are often synthesized with sulfonic acid functional groups so that they can act as acid dyes.

Basic dyes possess a basic functional group, usually an amine, which can combine with acidic sites on the fiber. The basic functional group is usually the amine group, $R—NH_2$, which is employed as an amine salt, $R—NH_3^+Cl^-$. Basic dyes will attach themselves to the acidic sites in wool, silk, leather, and acrylics by forming saltlike linkages. They will not dye cotton alone, but will dye the cellulosic fibers of cotton when they are first mordanted with tannin. The basic dyes tend to have a very high tinting power (usually orange and red), but comparatively poor light fastness. Note that in both the acidic and basic dyes an auxochrome group provides the site of attachment to the fiber.

Mordant dyes are based on dyes which have no natural affinity for the fiber being dyed. A substance called a mordant is first precipitated onto the fabric and the mordant instead of the fabric is dyed. There is a wide variety of mordants available. Salts of Fe^{2+}, Sn^{2+}, Cr^{3+}, and Cu^{2+} are often used in this role. The hydroxides $Al(OH)_3$, $Mg(OH)_2$, $Ca(OH)_2$, and $Ba(OH)_2$ and the acids tannic acid, phosphotungstomolybdic acid, and phosphomolybdic acids find common use as mordants. A metal complex of the water-soluble dye forms in the fabric and renders the dye insoluble. An example of a mordant dye complex attached to a fiber is as shown at the top of next page for the dye alizarin.

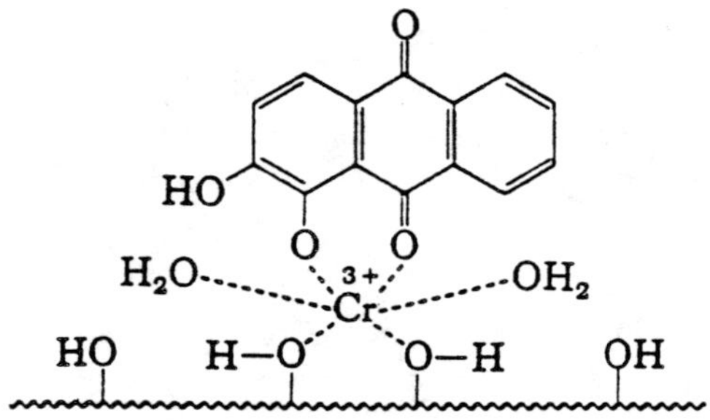

The dye–mordant complex is often called a *lake*. Madder forms a lake with an aluminum hydroxide mordant. The color of the dye often depends upon the mordant used. The choice of the mordant is made on the basis of the dye being used as well as the fiber being dyed. For example, Al^{3+} salts are needed when wool is to be dyed with alizarin, but Cr^{3+} salts must be used with azo dyes on wool.

The process of mordanting has been known for a very long time. Of the naturally occurring dyes used through the centuries, many must be used with mordants to successfully color the fiber. Madder, cochineal, the lac dyes, fustic, logwood, brazilwood, Persian berries, quercitron, cutch, and weld are all mordant dyes.

Direct dyes are dyes which are absorbed into a fiber simply by placing the fiber in a hot water solution of the dye. Cellulosic fibers of cotton can be dyed directly. Wool and silk can also be dyed in some cases by direct dyes, but inorganic salts may have to be present. Most direct dyes are synthetic, but a few, such as turmeric and safflower, are naturally occurring. Except for turmeric, direct dyes have rather good light fastness, but, because they are usually bond to the fiber by hydrogen bonding, they are not particularly fast to washing. An example of the linkage which attaches a direct dye to a fiber is as follows:

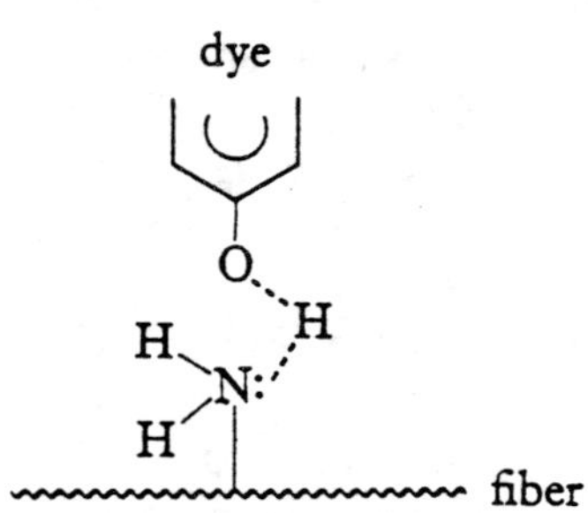

Disperse dyes have no functional groups to anchor the dye to the fiber. They simply disperse themselves in the polymer framework of the fiber. Unlike most other dyes, the disperse dyes must be quite insoluble in water to be of much value. Acetates, polyesters, nylon, and acrylics can be dyed with disperse dyes. These dyes were originally developed for use on cellulose acetate fibers. Azo and carbonyl dyes are commonly used as disperse dyes.

HISTORICAL CLASSIFICATION

A classification of dyes by the date of their availability is of special interest and utility in museum work. Perhaps the most significant demarcation point in this classification is 1856. In this year William Henry Perkin synthesized mauve (mauveine), and synthetic dyes began to be used in place of the naturally occurring materials. In the mid-nineteenth century many of the synthetic dyes which came onto the market were less expensive but inferior in quality to the natural ones. In the present day, however, this is not the case in regard to quality. Many of the synthetic dyes are vastly superior to the natural products.

In the pre-1856 era, natural dyes were either of animal or plant origin. A few of the most important dyes in each of these classes will be mentioned. Some of the dyes of *animal origin* are Tyrian Purple, kermes, cochineal, stick lac, and sepia. *Tyrian Purple* is also called Royal Blue because its price was so high that only royalty could afford to use it. In certain periods of history its possession by anyone except royal families has even been considered a criminal offense. Tyrian Purple is the origin of the term "born to the purple" occasionally applied to persons born to royalty. The cost of Tyrian Purple is one of the highest of all dyes because of the tedious task of obtaining it. The dye is found in mollusks (*Murex*) and it requires about 9000 live snails to obtain 1 g of the dye. A legend recorded on the coins of the ancient Phoenician city of Tyre says that Hercules discovered the dye when his dog was chewing snails. The dog's teeth became stained purple. Tyrian Purple was used primarily prior to the eighth century A.D. It found the most application on wool and silk. Chemically, it is an indigoid dye and has a very high stability toward decomposition by light, alkali, soap, and most acids. Nitric acid and chlorine will destroy it, however. The chemical name of Tyrian Purple is 6,6′-dibromoindigo and its structure is shown below:

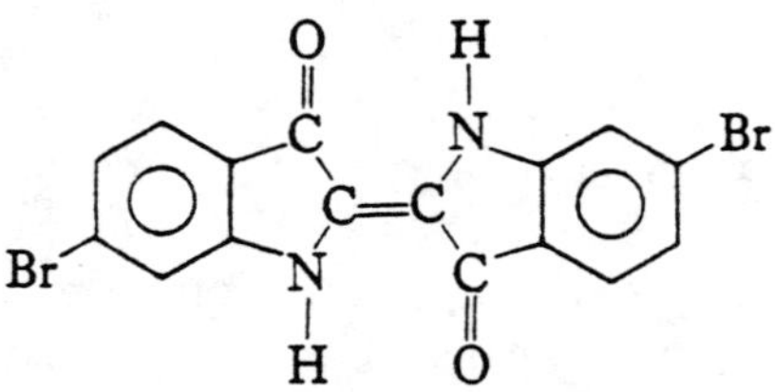

(Tyrian Purple)

The mollusks probably synthesize the monomer, which dimerizes when it is removed from the body and oxidized by air.

Kermes is one of the most ancient of all dyes, having known usage for 3500–4000 years. It is obtained from the dried bodies of female insects found in kermes oak, which grows in southern Europe. About 1 lb of dye can be obtained from 70,000 insects. It was used more in the fabrics of southern Europe than in middle Europe. Kermes is a red dye which was usually precipitated on a fabric with alum, $K_3Al(SO_4)_3$, to form a crimson lake. It is not brilliant in color and was replaced in the Middle Ages with dyes from India, and by *cochineal* from Mexico in about A.D. 1600. Cochineal was used in Mexico and Central America from very early times. Paintings on the pyramids of Teotihuacan were made in pre-Christian times and contained cochineal. Cochineal is very similar to kermes in chemistry, color, and origin. These two dyes are light-sensitive, particularly in fabrics and watercolors. The decomposition products are brown. They are more light-stable when used in oils, and found application for painting coaches. Their use in dyeing fibers has been on wool and silk to give a carmine red. Cochineal was also probably one of the first food colors. Kermes and cochineal are carbonyl dyes of the anthraquinone structure derived from kermesic acid, $C_{18}H_{12}O_9$, and carminic acid, $C_{22}H_{22}O_{13}$, respectively. *Stick lac* (lac, Indian lake, or lac dye) is another red dye of anthraquinone structure also obtained from insects. It is often found with cochineal. It is derived from laccaic acid, and was normally used as an alum lake. Laccaic acid is believed to have an hydroxyanthraquinone structure, the details of which are not known. Its chemical formula appears to be $C_{20}H_{14}O_{10}$. Lac dye is much like cochineal but its color is duller, and it is less sensitive to light. *Sepia* is a blue-black ink obtained from cuttle fish or squid. It has been used since early times but became quite significant in ink drawings and watercolors on paper only since the eighteenth century in Europe. Sepia is somewhat light-sensitive but less so than cochineal.

Dyes of *plant origin* are greater in number than those of animal origin. *Indigo* is the earliest known organic dye (about 2000 B.C.) and has been used throughout Europe and Asia. There are many locations in the world where indigo is grown and its quality and color vary widely from place to place. It can be used as a blue paint pigment when ground to a very fine powder in oil, but is better in tempera or frescoes and as a watercolor. Indigo can be confused with *woad*, which is less expensive than indigo and is not as high in quality. Indigo is only one of the coloring agents in woad. As a dye on fabrics, woad and indigo are often used with a mordant, and on skin they were used for tattooing. In fact, the Picts, who lived in England when the Romans invaded, often tattooed themselves with woad to make themselves appear more fierce in battle. The word

Briton is a Latin form of a Celtic word meaning "painted men." Woad formed the blue color on high priests' robes in Biblical times. As a dye it is used on wool, cotton, and silk. Indigo came to Europe in the sixteenth century A.D. and the woad growers fought its importation. At times the use of indigo was declared illegal. Chemically, indigo is insoluble in water, is quite light-stable unless applied thinly, and is unreactive toward non-oxidizing acids, such as HCl, and bases. Its chemical structure is very much like that found in Tyrian Purple:

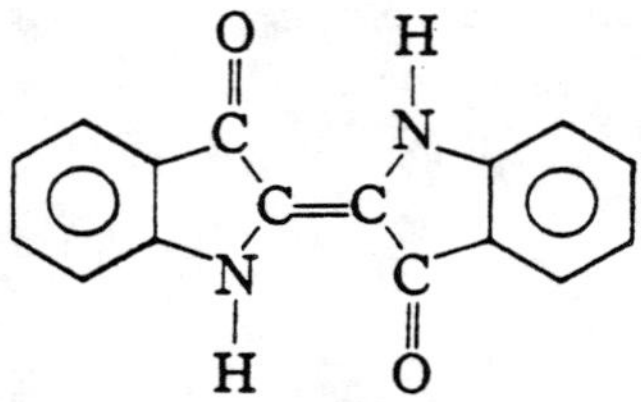

(Indigo)

Madder is obtained from the root of the madder plant found in Europe and Asia. It is sometimes known as Turkey Red when mordanted with Al^{3+}. It was most important in the seventeenth and eighteenth centuries, but has been used throughout the ages. Italian fabrics do not contain madder nearly as often as do materials from middle Europe. Its value as an artist's pigment is achieved by adding alum to the dye and precipitating it with base. When used as a dye, it is formed as a lake with $Al(OH)_3$. Historically, the "Redcoats" of the British Army were dyed with madder. Alexander the Great was reported to have splashed madder on his men's tunics and thereby tricked a much larger Persian army into thinking his men had been badly wounded in earlier battles. The Persian army then entered into battle with Alexander's Macedonians without due caution and this contributed to their defeat.

One of the chemicals found in madder root is ruberythric acid

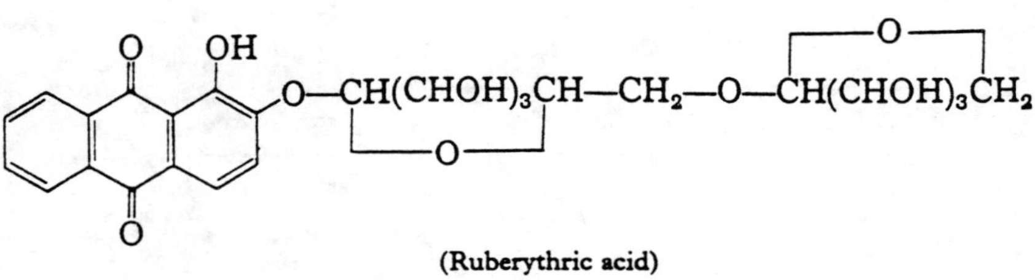

(Ruberythric acid)

which decomposes to alizarin, 1,2-dihydroxyanthraquinone,

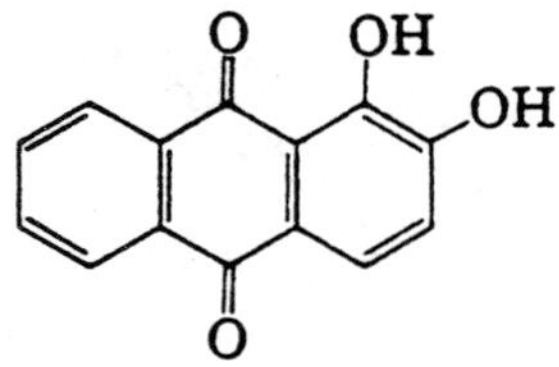

(Alizarin)

upon boiling in water. Madder gains its dye properties when the alizarin forms a lake with the alum. Madder also contains another material closely related to alizarin and known as *purpurin*. Purpurin is 1,2,4-trihydroxyanthraquinone which can also react as shown below with alum to form a colored lake:

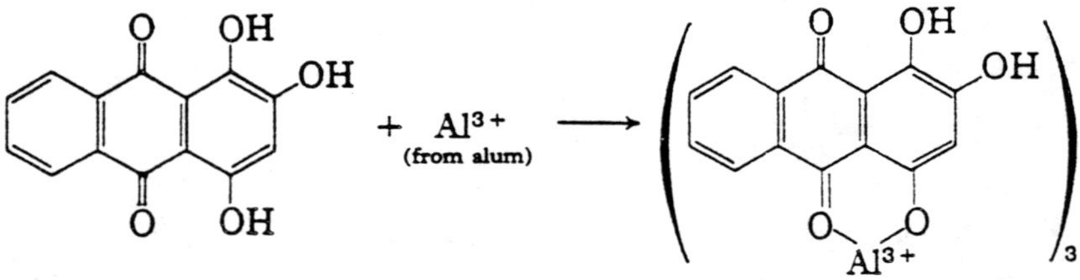

(Purpurin) (Purpurin lake)

The presence of purpurin can be used to distinguish the natural dye from the synthetic product and thereby establish the approximate time when the dye was applied. Synthetic alizarin was not known prior to 1875. Purpurin causes madder lake to fluoresce with a fiery, yellow-red color when exposed to ultraviolet radiation. Synthetic alizarin lakes show only a feeble, violet fluorescence. Alizarin forms mordant lakes with compounds other than $Al(OH)_3$. It is violet with $Mg(OH)_2$, purple red with $Ca(OH)_2$, and blue with $Ba(OH)_2$. Madder lake is stable toward light and dilute acids. It is rendered violet in color by dilute bases.

Logwood is a red dye which comes from trees indigenous to Mexico, Central America, and the West Indies. It was imported into Europe

Table VIII-1. Colors of Logwood Mordants

Mordant name	Mordant formula	Color with logwood
Tin	$SnCl_2$	Violet
Alum	$K_3Al(SO_4)_3$	Blue
Chrome	$K_2Cr_2O_7$	Blue-black
Iron	$FeSO_4$	Navy blue-lack
Copper	$CuSO_4$	Blue

during the time of Queen Elizabeth I and is one of few natural dyes still in use. When used as a mordant dyeing agent, it can give black on silk, cotton, and wool, but sometimes blue on wool. It has been used in making the dark background of calico. Some of the colors obtained with logwood mordants are shown in Table VIII-1. The chrome mordant with logwood was used in early writing inks. The coloring agent in logwood is haematein:

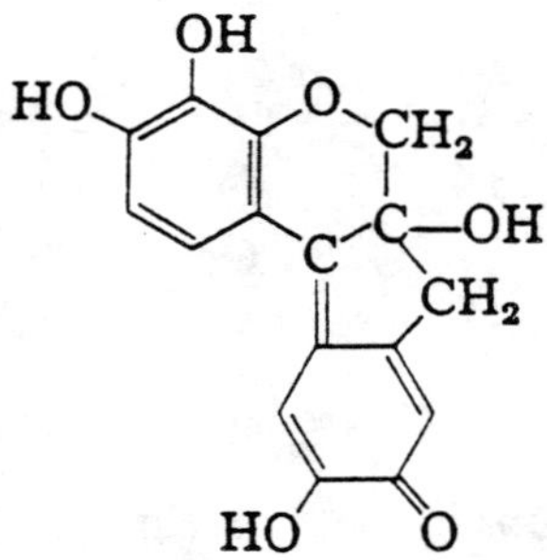

Haematein is light-sensitive and *p*H-sensitive. The *p*H sensitivity has made it useful in the manufacture of litmus paper.

Redwood is much like logwood, except that it is grown in Asia as well as Central America. It has been used primarily in cotton dyeing. As a chrome mordant it gives a red-brown color and with alumina, Al_2O_3, it gives blue and red. *Brazilwood* is a red dye which was imported into Europe from Ceylon during Medieval times and was more important than madder in this era. Redwood and brazilwood appear in most red fabrics in the 1000–1400 A.D. period. After about 1450 redwood and brazilwood greatly decreased in usage. Brazilwood was also used as an alum lake to give colors from light cherry to deep red. Chemically, brazilwood is very similar to logwood and redwood and probably has the following structure:

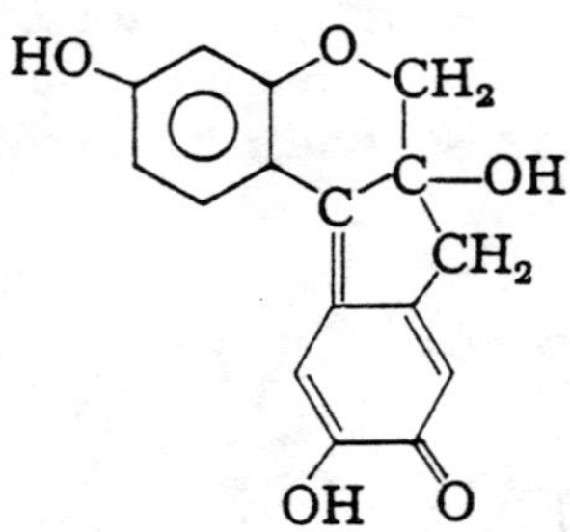

(Brazilwood)

Brazilwood is not stable in acids and is decomposed by strong light.

Quercitron bark was used in the eastern United States by colonists to dye cotton, wool, and linen. Its utility for dyeing wool was very good and it was not replaced until well into the synthetic dye era. Quercitron gives a yellow color as an alum lake, yellow and orange with tin mordant, and yellow-green colors with alumina and chrome mordants. Chemically, the dye is sensitive to acids and bases and fades in sunlight. The structure of the coloring agent quercitron is as follows:

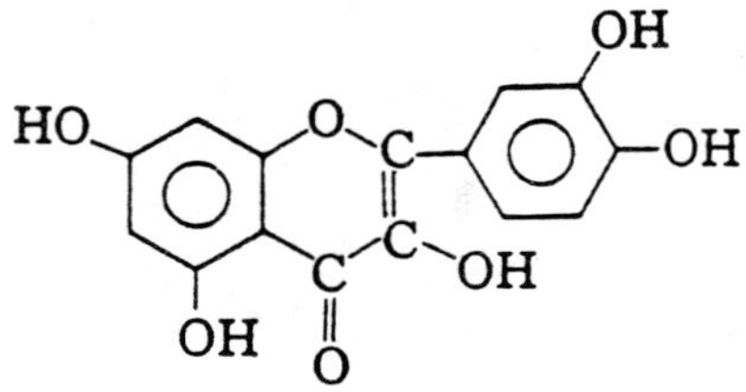

(Quercitron)

Persian berries from shrubs in southern Europe form a yellow lake with alum and yellow-orange colors as a tin mordant. The dye is somewhat sensitive to acids and bases as well as light, but it was competitive well into the twentieth century for calico printing. Persian berry dye is chemically similar to quercitron bark and the structure is as follows:

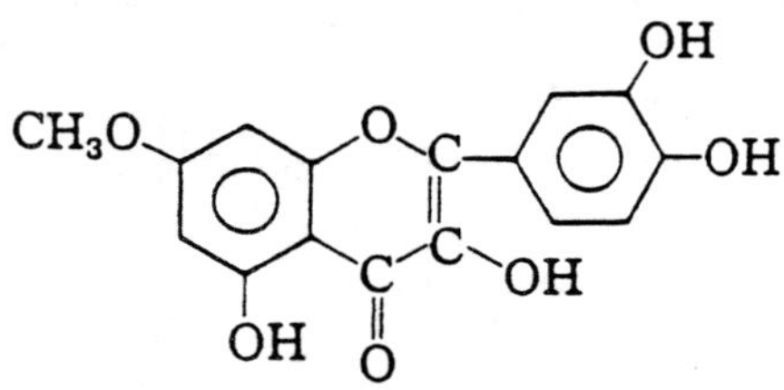

(Persian Berries)

Weld may be the oldest yellow dye. The plant was cultivated in central Europe. It can be formed as a lake and used as a pigment. It has lower tinting power than quercitron, but it is faster to light. Its chemical structure is closely related to that of quercitron and Persian berries:

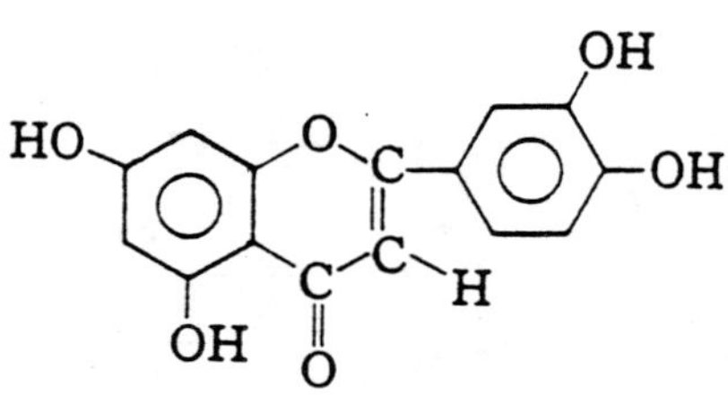

(Weld)

There are many other dyes of plant origin and briefly a few of these can be noted. *Cutch* was imported from India and gives various shades of brown used in calico printing. *Fustic* is derived from the mulberry family and gives yellow-brown and greens with certain mordants. *Turmeric* gives bluish-red on silk and wool with certain mordants but otherwise is yellow. Its coloring agent is curcumin. *Safflower* is obtained from the seeds of Dyers thistle and gives a water-soluble yellow dye and a less soluble pink-red dye. It has been used for thousands of years and is found in mummy cloths prepared by the Egyptians. *Annatto* is obtained from the seed of a small tropical shrub and is red in color when used on cotton, silk, and wool.

All of the natural product dyes and pigments are somewhat sensitive toward light. A qualitative ordering of light fastness for several of the dyes mentioned above can be made. In general, one has Persian berries < cochineal < madder < sepia < quercitron < indigo in light fastness. The fading rates also depend upon the amount of oxygen and moisture present, with the fading becoming worse as the oxygen and moisture levels increase.

After 1856 synthetic dyes gradually began to replace the natural dyes. The year 1856 was a landmark one because in that year an 18-year-old English chemist, William Henry Perkin, accidentally synthesized the gray-lavender dye, mauve, while attempting to prepare quinine. Mauve is a salt with the structural formula

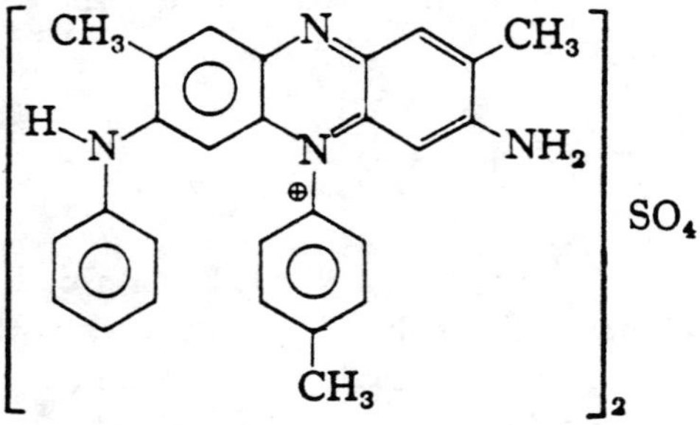

(Mauve)

His discovery revolutionalized dyeing as well as the field of organic chemistry to a considerable extent. Since dyes contain aromatic hydrocarbon rings, further developments required extensive research on the chemistry and commercial production of benzene and other unsaturated hydrocarbons. Scientific advances in chemistry had reached this level of sophistication and thus rapid advances could be made. In 1859 Verguin discovered fuchsin (magenta). Girard and DeLaire discovered roseaniline blue in 1860. By 1866 a whole series of brilliant but expensive artificial dyes had become available; however, some important colors, such as scarlet, were still missing. Finally in 1878, ponceaux, the first good scarlet, was discovered and it replaced cochineal.

Research and production of dyes rapidly fell primarily into German hands, probably because science and industry were more closely associated in Germany than in England. Many of the German chemical industries in existence today started as "color factories" and their names reflect this fact. Thousands of dyes with widely varying properties are now known. Most modern dyes bear little similarity in their chemical structure to the natural dyes, except in the case of indigo and alizarin.

CHEMICAL STRUCTURE

Classification of dyes according to their chemical structure is commonplace now because it is the approach the chemist must take. There may be thousands of dyes in each of the classes of structures given here. Within each class, most colors can be achieved merely by making subtle variations in the structure or composition of the molecules or by using different mordants.

Azo dyes were first synthesized by Griess, who in about 1863 prepared Bismarck brown. They did not come into extensive use until after 1875. Azo dyes are usually used in dyeing of cellulose fibers, such as cotton. Of all the dye groups, the azo dyes probably have the widest range of colors and physical properties. The chromophore is the —N=N group, as is present in azobenzene:

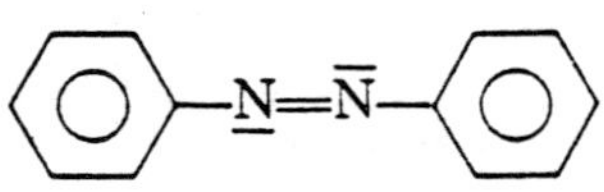

In fact, azobenzene is the nucleus of virtually all azo dyes. The light fastness of azo dyes is generally good, but is decreased by the presence of $—NH_2$ groups on the ring and increased by the presence of $—SO_2R$ and —OR groups. When an azo dye fades, it usually does so by oxidation of the azo group to the following azoxy compounds:

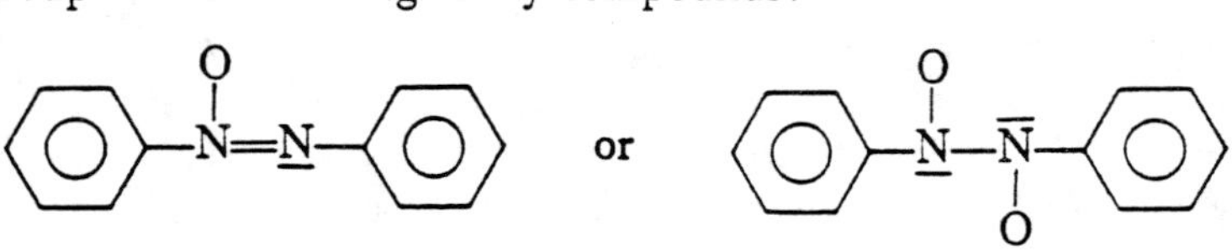

The color as well as the photodecomposition is activated by the $n \rightarrow \pi^*$ electronic transition involving the lone pair of electrons on nitrogen and the π^* molecular orbitals of the molecule. Total cleavage of the N=N bond also is known to occur in some instances.

Nitro and nitroso dyes usually produce yellow and yellow-brown colors. They became available first in 1864 as Martins yellow and Palatine orange. They formed the first set of yellow-to-orange colored dyes and were not rivaled until the azo dyes challenged them in the mid-1870s. Nitro and nitroso dyes possess as chromophores the $-NO_2$ or $-NO$ group bonded to an aromatic ring:

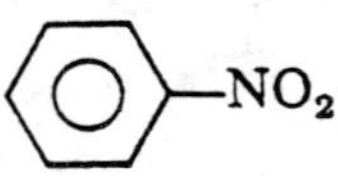

and

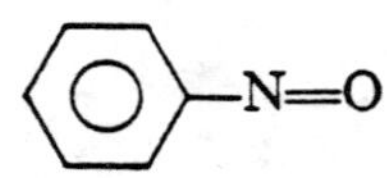

Their usual mechanism of fading is reduction to azoxy and azo compounds and finally to amine-containing products. Both the color and fading arise from the $n \rightarrow \pi^*$ electronic transitions of the molecule.

Cyanine dyes were first prepared in 1873. They have "resonance" structures of the type

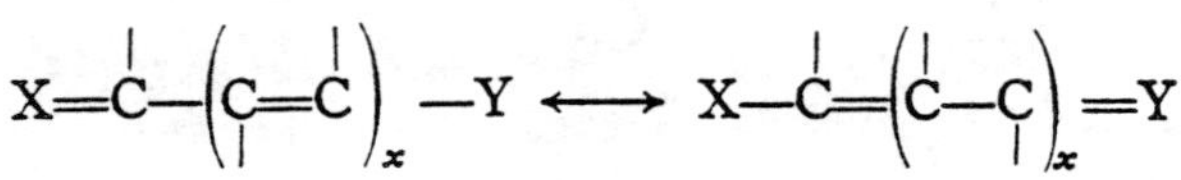

X and Y can be organic groups or fragments of an extensive variety which participate in the resonance structures. On occasion, these dyes are called polymethine dyes because of the presence of the conjugated carbon–carbon backbone in each system. They generally have relatively poor light fastness and are not used to any significant extent on textiles. They find their principal use as photographic sensitizers and in electrographic reproduction processes. As sensitizers, the polymethine dyes are unsurpassed in sensitizing the silver halides in wavelength regions where they are not otherwise sensitive (Chapter XII). Many natural coloring agents also belong to this class of dyes. For example, the carotenoids give color to some vegetables and are used to dye food. β-Carotene is an example:

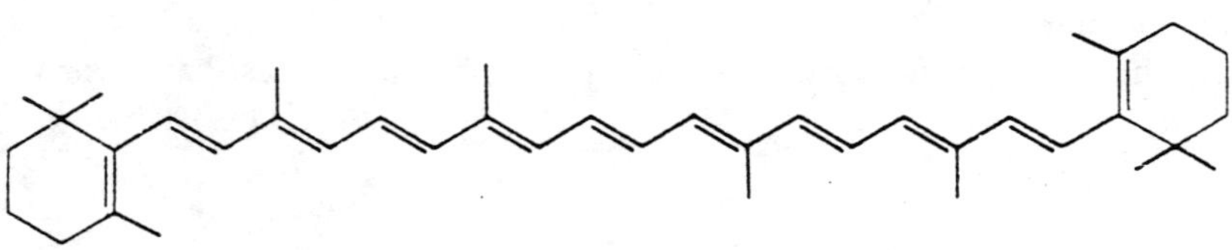

The *aza[18]annulenes* or *phthalocyanine dyes* are closely related to the cyanines, and are useful both as dyes and as pigments. They have a conjugated system of double bonds with 18 π electrons involved. Animal blood, both vertebrate and invertebrate, and chlorophyll fall into this

class of coloring agents. The phthalocyanine dyes were first prepared in the 1920s and became commercial in the 1930s. Their solubility is such that they could be classified as either dyes or pigments. They achieve their color by forming complexes with metals (usually from the transition elements). The source of color involves the electronic structure of the metal as well as the ring system. The tetrabenzoporphyrazine ring complexes a divalent metal M^{2+} as follows:

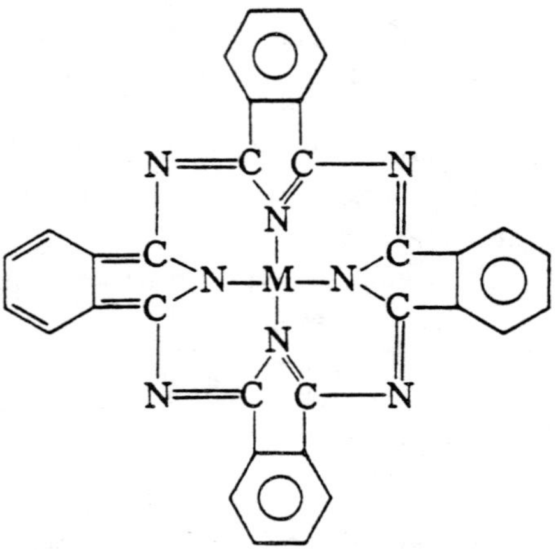

$M = Fe^{2+}, Mn^{2+}, Cu^{2+}$, etc.

These dyes are rather insoluble in water, brilliantly colored, and very stable to heat and light. In fact they are among the few truly light-stable organic-based dyes. This stability is very much tied to the presence of the metal and the lack of strain in the ring system. They have no affinity for the fiber and must be supported in some way. Usually an adhesive or resin is used for this purpose. The fastness of the dye is thus very dependent upon the binding agent. Phthalocyanine dyes have been used on cellulose, glass, and many man-made fibers.

Di- and *triarylcarbonium dyes* were first prepared in 1859 with the synthesis of fuchsin. They are basic dyes when used on wool, silk, and, a century or more ago, cotton with a tannic acid mordant. Structurally, they contain the carbonium ion group, which is stabilized by the resonance forms

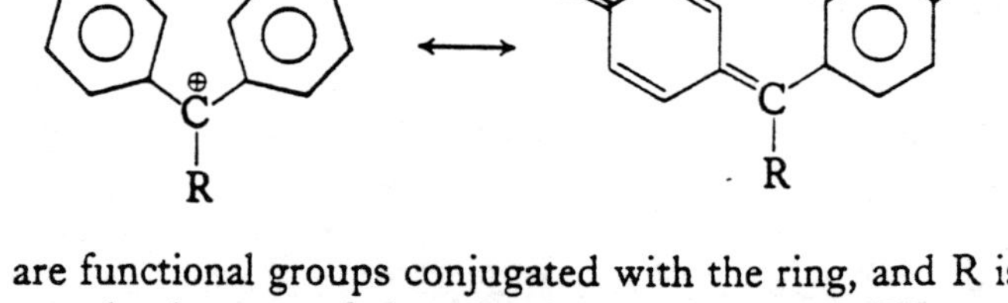

X and Y are functional groups conjugated with the ring, and R is another phenyl group in the case of the triarylcarbonium dyes. There are many

dyes in this class and they have brilliant colors, but they are not particularly fast toward light. One of the most important triarylcarbonium dyes is crystal violet. It has the strongest tinting strength of any synthetic dye, and its primary use is in the manufacture of inexpensive purple duplicating paper employed throughout the world. Mercurochrome, used as an antiseptic, and phenolphthalein, used as an indicator in acid–base titrations, fall into this class of dyes. The color of phenolphthalein is sensitive to *p*H. The following reaction illustrates how the molecule is altered by the *p*H:

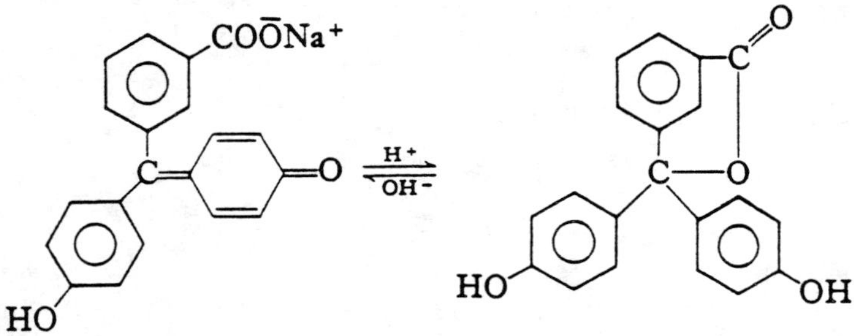

(pink) (colorless)

In acid solution the conjugation of the phenyl rings is disrupted.

Sulfur dyes were first prepared on a regular basis in 1893, although the process was known in France in 1873. The principle in using sulfur dyes is to mix sodium sulfide, Na_2S, with aromatic amines, phenols, etc., having dyeing potential. The dye fixes itself to the fabric but its structure is ill-defined and probably very complex. The dye starts out as water-soluble but as it dries, it becomes oxidized by oxygen from the air. The oxidized form is water-insoluble. Sulfur dyes are primarily used on cotton to produce greens, blacks, blues, and browns and they have good light fastness. They cannot be used on wool because the sodium sulfide damages the wool fiber.

Carbonyl dyes are sometimes called *vat dyes* because they are normally reduced in water solution to a closely related colorless or slightly colored compound called the "leuco" form of the dye. This was originally done in a special vat of reducing solution. The water-soluble form is bound to the fiber and is then oxidized to the original dye, usually by air oxidation. This process for indigo is as follows:

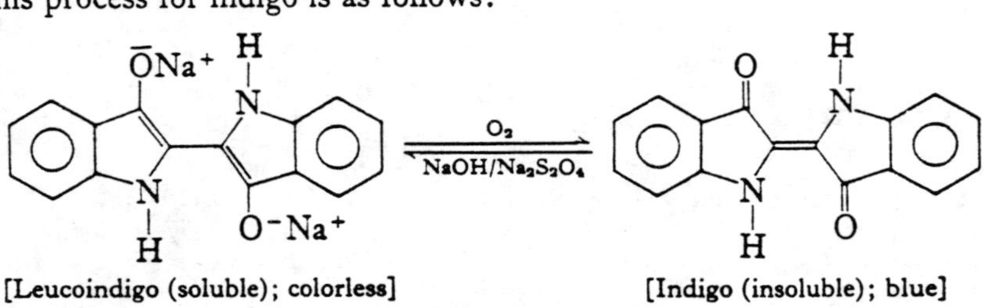

[Leucoindigo (soluble); colorless] [Indigo (insoluble); blue]

In the oxidized form, the dye is very insoluble in water and behaves as a pigment. However, vat dyes have little value as pigments, because of their low index of refraction.

The method of commercial vat dyeing has been known for over seventy years. This development was very important because carbonyl and vat dyes were among the first fast synthetic dyes for cotton. These dyes are most commonly used on cotton, but special methods have been developed to color other fabrics because of the outstanding light fastness. Carbonyl dyes contain the following structural unit:

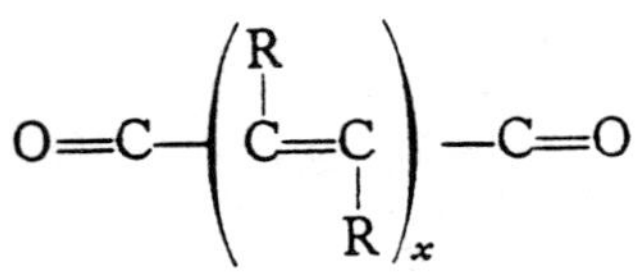

The parenthetical portion of the formula varies widely in form but is conjugated in all cases. Important examples of carbonyl or vat dyes include indigo, first prepared synthetically by Adolf Baeyer in 1879, and alizarin. Synthetic indigo was not introduced onto the market until about 1897. Alizarin was first prepared synthetically in 1879 but natural alizarin dyes were, of course, extensively used prior to this date. Alizarin is an anthraquinone compound used as a red mordant dye. A similar dye, alizarin yellow, which, as the name implies is yellow in color, eventually replaced fustic.

It is noteworthy that while most organic coloring agents are homogeneously dispersed in the substrate they dye and derive their colors from the particular molecular structure of the dye, some organic dyes are heterogeneously dispersed and thus are really pigments. In the pigment form, the molecular structure is still the fundamental source of the color, but wide variations in the depth, intensity, and brilliance of the color are often dependent upon how the particles are dispersed in the fiber. The effects of particle size on the appearance of surfaces were discussed in Chapter VI. Some further aspects of pigment particles are taken up in Chapter XI.

BIBLIOGRAPHY

Allen, R. L. M., *Color Chemistry*, Appleton-Century-Crofts, New York (1971).

Bird, C. L., and Boston, W. S., *The Theory of Coloration of Textiles*, Dyers Company Publications Trust, London (1975).

Brommelle, N., The Russell and Abney report on the action of light on watercolors, *Stud. Conserv.* **9**, 140 (1964).

Colour Index, Society of Dyers and Colourists, Bradford, England (1971).

Georgievics, G. V., and Grandmougin, E., *A Textbook of Dye Chemistry*, Scott, Greenwood and Sons, London (1920).

Gettens, R. J., and Stout, G. L., *Painting Materials, A Short Encyclopedia*, Dover, New York (1942).

Giles, C. H., *A Laboratory Course in Dyeing*, The Society of Dyers and Colourists, Bradford, England (1971).

Krochmal, A., and Krochmal, C., *The Complete Illustrated Book of Dyes from Natural Sources*, Doubleday, Garden City, New York (1974).

Mayer, R., *The Artist's Handbook of Materials and Techniques*, 3rd ed., Viking Press, New York (1970).

Mellor, C. M., Dyeing and dyestuffs 1750–1914, *Color Eng.* **7**, 50 (1969).

Pratt, L. S., *The Chemistry and Physics of Organic Pigments*, Wiley, New York (1947).

Rys, R., and Zollinger, H., *Fundamentals of the Chemistry and Application of Dyes*, Wiley–Interscience, New York (1972).

The Application of Vat Dyes, AATCC Monograph No. 2, American Association of Textile Chemists and Colorists, Lowell, Massachusetts (1953).

Trotman, E. R., *Dyeing and Chemical Technology of Textile Fibers*, Charles Griffin, London (1975).

Venkataraman, K., *The Chemistry of Synthetic Dyes*, Vols. I–V, Academic Press, New York (1952–71).

IX

Decomposition of Organic Materials by Light

In Chapter VII the origin of color in organic molecules was discussed from a molecular point of view. Color results from the excitation of electrons from the ground-state electronic structure of the molecule to an excited state. The electrons must rapidly return to the ground state in order to be active in color production.

In this chapter some of the elementary photochemical processes are considered both from a physical point of view and a chemical point of view. It is possible to apply the laws of photochemistry to colorless as well as colored molecules. Hence, in addition to the photochemistry of dyes, we can consider the photochemistry of paper, textiles, oils, and wood in this chapter. The rates and mechanisms of fading of dyes are discussed along with some precautionary measures that one might consider when working with valuable materials. Finally, a Spenglerian view toward the future of organic materials concludes the chapter.

The decomposition of dyes and organic materials is one of the most sinister aspects of the interaction of electromagnetic radiation with organic matter. Museum personnel, artists, and collectors must be constantly aware of this problem when displaying their possessions. A distressing illustration of the degradation of dyes by light can be seen in The Martyrdom of St. Paul, a fifteenth century European tapestry now housed in the Boston Museum of Fine Arts. The side of the tapestry facing the viewer has lost much of its brilliant coloration. However, when the linen backing of the tapestry was removed, the *verso* was found to have retained much of its original brilliance. Such degradation in exposed areas is difficult to avoid. This chapter deals with the mechanisms and the nature of this occurrence. These factors need to be recognized if deterioration is to be brought under control.

PHOTOCHEMICAL PRINCIPLES

Light-induced reactions are an aspect of chemistry known as *photochemistry*. The process of absorption of light by a molecule was discussed in Chapter VII. Figure IX-1 illustrates some of the ways in which a molecule which has just absorbed energy might relax back to the ground state. In most cases, the molecule rids itself of the absorbed energy by converting it to heat in about 10^{-8} sec. The molecule, having returned to the ground state, can absorb another photon of light to produce color, and the sequence is repeated. In the process, the material becomes slightly warmer than its surroundings. However, some molecules emit the absorbed energy in the form of visible light of longer wavelengths than the absorbed energy. If this light emission occurs within 1 sec or less, the process is generally called *fluorescence*. If the light emission occurs more than 1 sec after the photon is absorbed, then the process is known as *phosphorescence*. Many materials are found to fluoresce in the visible region when exposed to ultraviolet radiation. Included are varnishes, oils, minerals, gems, and dyes.

Whether the absorbed light is converted to heat (infrared energy) or emitted as visible light (phosphorescence and fluorescence), the excited state of the molecule can be a chemically different species from the ground state. The excited state is an energized form of the molecule and is inherently more reactive than the ground state. When the conditions are right, the excited state of the molecule may react with other materials in the vicinity, such as the fiber, another dye molecule, oxygen, or water, or it may simply destroy itself by bond rupture. Light-fastness, stability, and fading rate are terms that loosely refer to the relative reactivity of the excited states of dyes and other molecules. A light-fast dye has a comparatively unreactive ground and excited state, and relaxes by production of lower energy electromagnetic radiation. On the other hand, a fugitive dye has excited states which may chemically decompose while getting rid

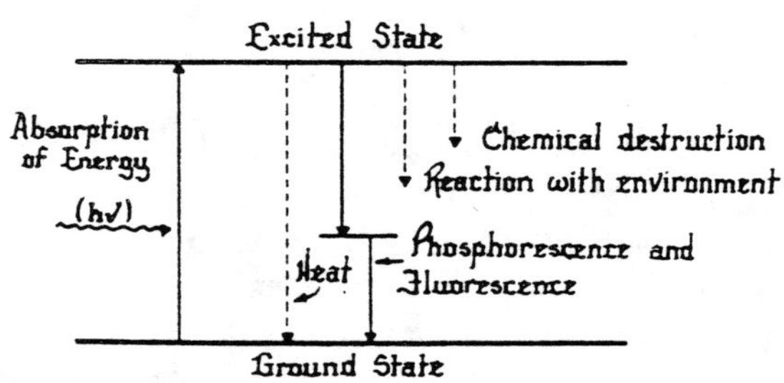

Figure IX-1. Absorption of light by a molecule and some mechanisms by which the molecule can relax back to the ground state.

of the absorbed energy. The degree of fastness depends not only upon the molecular structure of the dye, but also on the fabric dyed, the amount of moisture and oxygen present, the wavelength of light, the intensity of light, the temperature of the material, and the exposure time.

In discussing the fading of a dye, attention will be focused on how the excited state destroys the visible-light-absorbing properties of the ground state. Paradoxically, it is the mere fact that the dye has color that causes it to lose its color with time. Without color the compound would no doubt be longer-lived under the same conditions. Colorless compounds certainly can be decomposed by light, but most frequently the ultraviolet energies lead to the degradation. The photochemical concepts presented here, however, apply to colorless materials, such as the textile or paper being dyed, as well as to the dye itself.

There are several important rules of photochemistry.

1. A law proposed by Grotthus at the surprisingly early date of 1817 states that only light actually absorbed by the molecule can produce a chemical change. At a later date it was recognized that this law means the wavelengths of light must overlap an absorption band of the compound before it can produce a photochemical reaction. In Figure IX-2, wavelengths of light in region (b) will not be absorbed and thus will not produce a photochemical reaction. Any wavelength of light in region (a) can be absorbed and can produce photochemical effects.

2. *Einstein's law of photoequivalence* states that one photon of electromagnetic radiation can activate one molecule. Only one primary photochemical step and not the overall photochemical process is described by this law. Simple photoequivalence is rarely observed, because many molecules may be involved in subsequent steps after a single molecule becomes activated. The term *quantum efficiency* or *quantum yield* of a photochemical reaction refers to the number of molecules of reactant or product consumed or produced per quanta of light absorbed. The quantum yield for chemical reactions is usually in the range of 10^{-5}–10^{6}. Light-stable

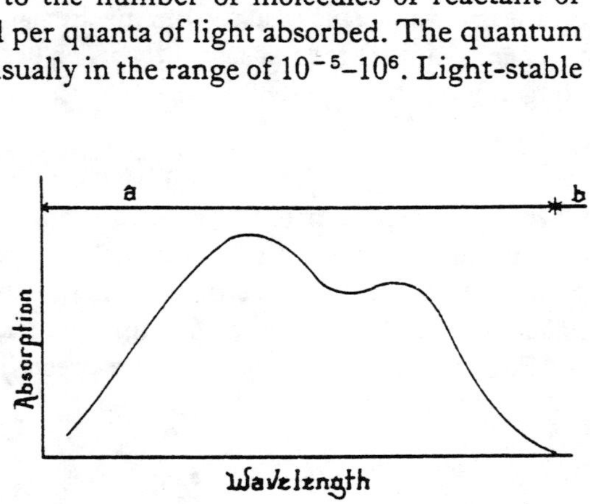

Figure IX-2. Light in region (a) will be absorbed and can produce a photochemical reaction. Light in region (b) will not produce a reaction.

dyes have very low values of quantum efficiency, while fugitive dyes have relatively higher values.

3. Wartburg in 1920 proposed that the number of absorbed photons (i.e., the intensity of light) and not the energy of the light determines the extent of the photochemical reaction. The photochemical reaction will be more extensive as the intensity of the incident light increases; however, the course of the reaction may vary with different wavelengths.

4. Turro in 1965 proposed that the lowest energy electronic transitions in a molecule determine the course of the photochemical reaction. One is mainly concerned with $\pi \rightarrow \pi^*$ and $n \rightarrow \pi^*$ electronic transitions in photodegradation reactions of colored molecules. These electronic transitions often tend to be the lowest energy ones in unsaturated molecules and therefore they control the course of the degradation reaction.

The subject of photochemistry can be carried on into the quantitative aspects of light absorption, the various pathways which the molecule follows in trying to rid itself of the absorbed energy, the reactions which take place in the excited state, and the structures which exist in the excited state. Such considerations rapidly lead into basic research in chemistry and physics. Rather than continue in that vein, a brief survey of the more common types of photoreactions that lead to the degradation of organic materials will be presented. The rates of these reactions are found to vary depending upon the nature of the reaction and how the dye is set into the fiber. To understand deterioration, one must consider both the reactions themselves and the rates of the reactions.

RATES OF PHOTOCHEMICAL REACTIONS

The *rate* of fading of a dye is not the same for all dyes and may occur in different ways, as is true for the mechanisms of fading. Most dyes in the presence of a fiber fade at a zeroth-order or first-order rate. Zeroth-order fading follows the equation

$$[\mathrm{D}] = D_0 - kt \qquad \text{(IX-1)}$$

[D] represents the concentration of dye after a particular time t; k is a rate constant characteristic of the dye; and D_0 is the initial concentration of the dye at the time that it was placed on the substrate. The number of dye molecules decreases linearly with time in zeroth-order fading.

First-order fading follows a rate prescribed by the equation

$$\frac{d[\mathrm{D}]}{dt} = kt \qquad \text{(IX-2)}$$

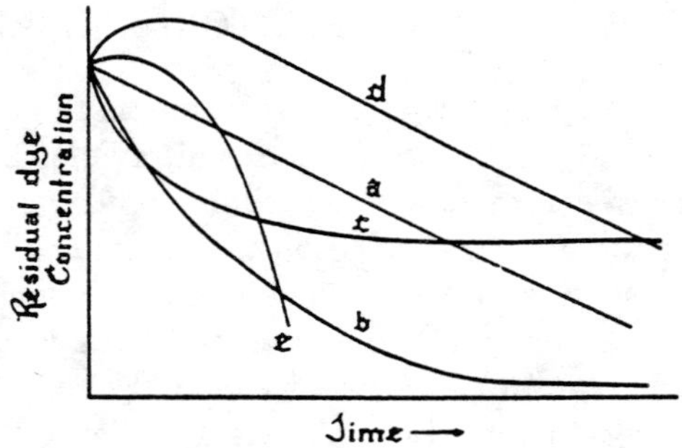

Figure IX-3. Some rates for dyes. (a) Zeroth-order, (b) first-order, (c) zeroth- and first-order, (d) negative initial fading, (e) increased fading rate with time.

Here $d[\mathrm{D}]/dt$ is a differential which measures the rate of change in the concentration of the dye with time. The dye concentration decreases logarithmically with time. Zeroth- and first-order fading rates describe many known chemical reactions, but with systems as complex as dyes dispersed inside of and on the surface of fibers, there are also other observed rates of degradation. Figure IX-3 shows some relative rates of fading expressed in terms of the concentration of the residual dye as a function of time.

Line (a) is the zeroth-order fading rate discussed above. It is often observed with large pigment particles and very fast dyes. The fading rate is linear in time according to equation (IX-1). Curve (b) represents first-order fading, which is characteristic of homogeneous, molecularly dispersed dyes. The concentration of dye decreases logarithmically with time according to equation (IX-2). Curve (c) is a combination of zeroth- and first-order fading and can be found when two types of dyes are mixed together. A dye molecularly dispersed in the fiber fades by a first-order rate, whereas an aggregated dye fades by a zeroth-order rate. This type of fading is very common in dyed fibers. Curve (d), with negative initial fading, is sometimes found if large particles of a coloring agent are initially present which disintegrate with time, perhaps due to the heat of illumination. The optical density of the dye increases during the time that the particles first disintegrate. However, a point in time is reached where the deterioration, which is occurring from time zero, becomes more important because the newly formed smaller dye particles fade at a faster rate. A zeroth- or first-order mechanism then takes over. The combination of initial particle disintegration and continual fading leads to a rate depicted by curve (d). Curve (e) illustrates the situation where the fading rate increases with time. Here the breakdown products of the dye may react with the dye and accelerate the rate of breakdown as more products form. This is known to occur with some insoluble azo dyes on cellulose.

MECHANISMS OF PHOTOCHEMICAL REACTIONS OF DYES

The energies of electromagnetic radiation of most concern in photochemical effects occur in the ultraviolet and visible regions. This is the highest energy radiation to which objects are likely to be routinely exposed. The radiation wavelengths of interest are in the 700–200 nm range, which corresponds to 40–140 kcal of energy (Chapter I). Solar energy below 270 nm does not reach the earth's surface in significant amounts, because of absorption by ozone and water in the earth's atmosphere. Most other light sources do not subject objects to radiation of wavelengths less than 270 nm either, if only because the glass envelope around the light source is opaque to these wavelengths. As a result, photofading in practical situations comes about because of radiation in the 300–700 nm region. The energies of radiation in this wavelength range can be compared to the bond energies typically found in organic compounds. Single carbon–carbon bonds have an energy of about 83 kcal/mole, C—O bonds are about 86 kcal/mole, C—H bonds are about 98 kcal/mole, and C—N bonds are about 73 kcal/mole. Electromagnetic energy must have wavelengths of 380 nm or less (Table I-1) to match these bond energies. Therefore, many degradation reactions in dyes and fibers are *not* a result of the direct rupture of covalent bonds by the light. The radiation simply does not have enough energy to cleave the bonds. Instead, the molecular excited states created by light absorption are more reactive than the ground state toward oxygen, moisture, the fiber, etc., and this causes most of the photodegradation.

It is interesting to note that organic materials are in general much more reactive than inorganic materials toward the action of light. This difference is not as much related to differences in bond energies between organic and inorganic compounds, as it is to the difference in pathways of decomposition of the excited states of the organic compounds compared to inorganic compounds. There are a greater number of pathways of decomposition and the products are often more volatile in organic compounds compared with inorganic compounds. On average, the bond energies of the two classes are quite similar to one another.

The most important general classes of photoreactions which are known to occur in organic materials are outlined here. It should be noted that in many cases the factors controlling the decomposition of dyes and organic materials are not well understood. It is not possible to identify with certainty a decomposition pathway of a compound except by doing careful research experiments. Most of the research in this area has been carried out under controlled idealized conditions and may not be indicative of the precise behavior of the material in real-use conditions. It is rarely satisfactory to determine only the straightforward variables such as the intensity of light, the wavelength of the light, and the environmental conditions when researching fading processes. One may also need to determine the energy, the lifetime, and the chemical nature of the excited state of the dye. These latter factors are much more complex to establish.

Intramolecular photoreactions are not particularly common in dyes of interest in art materials, but these reactions are very important in other applications. If D is a molecule of the coloring agent, then an excited state D* can be formed upon absorption of the appropriate energy of light:

$$D + h\nu \rightarrow D^*$$

D* then transforms to another product D′, which is similar to D but which may have a different color or optical density:

$$D^* \rightarrow D'$$

In many cases the overall reaction is readily reversible, and the amount of D or D′ depends upon the amount of light present:

$$D \rightleftharpoons D'$$

An intramolecular photoreaction is important in human vision. In the rod-type light receptors of the eye, the vitamin A aldehyde, retinal, is known to isomerize from *cis*-retinal to *trans*-retinal in the presence of light:

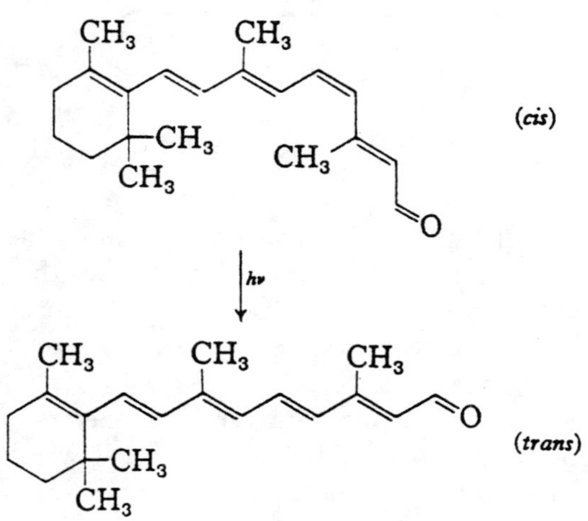

Colored intermediates between the *cis* and *trans* forms also exist, and thus a range of wavelength sensitivities can be obtained with only minor alterations to the retinal. If the reaction were not readily reversible, then vision might not continue beyond the first moments of eyesight.

Some dyes also reversibly change colors by intramolecular isomerization. This effect is called *photochromism*. Crystal violet, a synthetic triarylcarbonium dye, exhibits photochromism in solution by changing from violet to green upon exposure to light. In the dark the original crystal violet molecule is regenerated. The reaction known to occur is

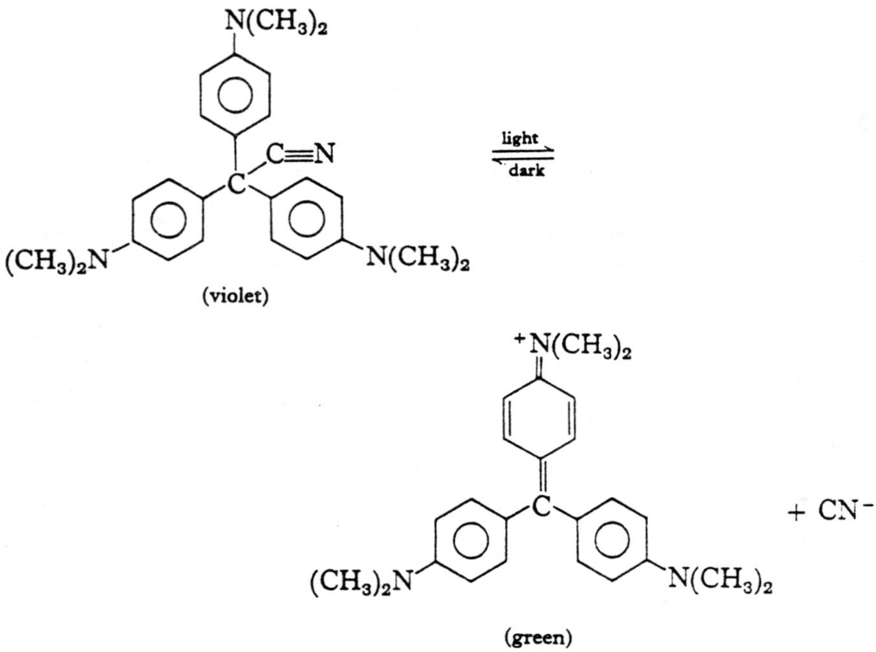

Variable-density light filters, in which a variation in optical density occurs depending upon the intensity of light, usually work on the principle of reversible intramolecular photoreactions. Strong light exposure causes the filter to darken. When the light intensity is reduced, the glass automatically becomes increasingly transparent.

Photoreduction is a mechanism whereby an energy-activated dye reacts with a neighboring molecule, such as the fiber to which it is bound, to reduce the dye and oxidize the fiber:

$$D + h\nu \rightarrow D^*$$

$$D^* + \text{fiber} \rightarrow D_{\text{reduced}} + \text{fiber}_{\text{oxidized}}$$

The dye and the fiber are both altered if not destroyed by the change. More specifically, the fiber can transfer electrons or hydrogen atoms to the dye in the course of the reduction process and produce radicals, which are molecules having unpaired electrons (here F = fiber):

$$D^* + FH \rightarrow D^- \cdot + \cdot FH^+$$

$$D^* + FH \rightarrow \cdot DH + F\cdot$$

Usually two electrons or two hydrogen atoms ultimately are transferred to a double bond in the dye, for example, as in the case of hydrogen atom transfer:

$$D^* + FH_2 \rightarrow DH_2 + F'$$

The end result might be the reduction (saturation) of a double bond which destroys the conjugation and thus the color of the dye. The overall chemical process is a multistep one and can be very complex. Nitro and nitroso dyes and carbonyl vat dyes are thought to fade by the photoreduction mechanism. Cellulose and wool can serve as the fiber which is oxidized in the process. In a few instances it may be possible to reverse the effects of photoreduction in a material by oxidizing the material back to its original state. For example, O_2 can react with a reduced radical intermediate $\cdot DH$ during the reduction step and oxidize it back to the original dye. The hydroperoxyl radical $HO_2\cdot$ forms in the process and it itself decomposes to H_2O_2 and O_2. These reactions are summarized as follows:

$$\cdot DH + O_2 \rightarrow D + HO_2\cdot$$

$$2HO_2\cdot \rightarrow H_2O_2 + O_2$$

Photooxidation is a very common pathway of degradation in dyes as well as fibers and other organic materials. Indigo is found to decompose by photooxidation. Cellulose fibers making up paper, wood, and some textiles also decompose by photooxidation. The common occurrence of this degradation mechanism stems from the fact that oxygen in the air is a strong oxidizing agent. Light-activated dye molecules routinely come into contact with oxygen and moisture and become chemically altered. The light-activated dye D* is usually oxidized by $HO_2\cdot$, which is itself formed by the light-induced reaction between O_2 and H_2O. The $HO_2\cdot$ is a very reactive, strong oxidizing agent. The reaction of the activated dye with moisture (H_2O) appears to follow the course

$$D + h\nu \rightarrow D^*$$

$$D^* + H_2O \rightarrow \cdot DOH + H\cdot$$

$$H\cdot + O_2 \rightarrow HO_2\cdot$$

The hydroperoxyl radical, $HO_2\cdot$, then oxidatively degrades the dye. Direct reaction of the dye with oxygen follows a different course, but with the same result:

$$D + h\nu \rightarrow D^*$$

$$D^* + O_2 \rightarrow D + O_2^*$$

$$O_2^* + H_2O \rightarrow \cdot OH + HO_2\cdot$$

The $HO_2\cdot$ radical formed in these reactions oxidizes the dye. The dye or fiber in its radical form can also react directly with O_2. Using the dye as an example, the following reactions are representative of this process:

$$D^* + O_2 \rightarrow DO_2^*$$

$$DO_2^* + \text{fiber} \rightarrow DO_2H + \text{fiber}'$$

A carboxylate group, —COOH, forms on the dye or fiber during this process. In addition, cross-linking as well as chain breaking may occur in fabrics, paper, hides, and oils. These chemical reactions also lead to mechanical changes, such as increased brittleness, which are rarely reversible.

A specific example of the pathway of oxidative degradation of a dye has been determined for the synthetic triarylcarbonium dye Cl-Basic Green-4 (also called malachite green) under controlled experimental conditions. The following sequence of reactions describes the process:

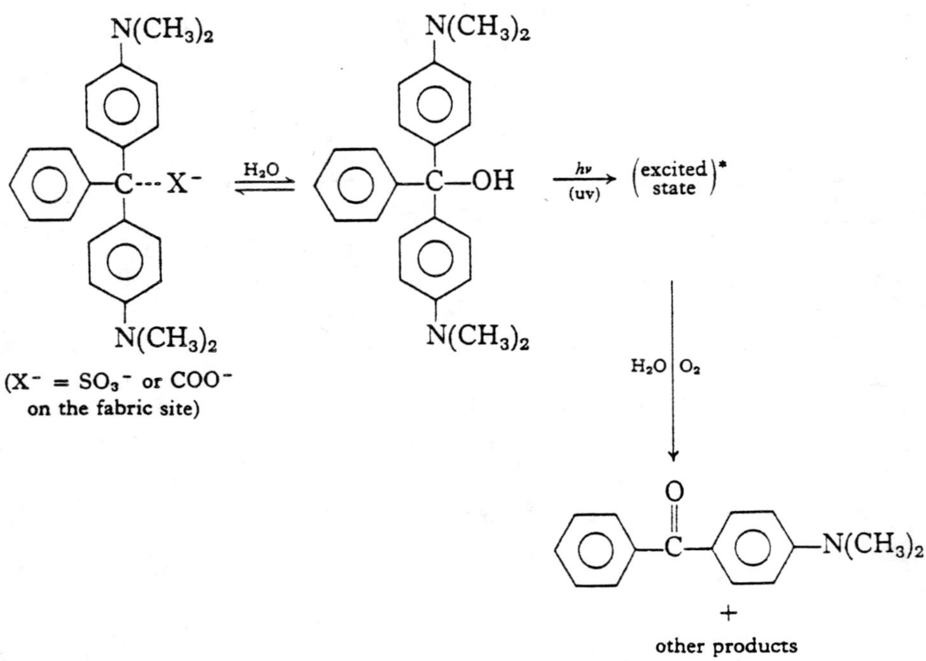

A salt form of the dye in water solution is in equilibrium with the hydroxide form. The hydroxide form reacts with ultraviolet radiation to produce an excited-state species of undetermined nature. Reaction with H_2O and O_2 in the solution destroys the dye molecules. Other primary products of the decomposition were not isolated because of further oxidation by oxygen.

In general there exist too many uncontrolled variables to carry out a detailed analysis of dye degradation in a fiber. However, general types of reactions can be classified by this type of experimental work.

Substrate activation or *phototendering* is a decomposition pathway characteristic of very light-fast dyes. Very light-fast dyes tend not to be significantly sensitive to light of wavelengths longer than 400 nm. What deterioration of the dye that does occur often results from decomposition of the fiber. The fiber rather than the dye absorbs light and causes reduction of the dye bound to the fiber, although in some cases the dye may act to transfer absorbed energy to the fiber. The fiber often acts as a hydrogen atom donor in these reactions. The process is basically the same as photoreduction described above, except that the substrate appears to decompose the dye rather than the dye decomposing the substrate. The dye is often reduced and the fiber is oxidized in the process:

$$\text{fiber} + h\nu \rightarrow \text{fiber}^*$$

$$\text{fiber}^* + \text{D} \rightarrow \text{fiber}' + \text{D}'$$

Both the fiber and the dye are decomposed as the end result. Examples of substrate activation occur in fabrics. Silk is especially susceptible to photoinduced deterioration in the dyed areas. Fiber-induced degradation of orange and yellow anthraquinone vat dyes occurs when they are used on cellulose. The N=N bond of azo dyes also can be ruptured by light-activated organic substrates with oxidizing power.

ENVIRONMENTAL FACTORS

Almost all of the fading mechanisms discussed so far involve the participation of other compounds in the light-induced decomposition of the dye. It is important to consolidate this point by identifying several of the common external influences on fading and organic degradation reactions. As noted by Russell and Abney in 1888, *it is rarely light alone that causes photochemical changes* in materials of artistic and historical interest.

Atmospheric oxygen and water actively participate in and accelerate fading in most cases. We know that oxygen reacts with the $\pi \rightarrow \pi^*$ excited state of the dye molecule and oxidizes the dye to a chemically reactive radical. Atmospheric moisture also accelerates most fading reaction rates. For example, dyes on cotton are more sensitive to humidity changes than are those on wool. The effect of H_2O is partly a result of the formation of the hydroperoxyl radical, $HO_2\cdot$, and hydrogen peroxide, H_2O_2, by the ultraviolet radiation-triggered reaction:

$$2H_2O + O_2 \xrightarrow{uv} 2H_2O_2$$

As a precaution against the destructive effects of oxygen and moisture, as early as 1893 paintings were being sealed under glass in a partial vacuum. An inert atmosphere, such as provided by helium, argon, or nitrogen gas, would be preferred to a vacuum. The United States Declaration of Independence is stored under a blanket of helium gas. In general, it is beneficial to store dyed materials in an oxygen-free environment having controlled humidity to reduce the occurrence of degradation reactions.

Around room temperature, temperature changes have relatively little effect on fading. This is because the activation energy of the fading process is usually small, being in the range of only a few kcal/mole. Materials taken from Egyptian tombs were frequently found to be in excellent condition because they had been spared from significant temperature changes, moisture, and oxygen. Moreover, light was absent. The oldest known wall paintings, the Altimira Cave paintings in Spain and those at Lascaux in France, were sealed by a landslide and when opened they were found to be in good condition. Since they have been opened, however, observers have noted a progressive deterioration due to changes in temperature and humidity. It is the organic binders and not the pigments that have been affected. The hazards of temperature, humidity, and oxygen thus cannot be stressed enough.

The wavelength of light has an affect on fading, depending upon whether the dye is fast or fugitive toward light. Ultraviolet radiation is hazardous toward all dyes, but visible light is comparatively worse toward fugitive dyes than it is toward fast dyes.

The physical state of the coloring agent in the fiber plays a role in the resistance of the dye to photodegradation. If the material is pigment-like and thus in the form of particles in the fabric, then a decrease in the size of the particle can cause an increase in the fading rate. If the dye is located only on the surface of the fiber, where it is constantly exposed to light, oxygen, and moisture, then it is apt to fade more rapidly than if it has been absorbed deeply into the fiber.

PHOTOREACTIONS OF PAPER, TEXTILES, WOOD, VARNISH, AND OILS

We know that the exposure of dyes to light can lead to fading and in some cases color changes in the dye. This effect of light causes immediate concern when it is observed in a valuable material. Deterioration of color is a dramatic form of photodegradation, but it is often not as potentially damaging as the photodegradation that occurs in the material forming the structure of the object. The materials of interest in this latter context are usually paper, textiles, and wood. Varnishes and oils will also be considered in this section, since they are traditionally not used for their coloring qualities. Paper, textiles, and wood may have color by themselves, but they are not thought of for their coloring attributes in the same way that dyes are.

Ultraviolet radiation is potentially the most damaging form of electromagnetic radiation to which an object is likely to receive significant exposure. Visible radiation and infrared radiation also have importance in causing photodegradation. These three wavelength regions do not create the same kinds of problems.

The degradation of polymeric organic materials can proceed on many fronts. Some of the routes were noted above for dyes. In addition, polymeric chains may be further polymerized (by cross-linking), depolymerized (by chain-breaking), oxidized, and reduced. Oxygen, moisture, and air pollutants enter into these reactions, as do dyes attached to the fiber. For example, dyes on paper can cause the paper to decompose as the dye degrades. Synergistically, as the paper degrades, its acidity may increase and this causes the dye to decompose even more rapidly. Therefore, the phenomena having to do with the photodegradation of polymeric fibers need to be considered.

The effect of light in bringing about the decomposition of cellulose concerns anyone interested in cotton, wood, and paper. Cellulose is a polymer of cyclic acetal rings bridged by oxygen atoms:

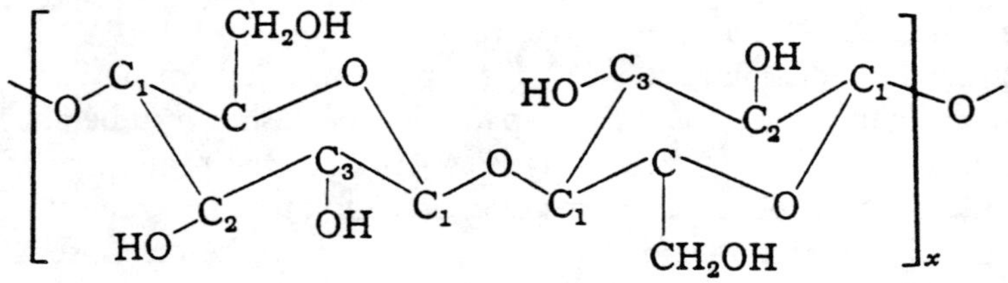

(Cellulose)

The subscript x can exceed 5000. Probably the first recorded example of the effect of sunlight on cellulose, if not on any organic material, can be found in *Exodus* 15:23. Moses led the Israelites into the Wilderness of Shur:

> When they came to Marah, they could not drink the water of Marah because it was bitter, therefore it was named Marah. And the people murmured against Moses, saying "What shall we drink?" And he cried to the Lord and the Lord showed him a tree, and he threw it into the water and the water became sweet.

Marah is on the Sinai Peninsula near the Red Sea. The area is desert, and desert water usually is highly alkaline, i.e., it is bitter. Bleaching of wood on the desert by the sun causes the alcohol groups, —COH, on the cellulose molecules to be oxidized to carboxylic acid groups, —COOH. By casting a piece of bleached wood into the alkaline water, an acid–base reaction occurs in which the alkalinity of the water is reduced. Thereby the water becomes drinkable (sweet).

The photooxidation of alcohol groups to carbonyl-containing functional groups, such as aldehydes and carboxylic acids, takes place at the C_2 and C_3 carbons on the acetal ring. The —CH_2OH group is not involved. It is believed that ultraviolet radiation of wavelengths less than 350 nm probably is responsible for the oxidation of alcohols to carbonyl groups. These carbonyl groups are chromophores and are in part responsible for the photochemical yellowing of the cellulose. The method of manufacture also plays a role in the rate of yellowing of paper. Newsprint contains larger amounts of lignin and rosin than does high-quality rag paper. Lignin and rosin yellow readily. All short wavelengths cause chain breaking in the fiber by oxidation of the acetal carbon (C_1). Some changes in conjugation occur and these also contribute to the formation of yellow color. The extent of chain rupturing affects the strength of the cellulose fiber. The shorter the chains are, the weaker is the fiber. Hence, paper, cotton textiles, and wood lose their strength when they are excessively exposed to electromagnetic radiation. In textiles, for example, the degree of deterioration can be ascertained by measuring the viscosity of the cellulose. Viscosity is a function of the chain length. When the chains are cleaved by light, the viscosity of the material decreases.

Ultraviolet radiation at wavelengths longer than 350 nm bleaches the cellulose to lighter color. Cellulose in paper has been shown to be affected by radiation with as long a wavelength as 460 nm even though cellulose does not absorb visible light. Other components of paper, such as glue and rosin, absorb this radiation and can deteriorate the cellulose fiber.

The acidity of paper is one of the main concerns of paper conservation. The acidity may be affected by the methods used in the manufacture of the paper. For example, alum-precipitated rosin used as a waterproofing agent produces hydrogen ions in the presence of moisture. Modern paper has an intrinsic *p*H of 4–5. However, an increase in the acidity is a commonly observed result of the aging of paper because the acidity can increase due to photoreactions. Photooxidation of an alcohol group in the subunits of the cellulose to a carboxylic acid group creates an acidic site in the fiber, particularly if moisture is present. The oxidation of a —COH group to a —COOH group requires the presence of oxygen, and it is a well-known fact that oxygen and moisture promote the photodecomposition of cellulose.

The restoration of paper or textiles which have been damaged by the effects of electromagnetic radiation is a delicate task. In paper, the reduction of acidity using buffered alkaline salts is a principal treatment. In an attempt to reverse the effects of light that cause yellow staining and embrittlement, some more drastic treatments have been tried. These call for chemically altering the chromophore causing the yellow color. Bleaching with hypochlorous acid, HClO, or its salts will oxidize all functional groups to the colorless carboxylic acid group, COOH. Hydrogen peroxide, H_2O_2, in a basic medium causes ring opening and breaks the conjugation in chromophores that produce the yellowing stains. Chlorine dioxide also functions as an oxidant in this way. Sodium borohydride, $NaBH_4$, used at $pH = 10$ will reduce the chromophores back to the —COH side chains originally present in the fiber. Not all of these methods are suitable for treatment of valuable materials and they should be used with caution and a trained hand.

It is important to note that exposure to light seems to cause a long-term activation of paper fiber. When the sample is removed from the light, it continues to react with atmospheric oxygen, water, and decomposition products, and degrades at a rate faster than if it had not been exposed in the first place. Storage of light-exposed paper in the dark should be done in anaerobic conditions. Nitrogen and helium are suitable atmospheres for this purpose.

Wood experiences oxidation of the lignin in addition to oxidation of the cellulose and hemicellulose. Most of this deterioration occurs on the surface of the wood. The interior regions do not undergo photooxidation. However, wood is damaged by the indirect effects of infrared radiation. Because infrared radiation is associated with heat, it causes loss of moisture, which leads to splitting and warping. Woods also contain natural dyes having red, yellow, and brown coloration. These dyes tend to be sensitive to light. Usually one observes a fading of the dyes in dark woods, but the light-colored woods are often observed to darken upon exposure to light.

Textiles made from wool and silk are subject to a different photochemistry from cellulose because the fibers are made of protein. The order of sensitivity of the fibers to light is silk > cotton > wool. However, the presence of moisture greatly increases the rate of fading in wool. Bleaching in wool appears to result from exposure to visible radiation, but yellowing in wool is caused by ultraviolet radiation in the 300–400 nm region. The yellow color is believed to result from the formation of carbonyl chromophores in the natural grease (suint) in wool. Silk also yellows in ultraviolet radiation and bleaches in visible light. The yellowing in silk results from oxidation of the amino acid, tryptophan.

Varnishes and oils are affected by light and produce complicated effects in paintings. Paints often increase in transparency with time because the index of refraction of the oil increases toward the value of the dispersed pigments. They can yellow in time and this leads to a change in the tonality of a painting. These effects were discussed in detail in Chapter VI. Deterioration of varnish and oil may also contribute to chalking in a paint layer. The course and extent of deterioration of a varnish or oil are affected by the pigments present and the purity of the original material. For example, the amount and purity of turpentine used as a thinner play a role in the yellowing characteristics of a varnish.

Yellowing in varnishes is due to oxidation reactions activated by light. Free radicals form in the varnish matrix, leading to the formation of ketones, aldehydes, and carboxylic acids, which function as chromophores and auxochromes. An increase in the number of unsaturated fatty acids occurs, which adds more chromophores and conjugation to those already present. The longer wavelengths of visible light (red) and the near-infrared region seem to be most responsible for the increase in the percentage of the unsaturated fatty acids. However, some natural varnishes are actually bleached by light and they tend to yellow faster in the dark.

One of the more troublesome problems in the cleaning and restoration of paintings is the change in the chemical nature of the varnish and oils from the time when they were first applied. Dammar and mastic contain alcohol and ketone groups when first applied. These functional groups become oxidized to acids with the passage of time. Hence the solvent needed to remove old varnish may be different from that needed for freshly applied varnish. Similarly, if the oil used as the paint vehicle polymerizes or depolymerizes over the years, then its solubility changes accordingly. Shrinking and cracking of the surface can result from the loss of volatile products formed in the degradation process.

The presence of pigments can influence the nature of photoinduced degradation of varnishes and oils. When inpainting, a common practice is to apply varnish before and after painting so that the new paint layer is sandwiched between two varnish layers. However, when zinc white and dammar are combined in this way, "chalking" is sometimes noted. This occurs because the vehicle degrades and leaves "unwetted" pigment particles behind to scatter light. In addition, the dammar is oxidized by ultraviolet radiation and becomes fractured. The whitish appearance is in part due to the fracture lines in the varnish. Zinc white, ZnO, hastens the ultraviolet-induced deterioration of the varnish because ZnO strongly absorbs ultraviolet radiation (Table VI-5). ZnO is also a basic anhydride and can react with the acids in dammar to further degrade the varnish. All of these effects produce chalking, but in different ways.

Synthetic varnishes are transparent in the visible and the near ultraviolet—at least more so than are the natural varnishes. An ideal situation would exist if the varnish were completely transparent in the visible region in order to have true color-rendering capability, but were completely opaque in the ultraviolet region to avoid the damaging effects of the higher energy radiation. The synthetics cannot provide this perfect ideal, but they are reasonably effective on both counts. The main concern with some synthetic varnishes (particularly Lucite) is their tendency to cross-link and form extensively polymerized chains upon exposure to ultraviolet radiation. This causes their solubility to decrease and makes them difficult to remove. Generally speaking, the high-viscosity synthetic varnishes are most apt to cross-link upon aging. However, aged linseed oil also exhibits the tendency to cross-link. Artists have known for centuries that a painting should be exposed to daylight to promote drying and oxidation of oils before it is varnished. Cross-linking is also responsible for the increased refractive index of the oil as it ages.

MINIMIZATION OF PHOTOCHEMICAL EFFECTS

The importance of light in the degradation of organic materials cannot be overestimated. Popular displays tend to be exposed to light for long periods of time. Many works have lasted for hundreds of years, but in most cases under far less illumination than is now used. As long as artifacts are to be displayed, there is no certain way to protect them from damage by light. The extent of exposure can be reduced by using the lighting only when a collection is open for viewing. A collection might be cycled so that the exposure time is minimized.

The most important method of minimizing the detrimental effects of light is, of course, to reduce the intensity of the most hazardous energies. Ultraviolet, visible, and infrared energies all are problematic, but some energies are more hazardous than others on selected objects. The intensity of visible light around oil and tempera paintings should not exceed 150 lx. For textiles, paper, and watercolors, the intensity should be held below 50 lx. These values are above the approximate value of 30 lx

Table IX-1. Damage Factors per Foot-candle of Zenith Skylight After Passing Through Various Filters[a]

Filter	Damage factor/foot-candle
Zenith skylight (ZSL)	4.80
ZSL through window glass	1.58
ZSL through window glass plus fresh dammar	1.45
ZSL through window glass plus aged mastic	0.91
ZSL through Plexiglas UF-1	0.41
Daylight fluorescent lamp	0.40
Incandescent lamp	0.14
Yellow NBS filter[b]	0.02

[a] From R. L. Feller, N. Stolow, and E. H. Jones, *On Picture Varnishes and Their Solvents*, The Press of Case Western Reserve University, Cleveland, Ohio (1971).
[b] Used on the U.S. Declaration of Independence.

needed for good color rendition. No object should ever be displayed in direct sunlight. It is probably wise to avoid, if possible, even indirect exposure to light coming from an unfiltered skylight or an exterior window. The risk of ultraviolet radiation from the sun is simply too great. Artificial light sources that may contain ultraviolet energies should be checked for their ultraviolet content. Ultraviolet filters, such as Plexiglas UF-3, should be installed on windows and worrisome light fixtures. They should be changed periodically as recommended by the manufacturer's specification.

The amount of damage per foot-candle of illumination for objects exposed through various coatings placed on ordinary glass and plexiglas is shown in Table IX-1. The damage factor for the daylight fluorescent lamp and that for the incandescent lamp are also shown for comparison. Zenith skylight (ZSL) has the potential for producing considerable damage by ultraviolet radiation. When ZSL is filtered by glass, glass with various coatings, or plexiglas, the damage factor is reduced. The high-energy radiation is filtered reasonably well by glass, but not well enough to protect highly sensitive materials. It can be seen that the presence of traditional varnish on a work will reduce the potential for damage but will not eliminate it completely. The limiting value for the damage factor of 0.46 is calculated for an ideal filter which removes all wavelengths of radiation below 400 nm and transmits all visible light above 400 nm.

Infrared radiation causes considerable damage to materials made of wood, ivory, paper, parchment, leather, and textiles. The most serious effects result from moisture loss as manifested in warping, cracking, and peeling. Color changes can occur in varnish and ivory. Incandescent lamps are an important source of infrared radiation in a museum. Spotlighting of particular objects is a practice that needs careful control. The temperature rise at an object's surface as a result of incandescent illumination can and should be measured if there is a chance that heating may occur. If heating is of concern, then glass infrared filters (KG-1 or KG-3) might be used between the source and the object. Also, air circulation or air conditioning systems might be employed to reduce the amount of heat.

For truly valuable objects, extreme protection should be brought into use. The Declaration of Independence of the United States is protected by a yellow-colored filter that absorbs ultraviolet and blue light. It is also housed in a hermetically sealed case under an atmosphere of helium. The purpose of the inert atmosphere is to eliminate the deleterious oxidizing effects of oxygen and the effects of humidity.

Control of the humidity will help stabilize many objects. An inert atmosphere will reduce the rate of, or eliminate altogether, many photochemical reactions. Light filters will absorb many harmful electromagnetic energies.

THE FUTURE OF NATURAL ORGANIC MATERIALS

To conclude this chapter, it is fitting to learn about the general instability of almost all carbon-containing materials, not just toward electromagnetic radiation, but also toward the more general laws of nature. Many objects and materials of artistic and historical value are composed of complex organic polymers originating from the plant and animal world. Paper, wood, tempera, dyes, oils, hides, textiles, and bone are examples of such materials.

A consequence of the law of conservation of matter is that once life ends for a creature, its comprising matter must convert back into primary building compounds to be recycled again. The primary compounds are CO_2, H_2O, O_2, carbon, and numerous other simple materials. The death of a living entity merely signifies a change in its chemistry. If an art object is partly or completely constructed from once-living materials, the recycling of matter beings to occur. An art conservator might be thought of as one who is fighting this law of nature. There is really no hope of victory, but there is opportunity to slow the advance. Unfortunately, nature has time on its side. The laws of thermodynamics provide the foundation for understanding, if not coming to grips with, this very real concern.

The second law of thermodynamics gives information about the spontaneity of chemical reactions. The reaction

$$\text{large organic molecules} \rightarrow \text{small organic molecules}$$

is spontaneous in nature but the rate of the reaction is slow. A thermodynamically favorable process is one in which the free energy change ΔG for the reaction is negative, i.e., $\Delta G < 0$. The change ΔG can be defined in terms of the temperature, the enthalpy change ΔH, and the entropy change ΔS. The ΔH measures the heat change during a reaction and is related to the energies of bonds broken and formed in the reaction. The ΔS measures the change in the amount of order in nature and is related to the change in the number of species present and/or the change in the ordering of the molecules relative to one another during a reaction. The free energy is related to entropy, enthalpy, and the temperature T by the equation

$$\Delta G = \Delta H - T\Delta S \tag{IX-3}$$

ΔG will always be negative in value if ΔH is less than zero (i.e., the reaction is exothermic) and ΔS is greater than zero (i.e., there is an increase in the disorder of the system). A higher temperature causes the $T\Delta S$ term to become still more negative and this favors spontaneity. We know from practical experience that an increase in temperature can cause chemical reactions to begin and will usually increase the rate of chemical reactions which are already occurring. There are other possible values for ΔG. If ΔG is greater than zero, then the reaction is not spontaneous at the specified temperature. If ΔG equals zero, then the reaction is in dynamic equilibrium.

The concern here is with the free energy of chemical reactions leading to degradation of organic materials. ΔG can be measured directly from the electrochemical potentials E in volts. The decomposition of organic materials can be discussed using these potentials. The following equation relates ΔG to the change in potential for a reaction ΔE:

$$\Delta G = -n\mathscr{f}\Delta E \tag{IX-4}$$

n is the number of electrons transferred in the reaction, $\mathscr{f}$ is Faraday's constant (23,060 cal/volt), and ΔE is the difference between the electrochemical potential of the reactants and products. For a hypothetical reaction $X \rightarrow Y$, $\Delta E = E_Y - E_X$. If $E_Y - E_X$ is less than zero volts, then ΔG will be greater than zero calories and the reaction will not occur spontaneously. If $E_Y - E_X$ is greater than zero volts, then $\Delta G < 0$ and the reaction is spontaneous at the particular temperature of interest.

Returning to the problem of art materials, there are two primary paths for decomposition of organic compounds, depending upon whether or not oxygen is present. In the presence of oxygen, complete degradation corresponds to the reaction

$$\text{organic compounds} + O_2 \rightarrow CO_2 + H_2O$$

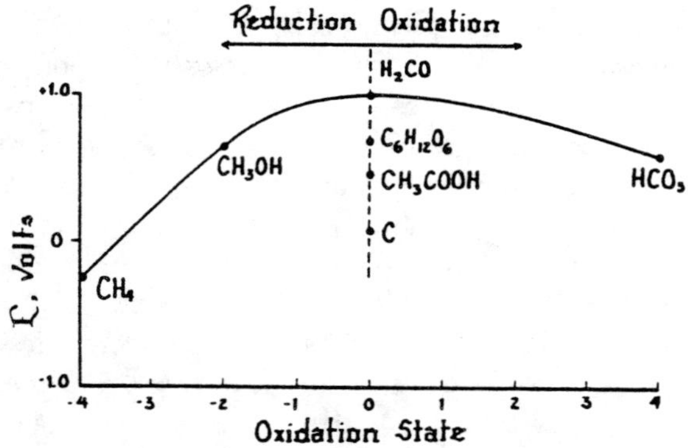

Figure IX-4. Oxidation state diagram for several carbon compounds.

This reaction completely or at least partly occurs in the presence of light, heat (fire), and some bacterial enzymes. It represents the oxidation of carbon to its highest oxidation state. In anaerobic conditions the following reaction may occur:

$$\text{organic compounds} \rightarrow CH_4 + \text{other products}$$

This reaction represents the reduction of carbon, usually by bacterial action, as in the formation of marsh gas or the formation of alcohol by fermentation. Intermediate reactions may occur which result in intermediate degrees of reduction of carbon, depending upon the conditions. Some of these reactions have produced coal and oil when the pressure is high and oxygen is deficient:

$$\text{organic hydrocarbons} \rightarrow C + H_2O + \text{other smaller hydrocarbons}$$

Yet a third mode of degradation is known to occur in organic molecules, but is not important here. On occasion large organic molecules are converted to smaller ones without oxidation or reduction. For example, sugar can be converted to vinegar and no oxidation or reduction occurs.

To model the oxidation and reduction reactions, one can consider the oxidation potentials of some reactants and the products which might be formed in decomposition reactions of organic compounds. A plot is made in Figure IX-4 of the oxidation state of carbon and the E values in volts for a series of compounds in a water solution buffered at $pH = 7$. Virtually all naturally occurring complex organic compounds lie on the dashed line because these compounds have an average oxidation number of zero for carbon. Several useful points can be obtained from the plot. First, the greater stability of elemental carbon (as occurs in some coals) relative to compounds of the same oxidation state (glucose, cellulose, acetic acid, formaldehyde, etc.) makes the decomposition of these latter naturally occurring compounds to coal, which has a higher carbon content, a thermodynamically favorable process. This fact has led to formation of oil and coal deposits and also the degradation of organic plant matter to humus-like material. Second, the convex upward shape of the line connecting lower oxidation states to the higher ones indicates that the oxidation states between the extremes of +4 and −4 are unstable with respect to disproportionation to CO_2 and CH_4. The CO_2 exists in neutral solution mostly as bicarbonate, HCO_3^-. Thus, all complex organic molecules having an oxidation state for carbon of or near zero are "energy-rich," and are thermodynamically unstable with respect to the oxidation to CO_2 or the reduction to CH_4. Many art materials are made of carbon-containing compounds with an average oxidation state for carbon of zero and with E values which are more positive than that of elemental carbon. Hence they are inherently unable to avoid inevitable oxidation or reduction to more stable compounds of carbon.

There are no known violations of the laws of thermodynamics. Thermodynamics, through the second law, dictates that complex carbon-containing molecules will eventually decompose no matter what precautions are taken. The second law is the bane of the conservator. A conservator slows the process and minimizes its effects, but can never expect to retard it completely.

BIBLIOGRAPHY

Balder, J. J., The Discoloration of Coloured Objects Under the Influence of Daylight, Incandescent Lamplight, and Fluorescent Lamplight, *De Museumdag*, The Netherlands Museums Association, Leiden (1956); *Lichttechnik* **8**(2), 57 (1956); see also R. L. Feller, *Stud. Conserv.* **4**, 155 (1959).

Barltrop, J. A., and Coyle, J. D., *Excited States in Organic Chemistry*, Wiley, New York (1975).

Barrow, W. J., *Manuscripts and Documents: Their Deterioration and Restoration*, 2nd ed., University of Virginia Press, Charlottesville, Virginia (1972).

Brommelle, N., The Russell and Abney report on the action of light on watercolors, *Stud. Conserv.* **9**, 140 (1964).

Brommelle, N., and Harris, J. F., Museum lighting, Part II, *Mus. J.* **61**, 1 (1961).

Cooper, B. S., Fluorescent lighting in museums, *Mus. J.* **53**, 279 (1953); see also S. Rees Jones, *Stud. Conserv.* **1**, 196 (1954).

Feller, R. L., Control of deteriorating effects of light on museum objects: Heating effects of illumination by incandescent lamps, *Mus. News Tech. Suppl.* **46**, 39 (1968).

Feller, R. L., Deteriorating effects of light on museum objects, *Mus. News Tech. Suppl.* (June 1964).

Feller, R. L., Problems in retouching: Chalking of intermediate layers, *Bull. IIC-AG* **7**, 32 (1966).

Feller, R. L., Studies in Photochemical Deterioration, in *ICOM 4th Triennial Meeting*, Venice (1975).

Feller, R. L., Stolow, N., and Jones, E. H., *On Picture Varnishes and Their Solvents*, The Press of Case Western Reserve University, Cleveland, Ohio (1971).

Genard, J., Extreme ultraviolet radiation from tubular fluorescent lamps and its effect on museum lighting, *Mus. J.* **52**(1), 5 (1952).

Hanlan, J. F., The effect of electronic photographic lamps on the materials of works of art, *Mus. News* **48**(10), 33 (1970).

Harrison, L. S., *Report on the Deteriorating Effects of Modern Light Sources*, Metropolitan Museum of Art (1954); *Illuminating Eng.* **49**, 253 (1953); See also B. S. Cooper, *Stud. Conserv.* **1**, 198 (1954).

Use of Fluorescent Light in Museums, ICOM, Unesco House, Paris (1953).

Kühn, H., The use of heat protection filters when works of art are filmed or televised, *Stud. Conserv.* **12**(3), 102 (1967).

Leene, J. E., Demeny, L., Elema, R. J., de Graaf, A. J., and Surtel, J. J., Artificial Ageing of Yarns in the Presence and in the Absence of Light and Under Different Atmospheric Conditions, in *ICOM 4th Triennial Meeting*, Venice (1975).

Radley, J. A., and Grant, J., *Fluorescence Analysis in Ultraviolet Light*, Chapman and Hall, London (1954).

Rawlins, F. I. G., and Rawson-Bottom, W. E., Ultraviolet radiation from fluorescent lamps, *Mus. J.* **52**(3), 84 (1952).

The Bleaching of Pulp, W. H. Rapson, ed., TAPPI Monograph Series #27, New York (1963).

Thomson, G., Visible and ultraviolet radiation, *Mus. J.* **57**, 27 (1957).

Turro, N. J., *Molecular Photochemistry*, Benjamin, New York (1965).

Van Beek, H. C. A., and Hertjes, P. M., Fading by light of organic dyes on textiles and other materials, *Stud. Conserv.* **11**, 145 (1966).

Venkataraman, K., *The Chemistry of Synthetic Dyes*, Vols. I–IV, Academic Press, New York (1952–71).

Williams, J. C., Chemistry of the deacidification of paper, *Bull. IIC-AG* **12**, 16 (1971).

Williams, J. C., *Preservation of Paper and Textiles of Historical and Artistic Value*, Advances in Chemistry No. 164, American Chemical Society, Washington, D.C. (1977).

Wringer, H., and Smith, S., *The Deterioration and Preservation of Library Materials*, University of Chicago Press, Chicago, Illinois (1970).

X

The Origin of Color in Inorganic Compounds

The color of organic compounds was shown in Chapter VII to be brought about by an excitation of the molecule from the ground electronic state to an excited state. For the excitation to occur in the visible region, a conjugated system of double bonds or an intense chromophore was necessary. Electronic transitions producing color were labeled as $n \rightarrow \pi^*$, $\pi \rightarrow \pi^*$, $\sigma \rightarrow \pi^*$ indicating the nature of the orbitals involved in the ground and excited states. Similarly for inorganic materials, electronic transitions result from the absorption of light of the appropriate energy and thus are responsible for the production of color. However, the terminology and bonding models used to describe the absorption process are somewhat different in inorganic as opposed to organic systems.

Of course electrons and nuclei compose inorganic compounds just as they compose organic compounds, but there are some practical differences in the chemistry of these two classes. Organic compounds are mostly made up of the lighter elements in nature and contain a carbon atom backbone. The molecules tend to be discrete species having covalently bonded atoms. A crystal lattice built from organic molecules often tends to be malleable, soft, and low melting compared to one built from inorganic compounds. Carbon atoms easily engage in π-bonding with themselves and other elements located near carbon in the periodic table. Inorganic compounds, on the other hand, are composed of elements which span the entire periodic table. Since the chemistry of the known elements varies widely, inorganic compounds can be expected to be chemically very different from one another. One finds crystal lattices made up of discrete, covalently bonded molecules, such as mercuric chloride, $HgCl_2$. Covalent crystal lattices, where a three-dimensional polymeric lattice forms without discrete molecules, are commonplace. Quartz (SiO_2) and diamond are examples of this type of lattice. One regularly encounters ionic lattices containing either simple or complex ions. Sodium chloride, containing Na^+ and Cl^- ions, and Prussian blue, $Fe_4[Fe(CN)_6]_3$, containing Fe^{3+} and $Fe(CN)_6{}^{4-}$ ions are examples of ionic lattices. The properties of these structures are very different from one another. They range from low to high melting, soft to hard, malleable to brittle, and nonmetallic to metallic. The objective is to describe the bonding the origin of color in these complex and very different types of compounds. The formalism needed is necessarily different from that put forth in Chapter VI for organic molecules.

A review of inorganic pigments reveals that virtually all contain either (1) transition metal ions (metal elements whose valence-electron configuration involves *d* electrons), or (2) high polarizable main-group-element anions, such as O^{2-} or S^{2-}. Transition metals and polarizable main-group elements may be present together, but most pigments contain one or the other plus additional elements. The metal ions in the pigment crystal lattice are bonded to anions or neutral molecules in what is usually the most electrostatically and structurally favorable arrangement. These anions and neutral molecules are referred to as *ligands* when used in this way. The variety of colors produced results from light-induced electronic transitions intimately associated with the metals and ligands. To obtain a thorough understanding of the electronic spectra of inorganic compounds, a comprehensive discussion of physical and quantum mechanical principles is required. In this chapter, instead, only a qualitative description of the light-absorbing processes of metals bonded to ligands will be presented. As was the case with organic compounds, one is rarely safe in trying to *predict* colors on the basis of elementary bonding models. Only with highly sophisticated bonding considerations can one begin to explain the details of why a certain color and not another is produced by a certain pigment.

A logical beginning to a discussion of color in inorganic pigments is to examine the periodic table of the elements and note the positioning of elements which make up useful artists pigments (Figure X-1). The transition elements found in pigments are located in the first row and in Group IIb. The metals occur as cations whose formal positive oxidation state can be represented parenthetically after the metal. In most instances in this chapter a charge is used to represent the oxidation state of the metal. Whether the charge or the oxidation state is shown, the formalism is merely a matter of "bookkeeping." As examples, titanium exists as Ti(IV) (or Ti^{4+}) in TiO_2, chromium is found as Cr(III) and Cr(VI) in pigments (3+ and 6+ oxidation states), manganese occurs as Mn(II) (or Mn^{2+}) in umbers, iron occurs as both Fe(II) and Fe(III) in pigments (2+ and 3+ oxidation states), cobalt is found in pigments as Co(II) (or Co^{2+}), and copper as Cu(II) (or Cu^{2+}). The Group IIb elements zinc,

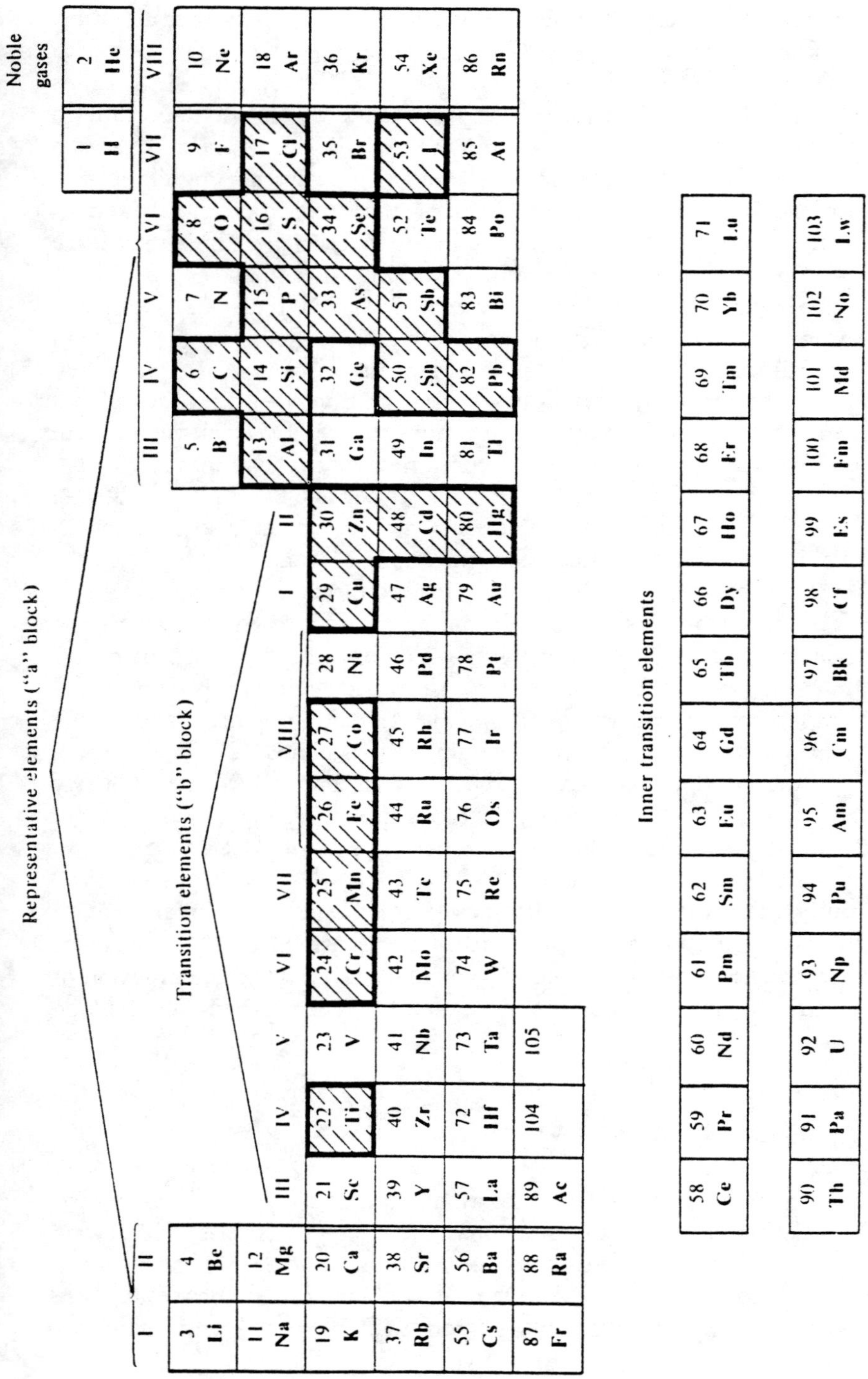

Figure X-1. Some elements found in artists' pigments.

cadmium, and mercury are not really transition elements, because they have a filled *d*-electron configuration ($3d^{10}$), but they can be grouped with the transition elements. They occur as Zn(II), Cd(II), and Hg(II) ions (M^{2+}) in pigments. The main-group elements which are important in pigments are widely spread throughout the periodic table. Several elements found in pigments, such as Na^+, K^+, and Mg^{2+}, are not highlighted in Figure X-1 because they do not play a primary role in color production. They are present in the crystal lattice in order to balance the electrostatic charge and provide structural stability by filling spaces.

The main-group elements shown in Figure X-1 can be present as anions or cations, depending upon the element. For example, oxygen (O^{2-}), sulfur (S^{2-}), iodine (I^-), and chlorine (Cl^-) are often found as simple anions and are important in color production when they are bonded to metals having a relatively large positive charge. Other elements form molecular anions, particularly by combining with O^{2-} to give OH^-, CO_3^{2-}, SO_4^{2-}, CH_3COO^-, SiO_3^{2-}, AsO_3^{3-}, SbO_4^{3-}, SnO_4^{4-}, PO_4^{3-}, AsO_4^{3-}. A few of the central elements in these anions also function as cations in the crystal lattice of pigments. For example, Al^{3+}, Pb^{2+}, Pb^{4+}, As^{3+}, and Sn^{4+} are found as monatomic cations in pigments.

The question is how do metals and metalloid ions combine with one another so as to produce stable colors which have been used in paintings for thousands of years? The answer has taken shape decisively only in the last sixty years. To discuss the origins of color in inorganic systems, one must consider three factors: (1) the covalency of bonds, (2) the electronic configuration of the ions, particularly the metal ion, and (3) the geometry of the anion arrangement about the cations. Fortunately, the discussion. does not require all three considerations to be involved at once, but in many cases two of them are necessary even for an elementary presentation. In combination with one another, these factors lead to energy absorption by the *charge transfer* and "*d–d*" electronic transitions. The importance of the covalency factor will be considered first and then the contribution of the geometry and electronic configurations to color production will be approached.

COVALENCY OF BONDS

A fundamental concept in chemical bonding of all systems is concerned with the distribution of the electrons in the bond. That is: (1) If two chemically different elements combine with one another, are the bonding electrons equally shared by the elements or does one element dominate the other in competition for the electrons? (2) If two elements bond together, does one element have an effect on the nonbonding electrons of the other? In other words, can one element "distort" another when they come into close contact? These questions have occupied the minds of some of the foremost scientists of the twentieth century.

As to question 1 of whether equal sharing of electrons exists when two different elements combine, the succinct answer is that it does not. It is an accepted fact that elements differ from one another in their demand for electrostatic charge. This causes a skewing of the lowest energy configuration of electron density in a bond toward one of the atoms and away from the other. The qualitative description of this effect has produced the terms *ionicity* and *covalency*. These terms are complementary. The more ionic a bond is, the lower is its covalency and vice versa. There have been numerous attempts to quantify ionicity and covalency. The most elementary attempts involve the concept of electronegativity of elements, but this is not the crucial point here. The important point is that a highly ionic bond is one in which there is a very unequal sharing of electron density between the atoms. For example, the crystal lattice of sodium chloride, NaCl, contains a high degree of ionicity. The sodium atoms are considered to be unipositive ions, Na^+, and the chlorine atoms become chloride ions, Cl^-, in order to more accurately depict the ionicity in the bonding. Actually, the Na^+ ion does not have a full unit of positive charge in the crystal lattice, nor does Cl^- have a full unit of negative charge. Rather, the sodium ion is about $Na^{+0.9}$ and the chloride ion is about $Cl^{-0.9}$ because some sharing of the electrons in the bonds exists. This fact is cited to emphasize that no bond can be completely ionic.

The opposite extreme is represented by a homonuclear diatomic molecule, such as H_2 or O_2, where both atoms have equal demand for the electron density in the bond. There is no electrostatic charge difference between the atoms and hence the bond can have no ionicity. Such a bond is then a completely covalent one. These facts serve to illustrate that covalency and ionicity are subjective terms referring to the extent of migration of electron density from one atom to another when a stable chemical bond forms.

Question 2 concerning the tendency of one atom in a bond to distort its neighboring atoms is really intimately tied to the first question. A foothold on this problem was attained by K. Fajans in the early 1920s. Fajans proposed a theory of deformation or *polarization* of ions which allowed him to predict, or at least justify, a large body of chemical facts. (Polarization of chemical bonds and polarization of light are different and should not be confused.) Fajans championed the idea that a bond could be considered ionic, and a variable amount of covalency could then be brought into the picture. A small, highly charged cation, when placed next to a large anion, should be capable of distorting or polarizing the anion. To express the polarizing power of a cation, the term ionic potential Φ was introduced. It expresses the charge–size relationship involved in polarizing power. The relation

$$\Phi = \frac{\text{charge of cation}}{\text{radius of cation}} \tag{X-1}$$

expresses the fact that the ionic potential of a cation increases with an increase in its charge and/or a decrease in its radius. In Figure X-2 an idealized ion pair with no polarization of the electron shells is shown along with a pictorial representation of a mutually polarized ion pair.

Polarization occurs most readily with anions that are large in size and carry a high negative charge. These factors make such anions "soft" and relatively distortable under the influence of a cation with a large ionic potential. The electronic configuration of the cation also plays a role in its ability to polarize an anion. It is found that transition metals tend to have greater polarizing power than main-group elements of the same size and charge. For example, Hg^{2+} is more able to polarize an anion than is Ca^{2+} even though both carry the same charge and both have an ionic radius of about 1.0 Å.

The effects of polarization in a closely related series of ionic compounds can be noted in their specific properties. Increasing covalency in a series of largely ionic compounds leads to decreasing melting points, a shift of color to longer wavelengths, and an increase in the intensity of the color. The following examples illustrate the trends in the melting point with changes in covalency (changes in polarization). For the same cation, but an increasingly easily polarized series of anions, we have a melting point for LiF of 870°C, for LiCl of 613°C, for LiBr of 547°C, and for LiI of 446°C. The bond covalency increases along this series. For the same anion but decreasingly polarizing cations, we have a melting point for $BeCl_2$ of 405°C, for $MgCl_2$ of 708°C, for $CaCl_2$ of 772°C, and for $SrCl_2$ of 873°C. The bond covalency decreases along this series.

Changes in color resulting from different degrees of bond covalency are of more interest than are melting points in pigments. Cadmium oxide, CdO, is colorless; cadmium sulfide, CdS is yellow; and cadmium selenide, CdSe, is red. The covalency increases along this series. Some metal halides also show the trend nicely: AgCl (white), AgBr (pale yellow),

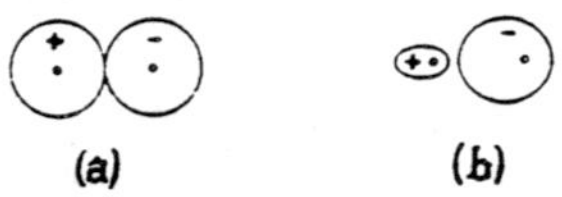

Figure X-2. (a) An idealized completely ionic bond; (b) a mutually polarized ion pair.

AgI (strong yellow). The question is, how does covalency operate to influence the color?

CHARGE TRANSFER TRANSITIONS

The origin of color in compounds is, of course, based upon more fundamental considerations than simple statements about the covalency of a bond. As with organic compounds discussed in Chapter VII, color results from the selective removal of certain wavelengths of light from the spectrum as a result of electronic excitations. In reality, what exists in the anion–cation pair of inorganic compounds is a series of electronic energy levels. These levels vary in energy, depending upon whether electrons are added to or removed from them as a result of differences in the degree of bond covalency. Light-induced electronic transitions can occur from the ground levels to the excited levels. As a rule, an increase in covalency of a bond tends to decrease the magnitude of the energy gap between the associated ground and excited states. In turn, this tends to shift the electronic absorption bands from the ultraviolet region into the visible range. Electronic transitions occurring between the anion and cation in inorganic compounds are known as *charge transfer transitions*. Charge transfer absorption bands have very great intensity and can be found somewhere in the visible or ultraviolet region in virtually all compounds containing metals and ligands. One might also consider the charge transfer band to be an "oxidation–reduction" electronic transition, because the distribution of electron density between two bonded atoms is altered. One atom momentarily loses some electron density (is oxidized) while the other momentarily gains it (is reduced).

For simplicity, a charge transfer transition can be imagined to occur when an electron on one atom or ion absorbs a quantum of visible light and is excited, not to a higher energy level of that atom, but to one of the vacant higher energy levels involving a neighboring atom or ion. The metal carries a net positive charge to begin with and in most cases retains a relative positive charge after the transition occurs. There are two possible

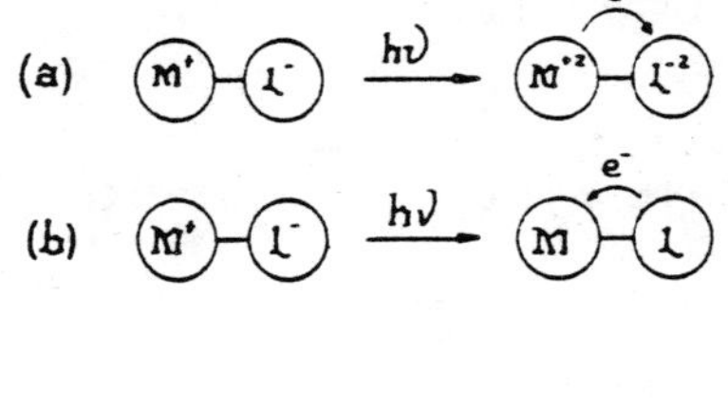

Figure X-3. Two possible light-induced charge transfer transitions; (a) metal-to-ligand, (b) ligand-to-metal.

types of charge transfer transitions as shown in Figure X-3. Case (a) is metal-to-ligand charge transfer and occurs when the ligand can function as an electron "sink." It is electrostatically less favorable than case (b). Case (b) is ligand-to-metal charge transfer, and is the transition most frequently encountered because the electron density is shifted in an electrostatically favorable manner. In order for ligand-to-metal charge transfer to occur, the ligand must have valence-electron density which is not used in bonding so that a shift in electron density can occur without disrupting the bond. The ligands oxide, sulfide, chloride, and iodide satisfy this electron density requirement. In case (b) the metal must have empty orbitals available to accept electron density from the ligand. The electron, shifted during the absorption of light, relaxes back into the ground state usually in a matter of microseconds, where it can participate again in the absorption process. Charge transfer absorption bands tend to be extremely intense because they represent allowed electronic processes. As a result, very intense color can be produced by a small amount of material. The actual color generated depends upon the ionization potential of the ligand from which the electron comes, the electron affinity of the metal where the electron goes, and the amount of electron–electron repulsion present. Charge transfer bands only appear in the ultraviolet range in many compounds. However, the band may overlap the higher energy side of the visible region (violet, blue, and green). The removal of violet, blue, and green energies from the reflected light renders pigments intensely yellow, orange, and red. Indeed, pigments achieving color by the charge transfer mechanism often have these colors. Some specific examples are given in Table X-1. The yellow and brown colors of the landscape, particularly in

Table X-1. Pigments in Which Color Is Produced by Charge Transfer from an Ionic Ground State

Pigment	Charge transfer reaction	Primary orbitals involved
Cadmium yellow (CdS)	$Cd^{2+}S^{2-} \xrightarrow{h\nu} Cd^{+}S^{-}$	Ligand $p_\pi \rightarrow$ metal $5s$
Vermilion (HgS)	$Hg^{2+}S^{2-} \xrightarrow{h\nu} Hg^{+}S^{-}$	Ligand $p_\pi \rightarrow$ metal $6s$
Orpiment and realgar (As_2S_3, As_4S_4)	$As^{3+}S^{2-} \xrightarrow{h\nu} As^{2+}S^{-}$	Ligand $p_\pi \rightarrow$ metal $4s$ or $4p$
Naples yellow ($Pb_3(SbO_4)_2$)	$Sb^{5+}O^{2-} \xrightarrow{h\nu} Sb^{4+}O^{-}$	Ligand $p_\pi \rightarrow$ metal $5s$ or $4p$
Massicot (PbO)	$Pb^{2+}O^{2-} \xrightarrow{h\nu} Pb^{+}O^{-}$	Ligand $p_\pi \rightarrow$ metal $6s$
Chrome yellow ($PbCrO_4$)	$Cr^{6+}O^{-2} \xrightarrow{h\nu} Cr^{5+}O^{-}$	Ligand $p_\pi \rightarrow$ metal $3d$
Red and yellow ochres (iron oxides)	$Fe^{3+}O^{2-} \xrightarrow{h\nu} Fe^{2+}O^{-}$	Ligand $p_\pi \rightarrow$ metal $3d$

the desert, and the reds of tropical soils are due in large measure to the charge transfer band of the Fe(III)—O bond.

In some cases color can be produced in a compound when ions of the *same* element are present with two different oxidation states. A light-induced resonance of an electron can occur between these two oxidation states. The absorption of this light leads to a *metal–metal charge transfer* reaction:

$$M_a^{1+}M_b^{2+} \xrightarrow{h\nu} M_a^{2+}M_b^{1+}$$

This metal–metal charge transfer transition is very strongly allowed and thus the color it produces is intense. However, in some cases the absorption band broadly covers the visible region, causing the compound to be black. Some metal–metal charge transfer processes are important in pigment chemistry. Prussian blue, $Fe_4[Fe(CN)_6]_3$, contains Fe(II) and Fe(III). The intense blue color is a result of a light-induced resonance between the two oxidation states of iron:

$$Fe_a(II), Fe_b(III) \xrightarrow{h\nu} Fe_a(III), Fe_b(II)$$

The color is so intense that one part Prussian blue in 500 parts of lead white imparts a definite blue tinge to the lead white.

Resonance involving Cu(II) and Cu(I) may be responsible for green and brown colors produced by copper in glass. The black color of U_3O_8 in glass results from resonance between the U(IV) and U(II) oxidation states of uranium. The modern pigments known as molybdenum blue and tungsten bronze appear to gain most of their color from resonance between ions of molybdenum and tungsten which have different oxidation states.

In general, it can be said that the greatest absorption of light in inorganic compounds deriving their color from ion polarizability (charge transfer) will occur when the electric vector of light vibrates along the directions in which the ions are most easily distorted. Distortion is usually greatest along a metal–ligand bond axis.

GEOMETRY AND ELECTRONIC CONFIGURATION OF IONS

The contributions of geometry and electronic configuration in producing color in inorganic systems are inseparably tied together. Pigments containing transition metals often attain their color from electronic transitions localized on the metal. It should be kept in mind, though, that charge transfer bands discussed above also can occur in transition metal systems. However, charge transfer transitions cannot account for all of the color in certain pigments. Absorption bands occurring within the transition metal ions make an important contribution.

The basis of color produced by transition metal ions began to take shape in the late nineteenth century with the Nobel-prize-winning work of Alfred Werner. Werner is notable for his research on the synthetic and structural chemistry of coordination complexes of transition metal ions. Many of the compounds he synthesized were brightly colored. Without any of the present knowledge of bonding and molecular structure upon which to work, Werner succeeded in correctly picturing the arrangement of anions about cations and ligands about metal ions in many of his complexes. In the period 1915–25, G. N. Lewis in the United States and N. V. Sidgwick in England developed simple bonding models which were the beginning of the modern concepts of electron distribution in molecules. Linus Pauling in the 1920s further extended these concepts in his valence bond theory. These bonding ideas were discussed in some detail in Chapter VII. The Nobel laureate physicists Hans Bethe and John Van Vleck made important contributions by developing the theory of anion–cation interactions known as the *crystal field theory*.

We can begin with the Bethe–Van Vleck model and amalgamate it with Pauling's ideas. The crystal field model and the valence bond description, while basically different, provide a good foundation for understanding the electronic structure and light absorption potential of transition metal ions. Many of the first-row transition metals of importance in pigments have a coordination number of four or six ligands about the metal when forming compounds. The *coordination number* refers to the number of electron-donor ligands, usually derived from the main-group elements, which can bond to the metal at any one time. These ligands are oxide, O^{2-}, or sulfide, S^{2-}, in many pigments. Figure X-4 shows tetrahedral coordination of four ligands and octahedral coordination of six ligands about a metal located at the origin of the axis system. These geometries represent the most stable configuration of the ligands about the metal in the majority of four- and six-coordinate structures.

The octahedral geometry will be considered first. From the electronic structure of elements discussed in Chapter VII, we know that first-row transition metals have five 3*d* orbitals, one 4*s* orbital, and three 4*p* orbitals which are energetically suitable for bonding. Not all of these valence orbitals will be used for bonding, but Pauling's valence bond model demands hybridization of two *d* orbitals, the *s* orbital, and three *p* orbitals (as d^2sp^2 hybrids) in order to obtain six equivalent, octahedrally disposed orbitals on the metal. Provided these six orbitals contain little electron density to start with, they can act as acceptor orbitals on the metal and bond

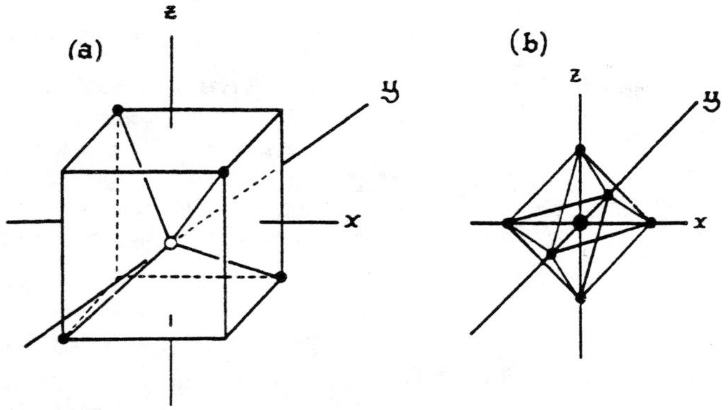

Figure X-4. (a) Tetrahedral coordination with the metal at the center of a cube and (b) octahedral coordination about the metal.

with the electron-donating ligands. As to which two 3*d* orbitals are used and what happens to the other three, we need to look at the shapes of the *d* orbitals and consider what will happen to their energies in the presence of the ligands. Figure VII-2 showed contour plots of the five *d* orbitals. These pictures show the shapes of hydrogenlike orbitals. An important feature to be noted is that two of the orbitals point along the *x*, *y*, *z* coordinate axes, while the other three have directional character which would concentrate electron density at 45° to these coordinate axes.

When a metal ion is isolated in the gas phase, the direction in which the *d* orbitals point is of no consequence. A single electron would have the same energy no matter which *d* orbital it chooses to occupy. Five lines, each line representing one of the five *d* orbitals of equivalent energy, can be drawn to illustrate the energy level scheme. This is shown in Figure X-5a. If the transition metal ion has more than one *d* electron, then all of the electrons would have the same energy. The complicating feature that the electrons tend to repel one another because of their negative charge is ignored here. However, because electrons do indeed repel one another, they will avoid pairing up in the same orbital until more than five *d* electrons are present. This occurrence is prescribed by Hund's rule as discussed in Chapter VII. For example, if an ion has five *d* electrons, as does Mn^{2+}, then each electron will occupy one of the energy levels shown in Figure X-5a.

If ligands are now brought in toward the metal along the *x*, *y*, *z* axes so that the ion is no longer an isolated gas-phase species, then it is distantly surrounded by electron-donating groups. The initial effect is to raise the energy of *all* the orbitals roughly equally because the electron pairs on the ligands begin to repel the electrons in the metal *d* orbitals (Figure X-5b).

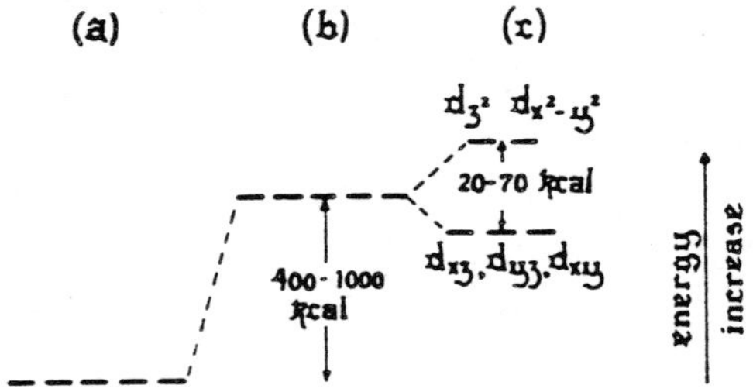

Figure X-5. (a) The five *d* orbitals on an isolated metal ion, (b) the effect on the *d* orbitals of a set of six ligands at a distance much greater than the bond distance, and (c) the effect on the *d* orbitals of six octahedrally disposed ligands bonded to the metal.

As the separation between the metal and ligands diminishes to the point where a bond can form, the *d* electrons on the metal ion experience repulsion by the electron pairs or negative charge on the ligands. As a result, the metal *d* electrons will preferentially locate in *d* orbitals which allow them to avoid, as much as possible, the approaching ligands. This means that the *d* electrons on the metal will concentrate as much as they can in the d_{xz}, d_{yz}, and d_{xy} orbitals, which are located at 45° to the x, y, z axes along which the ligands lie. The $d_{x^2-y^2}$ and d_{z^2} orbitals point directly at the ligands along the x, y, and z axes. Electron density in the $d_{x^2-y^2}$ and d_{z^2} orbitals will be repelled as the ligands approach the metal closely. It follows from this that the five *d* orbitals cannot have the same energy in the presence of the ligands. The d_{xy}, d_{xz}, and d_{yz} set of orbitals will be lower in energy than the d_{z^2} and $d_{x^2-y^2}$ set of orbitals. Figure X-5b illustrates this effect of the ligands on *d*-orbital energy levels. The energy lost due to repulsion between *d* electrons on the metal and the electron density on the ligands, as depicted in Figure X-5b, is less than the energy gained when metal–ligand bonds eventually form. Otherwise no bonds would form.

The splitting of the metal *d* orbitals into two sets by ligands, shown in Figure X-5c, was called *crystal field splitting* by Hans Bethe. The number of electrons available on the original metal ion now must be fed into the energy level diagram in Figure X-5c according to Hund's rule and to energy considerations. It is helpful to consider some specific electron configurations produced by an octahedral arrangement of ligands in order to illustrate the orbital description of transition metal ions. Chromium, Cr, has an atomic electron configuration of $[Ar]4s^23d^4$ ([Ar] represents the electron configuration for argon and is the configuration of the inner shells of chromium). As the Cr^{3+} ion, the configuration is $[Ar]3d^3$. The presence of an octahedral field of ligands about Cr^{3+} produces the following splitting and electron distribution according to Hund's rule:

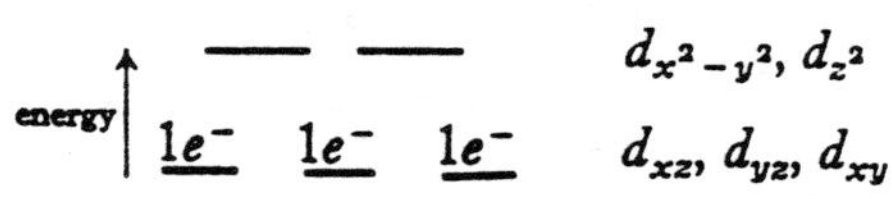

For iron (Fe) the electron configuration is [Ar]$4s^23d^6$. The Fe^{2+} ion has a configuration of [Ar]$3d^6$ and there are now two possible ways of distributing electrons. These two configurations are energetically different:

(a)			(b)		
	___	___		$1e^-$	$1e^-$
$2e^-$	$2e^-$	$2e^-$	$2e^-$	$1e^-$	$1e^-$

The arrangement (a) is called "low-spin" or "spin-paired" Fe^{2+}, while arrangement (b) is "high-spin" or "spin-free" Fe^{2+}. Whether the metal adopts the high-spin or low-spin electron configuration depends upon a number of factors, including the metal oxidation state, ligand electronegativity, metal–ligand bond length, and the amount of metal–ligand π bonding (double bonding) that exist. Without prior knowledge one cannot easily predict whether the high- or the low-spin state will be present in the ground state. The chemistry and properties of elements in the two electron configurations are quite different. Iron also has a 3+ oxidation state of importance in pigments and in the color of soil. Trivalent iron has the [Ar]$3d^5$ electron configuration, which also makes possible high- and low-spin states. The high-spin state is the most important in the ochres, where oxygen ligands coordinate the Fe^{3+} ions. The configurations are

(low spin)			(high spin)		
	___	___		$1e^-$	$1e^-$
$2e^-$	$2e^-$	$1e^-$	$1e^-$	$1e^-$	$1e^-$

Cobalt (Co) has an [Ar]$4s^23d^7$ configuration. The Co^{2+} ion is the most important cobalt ion in pigments, although Co^{3+} is a common oxidation state in the chemistry of cobalt. The high- and low-spin electron distributions for Co^{2+} are

(low spin)			(high spin)		
	$1e^-$	___		$1e^-$	$1e^-$
$2e^-$	$2e^-$	$2e^-$	$2e^-$	$2e^-$	$1e^-$

Copper, Cu, is usually found as Cu^{2+} in pigments and has an electron distribution

	$2e^-$	$1e^-$
$2e^-$	$2e^-$	$2e^-$

The possibility of high- and low-spin states does not exist for the Cu^{2+} ion.

THE "*d–d*" TRANSITION

In the preceding section a simplified picture of the relative energy ordering of transition metal *d* orbitals in an octahedral field of ligands has been developed. The difference in energy between the set of orbitals lying on the *x*, *y*, *z* axes and the set lying at 45° to these axes makes it possible for electronic transitions to occur between the two levels. Light-induced electronic transitions between the energetically different *d* orbital sets are called "*d–d*" *transitions*.

The 20–70 kcal gap in energy between the d_{xz}, d_{yz}, d_{xy} orbital set and the d_{z^2}, $d_{x^2-y^2}$ set falls in the range of the infrared and visible energies. A *d* electron at the lower energy level can absorb a quantum of energy from a photon of light and be excited into the higher energy set of orbitals. The absorption of the photons of a certain energy will remove that energy (wavelength) of light from the spectrum and produce color in the compound. The absorption of light by this mechanism depends upon the presence of at least one, but less than ten, *d* electrons on a metal, and the presence of the electron-repelling groups, i.e., ligands, bonded to that metal. The energy of light absorbed is controlled in part by the magnitude of the energy gap between the two *d*-orbital sets. Prediction of specific colors from this gap is rarely possible without also considering electron–electron repulsion. Electron–electron repulsion frequently produces several absorption bands in the spectrum. The color observed depends upon the wavelengths of these bands. The main point to be stressed is that selected wavelengths of light can be removed from the spectrum by "*d–d*" excitation.

It is worthwhile to note that correlations between color and absorption maxima can occasionally be made with success in a closely related series of compounds. A set of octahedral Co^{3+} complexes of the stoichiometry $Co(NH_3)_5X^{n+}$, where X = F^-, Cl^-, Br^-, I^-, H_2O, and NH_3, illustrates this point. These complexes are all water-soluble and have not been used as pigments. Figure X-6 shows the color rosette in the inner circle. In the middle circle appear the colors observed in these six compounds. In the outer circle appear the colors corresponding to the experimentally determined absorption maxima in each case. It must always be remembered that the observed color of the compound is the complement of the absorbed color. The colors observed for the compounds are indeed diametrically related to the absorption maxima. Thus, knowing the colors qualitatively

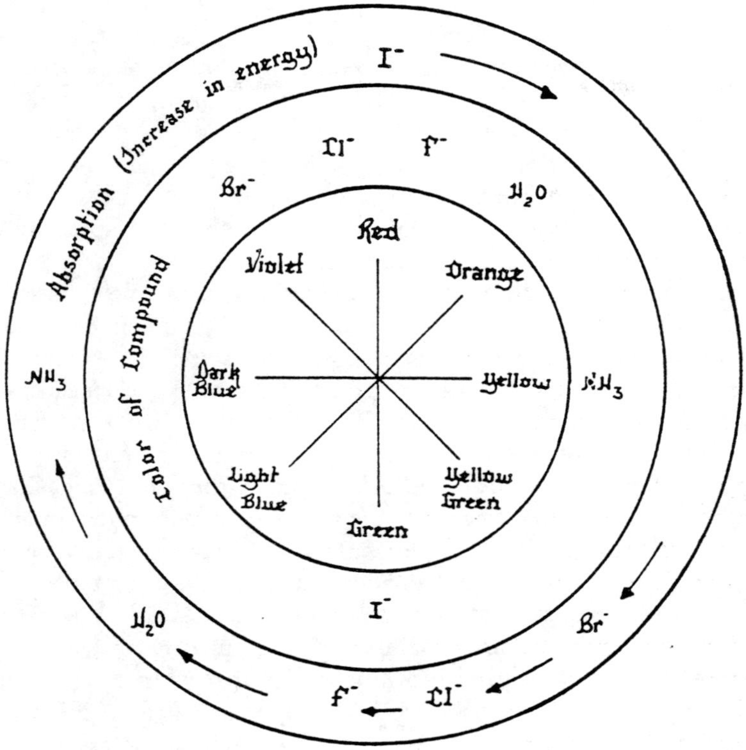

Figure X-6. The color rosette relating the observed color of $Co(NH_3)_5X^{n+}$ compounds, $X = F^-, Cl^-, Br^-, I^-, H_2O, NH_3$, to the experimentally determined absorption maxima of the compounds.

allows one to predict the wavelength of the main absorption bands of these compounds. Furthermore, the gap between the two sets of *d* orbitals in these compounds follows the trend $NH_3 > H_2O > F^- > Cl^- > Br^- > I^-$. This particular trend is well known to inorganic chemists. It represents the so-called spectrochemical series of ligands toward transition metal ions. That is, for many transition elements (not just Co^{3+}) this series of ligands produces a progressively decreasing energy gap between the two sets of *d* orbitals in an octahedral ligand field.

The *intensity* of the color produced by the "*d–d*" transition is usually comparatively low in octahedrally coordinated transition metal ions. In fact, the "*d–d*" transition can be shown to be theoretically forbidden and therefore should not appear at all. The selection rule which makes these bands forbidden is broken down to some extent by the vibrational motions of atoms in the complex. This breakdown leads to a small amount of absorption of light where otherwise none would occur. It is not necessary to focus on the theory of absorption band intensities. It suffices to note that there is a theoretical reason for the low intensity of "*d–d*" transitions which are experimentally observed for transition metal ions in an octahedral geometry of ligands. Per given quantity of colored material, the transition metal ions in octahedral sites produce much less color than do organic dyes and many other inorganic pigments.

Viridian, $Cr_2O_3 \cdot nH_2O$, is a pigment in which the Cr^{3+} ion is in an octahedral field of oxygen atoms. The green color results from a "*d–d*" transition. Umbers and ochres are iron oxides where the Fe^{3+} is in an octahedral field of oxygen ligands. The color of umber and ochre pigments is mainly a result of the charge transfer bands, but shading of the colors can arise from "*d–d*" transitions. Cobalt violet [$Co_3(PO_4)_2$], cerulean blue ($CoO \cdot nSnO_2$), Scheele's green ($CuHAsO_3$), cobalt green (CoO and ZnO), and emerald green [$3Cu(AsO_2)_2 \cdot Cu(CH_3COO)_2$] are pigments which achieve most of their color from "*d–d*" transitions on Co^{2+} and Cu^{2+}. The number of ligands coordinated to the metal ion is not known in all of these pigments. Several of these pigments may contain metal ions coordinated in a tetrahedral rather than an octahedral field of ligands.

All of the above statements about *d* orbitals in octahedral fields can be applied to deriving a similar crystal field splitting picture for tetrahedral coordination or any other ligand coordination geometry. In a tetrahedral arrangement of ligands about the metal, the ordering of the *d* orbitals is reversed from the octahedral case. Also, the magnitude of the crystal field splitting is not as large. For this reason green and blue colors tend to appear in metal complexes having tetrahedral geometry. More importantly, the "*d–d*" transitions are allowed in this geometry and thus the intensity of color is much greater for a given quantity of material than it is in the octahedral geometry. A crystal field splitting diagram for *d* orbitals in the tetrahedral field of ligands is shown in Figure X-7. An example of a pigment which achieves color from a metal ion in a tetrahedral ligand field is cobalt blue, $CoAl_2O_4$. A Co^{2+} ion is surrounded by four oxygen atoms in this pigment.

The CrO_4^{2-} ion in lead chromate gains its yellow color from charge transfer rather than "*d–d*" transitions even though chromium is a transition

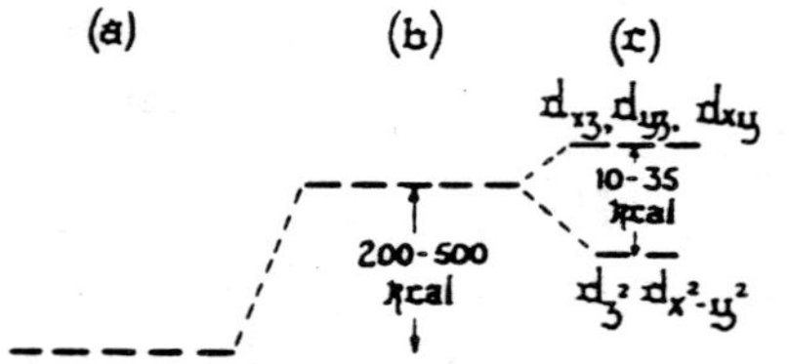

Figure X-7. (a) The *d* orbitals in an isolated metal ion, (b) the *d* orbitals in a "spherical" field from our ligands, (c) the splitting of the *d* orbitals in a four-coordinate tetrahedral ligand field.

metal. The Cr^{6+} ion has no *d* electrons. Hence, the color of CrO_4^{2-} cannot be due to a "*d–d*" transition. The yellow color results from a ligand-to-metal charge transfer. Likewise, the d^{10} metal electron configuration cannot produce color by the "*d–d*" transition mechanism, because the higher energy levels are completely filled with electrons. Yet, many pigments are colored and contain metals with a d^{10}-electron configuration. The primary source of this color is charge transfer transitions. In general, most main-group-element pigments derive their color from the charge transfer mechanism, as do a few transition-metal-containing pigments having no *d* electrons or no vacancies in the *d*-orbital levels. Most of the other transition-metal-containing pigments gain at least some of their color from the "*d–d*" mechanism.

MISCELLANEOUS SOURCES OF COLOR

In addition to charge transfer and "*d–d*" transitions, there are some other mechanisms which give color to important pigments. The presence of *color centers* in a crystal lattice gives color to certain pigments and glass. Some compounds contain cavities in their crystal lattice network in which otherwise unstable radicals (anions, cations, atoms, or molecules containing one or more unpaired electrons) can be trapped. These radicals are called color centers and they can absorb certain wavelengths of visible light. The pigment ultramarine blue, derived from the semiprecious gemstone *Lapis Lazuli*, is an example of a pigment which is colored by this mechanism. Sulfur appears to be trapped as S_3^- radicals in the silicate matrix. The S_3^- radical imparts a blue color. The S_2^- radical, which is yellow, is present in ultramarine in small amounts. The sulfur radicals S_2^- and S_3^- each possess one unpaired electron. The slight instability of ultramarine blue toward acid in varnish and atmospheric pollutants is due to the great reactivity of these sulfur radicals. Carbon-amber glass may also be colored by sulfur-containing color centers of this type.

The modern pigment known as monastral blue is a copper phthalocyanine complex with a structure shown in Chapter VIII. The intense blue color of this pigment results from allowed electronic transitions located mostly in the organic portion of the molecule.

EFFECTS OF CRYSTAL STRUCTURE VARIATION ON COLOR

Bonds and primary interactions between atoms are the main cause of color in inorganic compounds. However, variations in the crystal structure can perturb color-producing portions of the crystal lattice, particularly if several sources of color happen to be in delicate balance with one another. This perturbation arises because the electron orbitals of the color-producing compounds or ions can be distorted by small changes in the crystal lattice. Also, many lattices contain strongly associated molecules or molecular ions. Changes in the degree of association can result in a change of color. Several pigments exhibit these effects. The white pigment TiO_2 exists in two crystal forms called rutile and anastase. The rutile form is photostable and has a very faint yellowish tinge because the tail of a TiO charge transfer band in the ultraviolet very slightly overlaps the edge of the visible region. In the anastase form, this band no longer overlaps the visible region. Hence, anastase is pure white. Unfortunately, anastase is more photosensitive than rutile and therefore has found little application as a paint pigment. It had been used as a coating for paper when a very bright white copy is sought.

Lead chromate, $PbCrO_4$, has three different crystal forms, all slightly different in color. The orthorhombic crystals are greenish-yellow, the monoclinic crystals are yellow-orange, and the tetragonal crystals are orange to scarlet. Lead oxide, PbO, exists in yellow and red forms. Variations in color also occur in the copper–phthalocyanine complex, depending upon the crystal structure.

EFFECTS OF PARTICLE SIZE AND CONCENTRATION ON COLOR: THE "ALEXANDRITE EFFECT"

Particle size, in an indirect way, causes coloration by the Rayleigh effect if the particles are small enough (Chapter VI). The larger particles of pigments take on color due to selective absorption of light. In most cases, a general desaturation of the color is observed as the crystals are ground to finer sizes. Grinding has the effect of increasing the number of crystal faces. This results in an increase in the amount of scattered white light and reduces the amount of the light available for absorption.

A phenomenon occasionally observed in pigment crystals is a change of the color accompanying various degrees of pulverization of the crystals. When an actual color change occurs in the crystals, it does not result from an alteration in the number of absorption bands or their wavelengths. The spectrum cannot be altered simply by changing the number of unit cells which are layered on one another. However, the intensities of the electronic absorption bands can change and this is usually the source of the change in color with grinding. For example, crystals of α-FeOOH are yellow when finely powdered. As the crystals grow in size, they progressively darken and eventually become black when the particles are very large.

The source of this apparent change in color is found in the nature of the absorption bands. Strong absorption bands occur in the blue region of the visible range, but the bands have shallow minima between them. That is, light is absorbed broadly across the visible range. When the crystals are small, the intense, yellow-producing absorption bands dominate and give a yellow color. As the crystal size grows, broadband absorption of light throughout the visible range plays an increasingly important role because of the shallow minima between bands. In the extreme the crystals no longer transmit light and become black.

Orange crystals often become yellow with grinding. Cadmium yellow and cadmium orange illustrate this effect in pigments. Both pigments are cadmium sulfide, but cadmium yellow contains particles which are only one-fiftieth the size of those in cadmium orange. The orange crystals contain more red color than the yellow crystals. The tail of a strong, blue-absorbing electronic transition which produces yellow extends into the green and thereby gives an orange tint to the color. As the crystal size is reduced, the intensity of this band diminishes and the tail of the band ceases to play a role in the production of color. Consequently, the crystal becomes more of a pure yellow.

Elegant examples of color change with thickness of the absorbing media occur in many chromium salts containing the Cr^{3+} ion. In addition, many of these salts show a marked color change depending upon the type of illumination. The mineral *alexandrite* is a variety of chrysoberyl with a chemical composition of $Al_{2-x}Cr_xBeO_4$, and vividly illustrates this phenomenon. In fact the term *alexandrite effect* has been applied to materials which show a very marked color change depending upon the type of

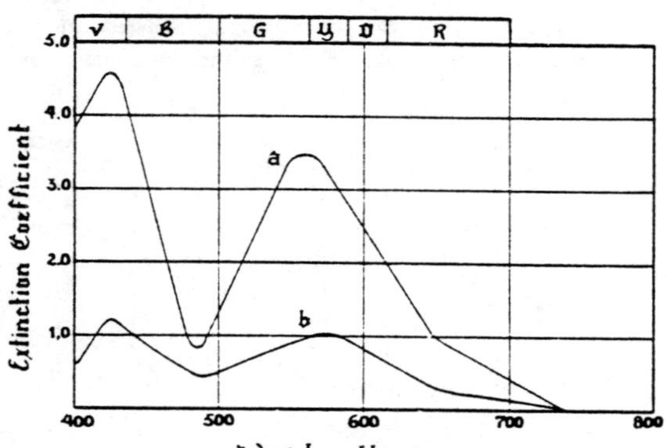

Figure X-8. The absorption spectrum of an aqueous solution of $K_3Cr(C_2C_4)_3$. (a) High concentration, red; (b) low concentration, green.

illumination. Chromium salts that exhibit the alexandrite effect are green in daylight illumination and/or in thin layers. In incandescent illumination and/or in thick layers, they are red in color. The origin of this difference in color can be understood by examining the electronic absorption bands in the visible range. Figure X-8 shows the absorption band trace for a solution of the chromium oxalate salt $K_3Cr(C_2O_4)_3$ containing Cr^{3+}. Two strong absorption bands are present which remove the blue and the yellow wavelengths from the visible region. The green wavelengths are weakened but not removed entirely. The red wavelengths are not removed at all. Hence concentrated solutions will transmit only red light. However, as the solution becomes more dilute, the intensity of light transmitted at all wavelengths increases. Since the eye is more sensitive to green than to red (Figure V-5), a green color takes over as the visual light stimulus.

The $K_3Cr(C_2O_4)_3$ solution which is green in daylight is red in incandescent light. This is because incandescent lamps produce a great deal more red light than they do green light. Daylight contains approximately equal amounts of red and green light. Hence the intensity of red color relative to green color reflected from the solution differs in the two sources of illumination. Glass containing neodymium oxide, Nd_2O_3, also shows the alexandrite effect.

BIBLIOGRAPHY

Fajans, K., *Chemical Forces and Optical Properties of Substances*, McGraw-Hill, New York (1931).

Figgis, B. N., *Introduction to Ligand Fields*, Interscience, New York (1966).

Huheey, J. E., *Inorganic Chemistry*, Harper and Row, New York (1978).

Jorgensen, C. K., *Inorganic Complexes*, Academic Press, London (1963).

Lever, A. P. B., *Inorganic Electronic Spectroscopy*, Elsevier, Amsterdam (1968).

White, W. B., Roy, R., and Crichton, J. M., The "alexandrite effect": an optical study, *Am. Mineral.* **52**, 867 (1967).

XI

Pigments and Colored Glass

In this chapter attention will be focused on the properties and usage of artist's pigments. The origin of color in the glass and glaze matrices will also be considered. Both painting pigments and colored glass have in common the fact that they are inorganic materials and therefore derive their color from the mechanisms discussed in the preceding chapter. The structure and some properties of glass and ceramic glaze were described in Chapter VI, where the physical appearance of materials was the subject.

It has been necessary to consider certain aspects of pigments throughout the earlier chapters of this book, particularly in Chapter VI. There the topics of hiding power, index of refraction, glazing, and particle size were covered. With these concepts in mind, the properties of individual pigments can be confidently understood. Artist's pigments and their usage are perhaps the most thoroughly studied and described areas of science related to art. It would be formidable to attempt to present the subject in a more useful way than have such standard authorities as Gettens and Stout, Doerner, Mayer, and Laurie, among others. What is intended in this chapter is an introduction to some of the most important artist's pigments used throughout the centuries. The Bibliography at the end of this chapter can be consulted for a more in-depth exposition. The methods of identification of pigments are not especially within the scope of the interaction of light with art and historical materials and will not be discussed here.

ARTIST'S PIGMENTS

Perhaps the most obvious method of classifying pigments is by their color. Color impressions can be subjective and thus only six intervals,

Table XI-1. Dominant Reflectance Wavelengths of Some Materials Used as Pigments

400–487 nm	488–515 nm	516–570 nm	571–700 nm
Cobalt blue	Malachite	Terre verte	Red lead
Ultramarine (natural)	Scheele's green		Vermilion
Cerulean blue	Verdigris		Cadmium red
Azurite	Emerald green		Madder lake
Smalt	Cobalt green		Alizarin crimson
Verditer	Viridian		Realgar
Indigo	Chrysocolla		Indian red
Prussian blue			Cadmium orange
Egyptian blue			Yellow ochre
			Yellow lake
			Strontium yellow
			Chrome yellow
			Cadmium yellow
			Orpiment
			Lead–tin yellow
			Raw and burnt sienna
			Raw and burnt umber

white, blue, green, yellow-orange-brown, red, and black, will be used. Metallic leaf and paint will also be briefly discussed. Table XI-1 contains the dominant reflectance wavelengths for most of the pigments considered here. White and black, of course, represent uniform reflectance and absorbance, respectively, across the visible region. The pigments listed all have been known for at least 100 years and a few have been used for 6000–7000 years.

The artist's palette has constantly changed. Peter Paul Rubens was a master colorist of the early seventeenth century, yet he used less than fifteen pigments in producing his great works. These were vermilion, madder lake, burnt sienna, red ochre, orpiment, yellow ochre, yellow lake, cobalt oxides, malachite, natural ultramarine, white lead, and carbon black. More recent painters had as many as forty or fifty pigments on their palette. Some of these pigments are naturally occurring and some are synthetic or artificial. After about 1900, however, the actual paints made from the pigments began to change.

Pigments and paints of the pre-industrial era tended to be simple in formulation. They were mainly composed of the pigment and a vehicle or binder. Many early pigments are silicates; in fact, most early pigments were naturally occurring and were used in this form without alteration except for a few purification steps. On the other hand, modern paints of the industrial era are usually more complex in formulation. It may be very difficult to determine what the composition of a modern paint actually is. Conservators in the years to come who might be concerned with these paints are likely to face a difficult problem identifying the reasons for deterioration of the paint and painted materials. In addition to synthetic pigments and vehicles, modern commercial paints may contain plasticizers, antifoaming agents, dispersing agents, fire retardants, antioxidants, deodorants, gloss controllers, viscosity controllers, curing agents, driers, emulsifiers, heat stabilizers, and antiskinning agents. The composition varies with the intended use. No space will be devoted here to the discussion of modern paints.

White Pigments

In Christian art, white has been accepted as a symbol of innocence of soul, of purity, and of holiness of life. In practice, white pigments can serve as more than simple pigments. They can act as bases for lakes and can be used as extenders.

The frequency patterns of pigments in a number of European easel paintings as based on chemical analyses carried out at the Doerner Institute will be given in this and the following sections. Figure XI-1 shows the results for several whites.

Lead white, formulated as $2Pb(CO_3)_2 \cdot Pb(OH)_2$, is a synthetic pigment first prepared in the fourth century B.C. It is the most common lead pigment and is, perhaps, the most commonly encountered pigment of all. It is present in almost all Western paintings. Lead white is used in tempera, oil, and water-base paintings, but is most satisfactory in oil. It is often found

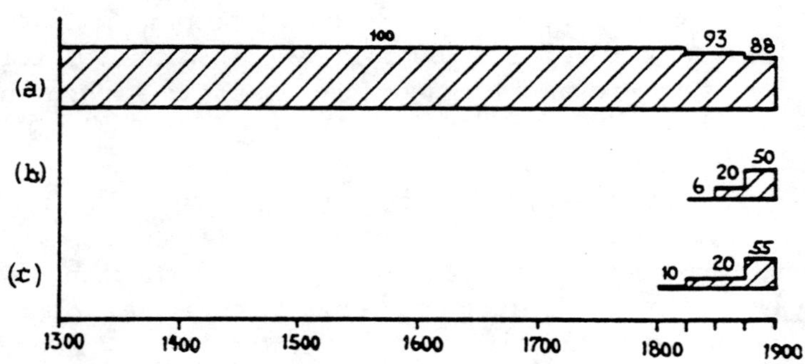

Figure XI-1. The occurrence of some whites in European easel paintings. (a) Lead white; (b) zinc white; (c) barium sulfate as an extender in white pigments. [Courtesy of H. Kühn, Terminal dates for paintings derived from pigment analysis, in *Application of Science in the Examination of Art Works*, Museum of Fine Arts, Boston, Massachusetts (1973), p. 199.]

in early paintings mixed with ochres, cinnabar, and terre verte to give flesh tones. The extensive absorption of X-rays by lead permits radiography to be used on most paintings. Lead white is darkened by H_2S in the atmosphere because of the formation of PbS; it reacts with HCl to precipitate $PbCl_2$, and reacts with NaOH to give Na_2PbO_2. With HCl, CO_2 is also released. Lead white is not light-sensitive and has good opacity ($\eta = 2.0$). It dries quickly in oil because the $Pb(OH)_2$ portion forms lead soaps (lead linoleate and other salts) and this fact is exploited nowadays to make quick-dry paints. Soap formation also contributes to the occurrence of pentimenti. Lead white is unsuitable in fresco painting because frescoes are alkaline due to the presence of $Ca(OH)_2$. Lead white will darken as a result of the reaction of $Ca(OH)_2$ with $PbCO_3$ to form brown-black PbO_2.

Chalk, $CaCO_3$, is a naturally occurring pigment and extender sometimes called "whiting." It has been used since very early times and is frequently found as a priming layer. It dissolves in HCl to release CO_2 gas and does not react with NaOH. It turns somewhat yellow when mixed with oils and has rather poor covering power ($\eta = 1.5$–1.65). In the same class with chalk can be included old sea sediments (mixtures of carbonates, silicates, and phosphates) and white clays sometimes called *creta*, which are silicate minerals with a crystal lattice composed of K^+, Ca^{2+}, Al^{3+}, Si^{4+}, and O^{2-} ions.

Gypsum, $CaSO_4 \cdot 2H_2O$, is a naturally occurring material which, when mixed with water and rabbit skin glue, is called *gesso*. Gesso is the Italian word for gypsum. Today the word gesso has taken on wider meaning and may include grounds made from chalk (whiting), zinc oxide, or any other inert white. It has rather poor hiding power ($\eta = 1.5$) and is frequently employed as an extender or a base. It is slightly soluble in HCl and insoluble in NaOH. *Baryta*, $BaSO_4$, has been used both in its naturally occurring and synthetic forms as an extender and a base. It is a common pigment in watercolor, and came into use in the early part of the nineteenth century. It has rather poor hiding power ($\eta = 1.637$). With both HCl and NaOH, there is no reaction of barium sulfate. *Blanc fixe* is another name applied to synthetic $BaSO_4$.

Zinc white, ZnO, is a synthetic pigment first used in 1834 and widely used since that time. At present, it is about as important as lead white in artists' pigments. Its covering power, however, is not quite as good as lead white, even though they have similar indices of refraction. This is probably because the oil absorption value of zinc white is greater than that of lead white and less pigment is therefore present in the paint. In practice zinc white and lead white are frequently found mixed together. Chemically, zinc oxide can be attacked by H_2S to form ZnS. It has a tendency to yellow and to chalk at times when it is exposed to ultraviolet radiation. The chalking has been attributed to reaction by ultraviolet-generated H_2O_2 (Chapter IX) to produce Na_2ZnO_2. It is soluble in HCl and NaOH. In NaOH, zinc oxide produces Na_2ZnO_2.

Lithopone, 30% ZnS–70% $BaSO_4$, is a synthetic pigment first used in 1874. It is quite common now as a poster paint, but has not found extensive use among artists' pigments. It has greater hiding power than does zinc white. In HCl it is somewhat soluble, releasing H_2S. It does not react with NaOH. Early samples tended to turn gray in light, probably because of the presence of trace metals, but modern lithopone does not discolor in light. Ultraviolet radiation can produce chalking because of oxidation of sulfur to form sulfates:

$$ZnS + 2O_2 \xrightarrow{h\nu} ZnSO_4$$

Titanium dioxide, TiO_2, is a synthetic pigment first produced in 1920. Other names for TiO_2 are titanium white and titanox. As a commercial paint, it is frequently mixed with as much as 70% $BaSO_4$. It is also used as a pigment in enamels and ceramic glazes. Chemically, it is inert to both HCl and NaOH. It has the best hiding power of all whites because $\eta = 2.7$ in the rutile form. An absorption band in the ultraviolet very slightly overlaps the visible region, giving a faint cream color to the rutile form. It also chalks on occasion because of simultaneous reduction of TiO_2 and oxidation of the vehicle. The reduced form of the pigment can be oxidized back to TiO_2 by atmospheric oxygen.

Blue Pigments

The color blue is often referred to as a retiring and cool color. It symbolizes truth, Heaven, and heavenly love in Christian art. Blue is achieved in inorganic compounds by a variety of mechanisms. Prussian blue is colored by charge transfer bands, cobalt blue by "*d–d*" transitions, and ultramarine by color centers. Although many blue pigments are rather transparent, their indices of refraction are increased somewhat by the effects of anomalous dispersion (Chapter VI).

Perhaps the purest natural blue is natural ultramarine. Cobalt blue reflects much green and violet light, making it a less pure blue. These purplish blues lose much of their blue tint when viewed in warm lighting. They tend to look more red.

The occurrence of blue pigments in European easel paints is shown in Figure XI-2.

Egyptian blue is a copper silicate having the formula $CaCuSi_4O_{10}$. It was probably the first synthetic pigment, and was prepared by heating

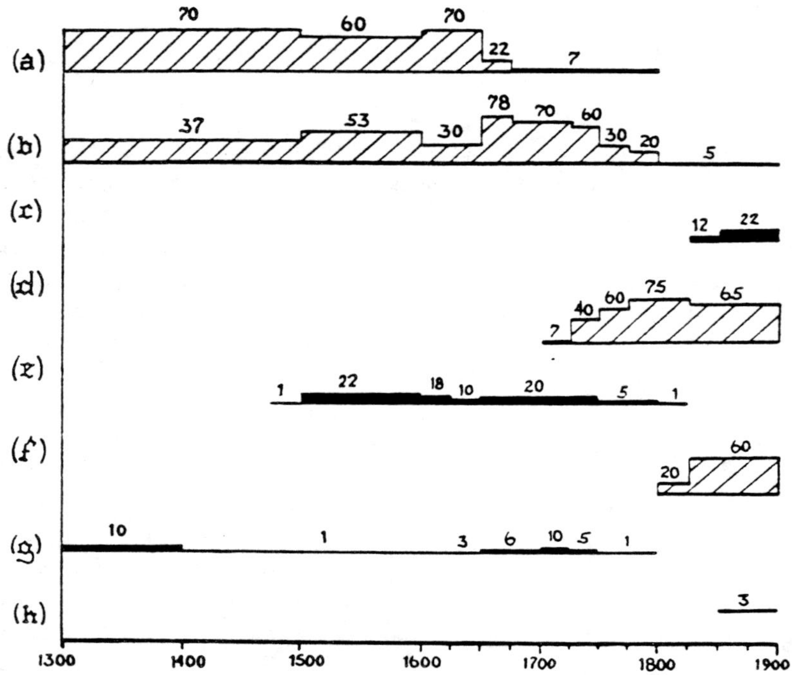

Figure XI-2. The occurrence of some blue pigments in European easel paintings. (a) Azurite; (b) natural ultramarine blue; (c) synthetic ultramarine blue; (d) Prussian blue; (e) smalt; (f) cobalt blue; (g) indigo/woad; (h) cerulean blue. [Courtesy of H. Kühn, Terminal dates for paintings derived from pigment analysis, in *Application of Science in the Examination of Art Works*, Museum of Fine Arts, Boston, Massachusetts (1973), p. 199.]

Cu_2O and sand to give a blue glass and/or malachite. Further reaction by heating with $CaCO_3$ gives all blue glass. This glass is then pulverized and used as a pigment. It seems to have been used only in very early times (at least before 2000 B.C.). Egyptian blue is chemically inert and has rather poor hiding power (η = 1.6). It does not react with HCl or NaOH.

Natural ultramarine is a silicate with a somewhat variable formula best represented as $Na_{8-10}Al_6Si_6O_{24}S_{2-4}$. Some chloride is often present in the crystal lattice as well. This blue is obtained from the semiprecious ore Lapis Lazuli and has a very complex structure. It has been used throughout the ages, but mostly in the sixth century A.D. and later. It was prized by Renaissance painters but decreased in usage after about 1800. Its earliest general use was in jewelry, although an ultramarine bowl from Ur, dated 2400 B.C., is known. Ultramarine was very expensive and it found use only in works intended to be of great value. It is frequently found to contain calcspar flakes in early works. Its covering power is very low (η = 1.5) and it is better when used in tempera than in oil. In acid it turns white, with the loss of H_2S. There is no reaction with base. Ultramarine is occasionally subject to "ultramarine sickness," in which the blue color turns to a gray-blue with time. Perhaps the deterioration is a result of reaction with SO_2 in the atmosphere, but an acceptable explanation is that acid in the varnish attacks the surface of the pigment particles and destroys the trisulfide radical color centers (Chapter X). It should be kept in mind that the green color of ultramarine paint in old paintings can also be due to yellowing of the varnish layer and the oil vehicle. The yellow combines subtractively to produce green.

Azurite, $2CuCO_3 \cdot Cu(OH)_2$, is a naturally occurring pigment and its purity is variable. It has been found in Egyptian paintings and was the most important blue used in the Far East. It is referred to by other names, such as blue ash, sky blue, mountain blue, and German ash. In HCl, azurite is soluble and effervesces by giving off CO_2. In base it slowly hydrolyzes to give black CuO on the surface of the particles. It is darkened by hydrogen sulfide from air pollution. Its crystals lose their color upon grinding so it must be used as a coarse-grained, almost sandy, pigment. Azurite has reasonably good covering power (η = 1.73–1.83) and is used in tempera instead of oil. In frescoes, azurite often converts to green malachite. *Blue verditer* or *blue bice* is synthetic azurite. It was most important as a pale, greenish blue in the fifteenth through seventeenth centuries in Europe and China. In has the same chemical reaction behavior as natural azurite toward HCl and NaOH.

Smalt refers to CoAsS (cobaltite), $CoAs_2$ (smaltite), or some natural minerals containing the ions K^+, Co^{2+}, Al^{3+}, Si^{4+}, and O^{2-}. Smalt was used in twelfth and thirteenth century wall paintings in Inner Mongolia. It found use in the coloring of glass and is present in early Venetian glass. The glass can be ground up and used as a pigment. In the sixteenth and seventeenth centuries, smalt was used in this way as a pigment. It was the earliest cobalt-containing pigment, but is difficult to paint with because large particle size is needed to retain the color. Smalt is chemically unreactive toward acid and base, and it does not have good covering power (η = 1.49–1.52).

Prussian blue, $Fe_4[Fe(CN)_6]_3$, was synthesized first in 1704 and is as intensely colored as indigo. It was the first truly artificial pigment. Prussian blue can be used in very small particle sizes. It does not react with HCl but dissolves in NaOH and precipitates orange-brown $Fe(OH)_3$. Its reactivity toward base means that it cannot be used in fresco painting. Its covering power as a pigment is not too large, because η = 1.56, but it has high tinting power. Prussian blue is the blue color characteristic of architects' blueprints.

Synthetic ultramarine has a variable composition. Some of the compositions and colors that are known are

$Na_{6.88}Al_{5.63}Si_{6.35}O_{24}S_{2.4}$	blue
$Na_{6.50}Al_{6.30}Si_{5.70}O_{24}S_{3.5}$	green
$Na_{5.88}Al_{5.76}Si_{6.21}O_{24}S_{2.52}$	rose

These compounds were first prepared in 1814 and have been used from 1831 to the present. Synthetic ultramarine is not as bright a blue as natural ultramarine, but it is much less expensive. The main reason for the dullness is the fact that the synthetic material has small, uniform particles which evenly diffuse light. On the other hand, the natural material has irregular particles with widely varying dimensions and shapes. The larger particles add brilliance to the pigment. The impurities present in natural ultramarine also allow it to be distinguished from the synthetic product under a microscope. The reactions with HCl and NaOH are the same as those of natural ultramarine. The presence of SO_2 causes slight fading of ultramarine. The hiding power of the pigment is not great ($\eta = 1.5$).

Cobalt blue, $CoAl_2O_4$, is a synthetic pigment prepared first in 1802 by Thénard and is also called Thénard's blue. It essentially replaced smalt as a blue, cobalt-containing pigment. It is chemically unreactive toward acids and bases and has been used in coloring glasses, ceramics, oils, and watercolors. Its index of refraction is 1.74. *Cerulean blue*, $CoO \cdot n(SnO_2)_x$, is another synthetic cobalt-containing pigment, and was first used in 1860. It is chemically stable and inert toward acids and bases. Its index of refraction is 1.84, giving it good hiding power.

Green Pigments

Green, the color of vegetation and spring, symbolizes the triumph of life over death in Christian art. On the whole, greens tend to be comparatively transparent pigments and are generally not very bright in color. A number of greens have been produced by mixing blue and yellow pigments or by glazing a transparent yellow paint over an opaque blue. Some of these methods are discussed later in this chapter and some were discussed in Chapter VI. Artists sometimes find it difficult to bring strong greens into a pleasing color scheme because of saturation of the eye by the green. Figure V-5 shows that the eye is most sensitive to green wavelengths, and a strong green can dominate other colors when viewed by the normal eye. One of the purest greens is emerald green. Viridian tends to be bluer

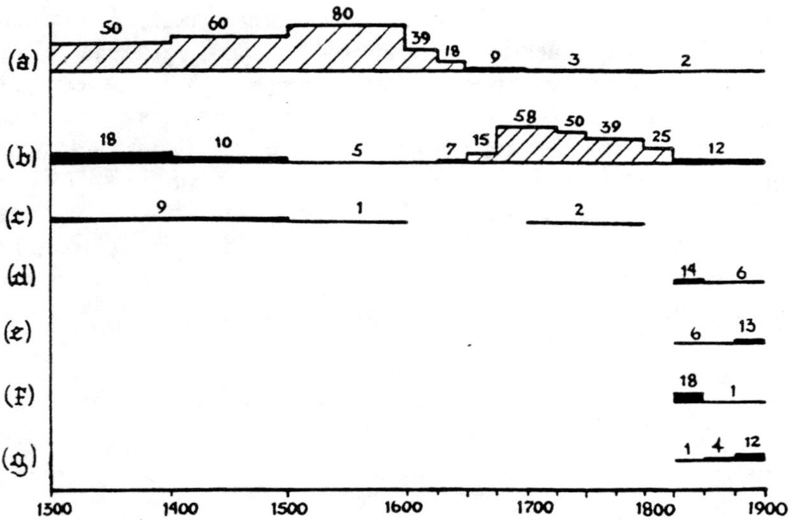

Figure XI-3. The occurrence of some green pigments in European easel paintings. (a) Verdigris; (b) green earth; (c) malachite; (d) chromium oxide green (opaque); (e) chromium oxide green (transparent); (f) cobalt green; (g) emerald green. [Courtesy of H. Kühn, Terminal dates for paintings derived from pigment analysis, in *Application of Science in the Examination of Art Works*, Museum of Fine Arts, Boston, Massachusetts (1973), p. 199.]

and more transparent. The occurrence of green pigments in European easel paintings is shown in Figure XI-3.

Malachite, $CuCO_3 \cdot Cu(OH)_2$, is a naturally occurring pigment and was the first green. It found use in Egypt and China and was used in Europe up to about 1800, when it was replaced by synthetic greens. Malachite is better when used in tempera than in oil. In acid solution it is soluble and releases CO_2, but remains green. In cold NaOH there is no reaction, but with hot NaOH, black CuO forms on the surface. It is similar in composition to patina found in copper and bronze metal corrosion.

Green earth or *terre verte* is a silicate containing Fe^{2+}, Mg^{2+}, Al^{3+}, K^+, Si^{4+}, and O^{2-}. Terre verte is actually an ochre and mica silicate which occurs naturally and was used by the early Romans and throughout the ages. In acid, it gives a green solution. In NaOH it is partly soluble, giving a gray precipitate of $Fe(OH)_2$ which turns brown on standing. Green earth is rather transparent in oil ($\eta = 1.62$). *Chrysocolla*, $CuSiO_3 \cdot nH_2O$, is a naturally occurring mineral used since very early times, particularly in watercolors, but it is not generally common as a pigment. Its appearance is similar to that of malachite but it is slightly more blue. It is rather transparent in oil because its index of refraction is 1.6. It can be ground finely and still retain its color.

Verdigris, $Cu(CH_3COO)_2 \cdot Cu(OH)_2$, is a synthetic pigment that has been made since early times by corroding copper metal with acetic acid, CH_3COOH, from vinegar. It was most commonly used between the fifteenth and seventeenth centuries. It has high transparency and was used as a glaze on paintings. Verdigris is chemically the most unstable of any pigment in use. It tends to turn brown or black unless carefully isolated by varnish from other pigments and air. A number of different compositions of copper acetates are known and they have slightly different colors. The first three compositions shown below can occur in mixtures with one another, while the last formula is only found in pure form:

$[Cu(CH_3COO)_2]_2 \cdot Cu(OH)_2 \cdot 5H_2O$	blue
$Cu(CH_3COO)_2 \cdot Cu(OH)_2 \cdot 5H_2O$	blue
$Cu(CH_3COO)_2 \cdot Cu(OH)_2 \cdot 2H_2O$	green
$Cu(CH_3COO)_2 \cdot Cu(OH)_2$	blue and green

In HCl, verdigris is soluble and produces a green solution. In NaOH it is soluble and precipitates bluish $Cu(OH)_2$. All of these verdigris formulations are darkened by atmospheric sulfur compounds due to the production of Cu_2S and CuS. They react with CO_2 in the air to form compounds containing $CuCO_3$ and also some Cu_2O. In the first three months after their use, the verdigris formulations can undergo a color change from blue-green to green. All react with resins to form copper resinates and this promotes drying. Many copper resinates may have been verdigris which reacted with the resins. *Copper resinate* is copper salts of resin acids. It is a synthetic pigment material that appears in eighth through sixteenth century paintings and manuscripts. Because it is rather transparent, it is often found as an overpaint to increase the depth (saturation) of an opaque green. It tends to be homogeneous with the vehicle and, in this sense, is not really a pigment. It undergoes the same acid–base reactions as verdigris does.

Scheele's green is $CuHAsO_3$ or $Cu(AsO_2)_2$. This is a synthetic pigment which was prepared first in 1778 and was used occasionally in and after the eighteenth century. Scheele's green is not a high-quality pigment and was rather unpopular because of its toxicity. Many arsenic poisonings are believed to have resulted in the nineteenth and twentieth centuries from the use of copper arsenate as a pigment in wallpaper. Extremely toxic and gaseous trimethylarsine, $(CH_3)_3As$, is formed by the action of fungus on copper arsenate in damp wallpaper. If the room is poorly ventilated, the arsenic levels become very high. Scheele's green is blackened by sulfur-containing air as a result of the formation of CuS, and reacts with HCl and NaOH in the same way that verdigris does.

Emerald green, $3Cu(AsO_2)_2 \cdot Cu(CH_3COO)_2$, is a synthetic pigment first prepared in 1814. It is the brightest green and it essentially replaced verdigris and Scheele's green. It reacts with HCl and NaOH in the same way as verdigris does and is rather toxic.

Viridian, $Cr_2O_3 \cdot 2H_2O$ or $Cr_2O_3 \cdot H_2O$, is a synthetic pigment prepared in 1838 but used first in 1859. It is in common use at present and replaced emerald green in many applications. Upon dehydration it gives opaque green chromic oxide, Cr_2O_3. It does not react with HCl or NaOH.

Cobalt green is a mixture of cobalt oxide (CoO) and zinc oxide (ZnO). It was primarily used in the nineteenth century and was considered to be of high quality. It is slightly soluble in acid, giving a pale pink color, but does not react with base.

Yellow, Orange, and Brown Pigments

Yellow is the most luminous of all colors. Chrome yellow is one of the brightest of the yellow pigments. Shades of orange from yellow-orange to red-orange can be made with certain pigments. Depending upon the size of the particles, cadmium yellow can range in color from a strong yellow to a strong orange. Browns are obtained from yellow and orange pigments when significant secondary absorption at other wavelengths also occurs. The use of yellow pigments in European easel paintings is shown in Figure XI-4.

Ochre, sienna, and umber are iron oxides and clays having various degrees of hydration. They are natural pigments whose composition, structure, and transparency vary widely. They are among the most important pigments and were used in times dating from the Egyptians up to the present. Yellow ochre is $Fe_2O_3 \cdot H_2O$ (goethite). Some samples are slightly soluble in HCl but are unreactive toward NaOH. Brown ochre, sometimes called raw sienna, is the limonite structure of $Fe_2O_3 \cdot H_2O$. It is not very reactive toward acid or base. Burnt sienna is made by heating raw sienna, $Fe_2O_3 \cdot H_2O$, to form red Fe_2O_3. This pigment is redder and darker than raw sienna. Raw umber is an ochre containing 8–15% manganese dioxide. Raw umber is important in wood finishing and has been used in all European periods since the fifteenth century. Burnt umber is made by heating raw umber. Burnt umbers range from dark brown to red in color. In general, the higher the Fe_2O_3 content, the more transparent is the pigment. Sienna, which can be entirely Fe_2O_3, has been used as a glaze in painting. All ochres are light-fast.

Orpiment, As_2S_3, is a natural pigment widely used in the Far East up

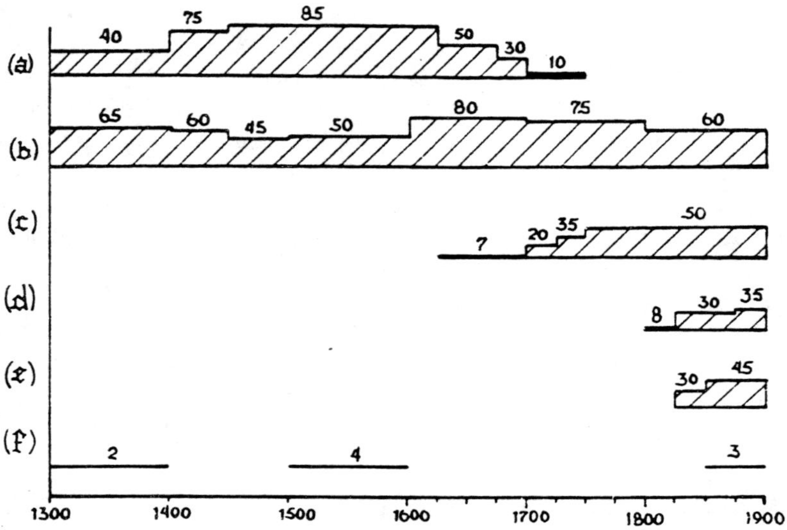

Figure XI-4. The occurrence of some yellow pigments in European easel paintings. (a) Lead–tin yellow; (b) yellow ochre; (c) Naples yellow; (d) chromium yellow; (e) cadmium yellow; (f) orpiment. [Courtesy of H. Kühn, Terminal dates for paintings derived from pigment analysis, in *Application of Science in the Examination of Art Works*, Museum of Fine Arts, Boston, Massachusetts (1973), p. 199.]

to the thirteenth century but not used very extensively since that time. It gives yellow to orange colors, depending upon the particle size. In acid it is sparingly soluble, giving off H_2S. In NaOH it dissolves slightly to give Na_3AsO_3. Orpiment is toxic, and is not compatible with copper pigments, because of the formation of black CuS.

Massicot, PbO, although naturally occurring, is usually a synthetic pigment also called "litharge." It is made by heating white lead to 620°C. Massicot is essentially absent from paintings after the thirteenth century but was used in very early times (400 B.C. in Egypt, for example). In HCl solution it precipitates white $PbCl_2$, and is soluble in NaOH, forming Na_2PbO_2. It was mixed with blues to make greens in early times, but it is somewhat light-sensitive. For example, the modern iron arc test reveals that massicot darkens in ultraviolet radiation in 45 sec of exposure. It reacts with CO_2 from the atmosphere to give white $PbCO_3$. Massicot has good opacity and dries quickly because of the formation of lead "soaps" with the vehicle. Another lead-containing yellow pigment is lead–tin yellow. *Lead–tin yellow*, Pb_2SnO_4, frequently contains some silica. It was a synthetic pigment used in the era 1300–1750. *Naples yellow*, $Pb_3(SbO_4)_2$, is a synthetic pigment first prepared in 1758. However, it occurs naturally on the slopes of Mount Vesuvius and this material has been used since the fourteenth century. A yellow of similar composition is found in glaze on ceramics from the Middle East. Naples yellow is very similar in appearance to massicot. It began to replace lead–tin yellow as a yellow pigment in the late seventeenth century. It is darkened by H_2S because of the formation of PbS and precipitates $PbCl_2$ from HCl solution. In NaOH it is soluble and gives Na_2PbO_2. *Turner's patent yellow*, $PbCl_2 \cdot 5\text{–}7PbO$, is a synthetic pigment first prepared in 1775. It is blackened by water and sunlight but works well in oil and dries quickly.

Chrome yellow is the monoclinic polymorph of $PbCrO_4$. Both $SrCrO_4$ (strontium yellow) and $BaCrO_4$ (barium yellow) have a similar appearance to it. Chrome yellow is a synthetic pigment first prepared in 1809. It was fairly commonly used after 1818 and was an important commercial pigment. It was inexpensive and superseded patent yellow. Chemically, it is not highly stable. In HCl solution it precipitates $PbCl_2$ and gives a yellow solution due to the formation of $H_2Cr_2O_7$. In NaOH it gives a yellow solution of CrO_4^{2-}. It is not light-fast. In the iron arc test, it darkens in 1500 sec of ultraviolet radiation. The presence of H_2SO_3 from air pollution reduces the CrO_4^{2-} ion to yellow-green Cr_2O_3. The presence of H_2S turns $PbCrO_4$ brown due to the formation of PbS. There are three polymorphs of $PbCrO_4$. These are the orange-red tetragonal form, the yellow monoclinic form, and the greenish-yellow orthorhombic form. Lead chromate also produces mixed crystals with $PbSO_4$. These facts allow one to produce colors from pale primrose to scarlet, using mixtures with slightly different crystal composition.

Yellow lake is quercitron in alum on chalk. Quercitron dye was discussed in Chapter VIII. This pigment was used in the seventeenth century. It is sensitive to ultraviolet radiation. Vermeer used it to mix with blue pigments to obtain green coloration. His green foliage has now turned blue, however, because the yellow lake has faded. In HCl solution it decolorizes, and in base it gives a yellow-brown solution.

Cadmium yellow, CdS, is a synthetic pigment first prepared in 1829. It was expensive but very important as a yellow and quickly replaced chrome yellow. Colors ranging from yellow to orange can be obtained, depending upon the particle size. It has good opacity but is bleached by H_2O and O_2 from the atmosphere because of the formation of white $Cd(OH)_2$. In HCl it is fairly insoluble and gives off some H_2S. It does not react with NaOH.

Van Dyke brown is an organic bituminous material. It is naturally occurring and was used mainly in the sixteenth through seventeenth centuries but is still available today. It fades to gray in sunlight and the fading is worse in water-base paints than in oil-base paints. Because of its high transparency, it has been used for staining wood and glazing in paintings.

Red and Violet Pigments

Red is associated with emotions and is symbolic of both love and hate in Christian art. Of the reds, vermilion provides a strong opaque red. Madder and cochineal lakes are more purple-red than is vermilion by virtue of having a slightly different reflectance spectrum. The frequency of occurrence of reds in European easel paintings is shown in Figure XI-5.

Vermilion is the red form of HgS. Vermilion can be obtained from the mineral cinnabar, but has also been prepared synthetically since about the fifteenth century. It was used as early as 3000 B.C. and has been the most important red throughout the ages. It was very expensive and thus valued as a pigment. The color depends somewhat on the particle size. Orange to deep red colors are obtained as the particle size is increased. The large index of refraction of vermilion causes it to be highly opaque. It is often found mixed with zinc white or lead white to give flesh tones. Some darkening is occasionally observed after ten or more years of exposure to light. The reason appears to be a light-induced transition to a metastable black polymorph of HgS as discussed later in this chapter. Chemically, HgS is unreactive toward HCl and NaOH.

Red ochre, Fe_2O_3, can be synthesized by heating other ochres. It was used in very early times and was common in works of the American

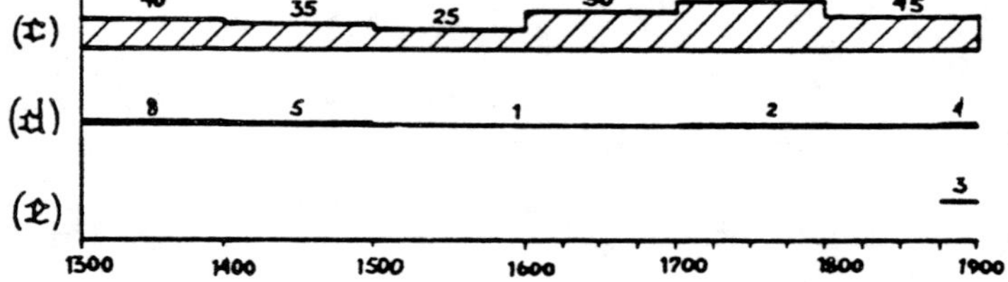

Figure XI-5. The occurrence of some red pigments in European easel paintings. (a) Vermilion; (b) red lake pigments; (c) red ochre; (d) minium; (e) cadmium red. [Courtesy of H. Kühn, Terminal dates for paintings derived from pigment analysis, in *Application of Science in the Examination of Art Works*, Museum of Fine Arts, Boston, Massachusetts (1973), p. 199.]

Indians. It is very stable, but dissolves slightly in HCl and does not react with NaOH.

Realgar, As_4S_4, is a naturally occurring mineral that has been used since very early times to give reddish-orange coloration. It is not as commonly used a pigment as orpiment, and is toxic. Realgar reacts with acids and bases in the same way as orpiment does.

Minium or *red lead* is Pb_3O_4. Minium is a synthetic pigment prepared by heating white lead at 480°C. It has been used since very early times and found use in the Middle Ages in illuminated manuscripts. Nowadays it is important because of its anticorrosive properties when applied to iron and steel. It is somewhat unstable toward light. Light causes the formation of brown PbO_2. Minium lightens upon reaction with CO_2 because of the formation of $PbCO_3$. In acid solution, minium dissolves, with the formation of $PbCl_2$, but it does not react with NaOH.

The *red lakes* were discussed in some detail in Chapter VIII. Madder, alizarin, and cochineal can be precipitated on clays and used as pigments. However, because they are organic colorants, they are less stable toward light than are inorganic pigments. In HCl they turn orange and in NaOH they turn violet.

Cobalt violet, $Co_3(PO_4)_2$ or $Co_3(AsO_4)_2$, is not too commonly found in paintings. Cobalt violet as a phosphate was prepared first in 1859 and is darker than the arsenate, which was prepared first in 1880. They have rather weak tinting strengths and are unreactive toward HCl and NaOH.

Cadmium red, CdS/Se, is a mixed crystal of cadmium sulfide and selenide. By variation in the ratios of the components and control of the synthetic conditions, shades of color from vermilion to maroon can be realized. It is a popular modern pigment and has essentially replaced vermilion. Cadmium red was introduced in about 1910. Its chemical reactions are similar to those of cadmium yellow.

Black Pigments

Even before Christianity, black was symbolic of death and the underworld. The use of blacks in European easel paintings is summarized in Figure XI-6.

Asphaltum and *mummy* are tarry complex organic–inorganic materials. Mummy was obtained from the remains of Egyptian mummies embalmed with asphaltum. These blacks are plastic in character, and, when used as undercoating on a painting, they can chip and peel as they age. They do not react with HCl, but turn brown in NaOH.

Carbon black is mostly elemental carbon and is sometimes called *lamp black*, *charcoal*, or *bone black*. These blacks have been used since very

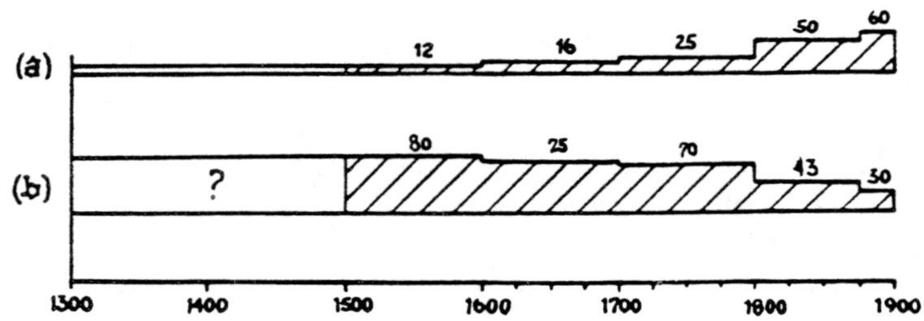

Figure XI-6. The occurrence of black pigments in European easel paintings. (a) Bone black; (b) charcoal black. [Courtesy of H. Kühn, Terminal dates for paintings derived from pigment analysis, in *Application of Science in the Examination of Art Works*, Museum of Fine Arts, Boston, Massachusetts (1973), p. 199.]

early times (1500 B.C.). Old India inks were made from carbon black. They do not react with acid or base. Bone black contains a large amount of colorless $Ca_3(PO_4)_2$ and $CaCO_3$. The forms of these blacks can be distinguished under a microscope. Charcoal looks like tiny wood splinters, while bone black contains irregular particles, some of which are brown and transparent. Lamp black is homogeneous, particulate carbon.

Metallic Paints and Coatings

An important property of metals, particularly gold, is that of harmonizing and setting off ordinary colors. This is mostly due to the bright luster of the metals, which brings up the luminosity of a weak color.

Gold leaf and *powder* have been used for centuries in valued works. Gold leaf was made by beating gold foil into extremely thin sheets. The leaf was then made to adhere to the surface of the work. Gold powder has been used in writing inks and panel paintings since Medieval times. The powder was applied with egg white or gum and then sometimes burnished to obtain high luminosity.

Silver leaf and *powder* have been used in the same way as gold, but their tendency to tarnish by formation of silver sulfide has limited their utility.

Bronze powders are copper–zinc and copper–tin alloys and have been used to imitate gold paint. Colors from light yellow to orange can be achieved by varying the alloy composition. The vehicle for the bronze powders is usually nitrocellulose in an organic solvent, such as amyl acetate. Nitrocellulose is rather sensitive to light. It becomes brittle and yellows in ultraviolet radiation. It is also unwise to use bronze powders for fine art retouching, as they tarnish and discolor rapidly.

The Mixing of Pigments

In Chapter V some of the problems and considerations were discussed that arise in the mixing of colors. The actual mixing of pigments can be taken up in more detail now that specific pigments have been presented.

The subtractive color process is used when mixing colors in opaque media, such as paints. However, it is always wise to try an actual mixture rather than to accept a prediction. Painters know from experience that it is difficult to foretell whether a green paint produced by mixing a certain yellow with a certain blue will give the desired green or perhaps one with a brownish or bluish tinge.

In Figure XI-7, the reflectance pattern of a mixture of two parts cadmium sulfide (cadmium yellow) and one part ultramarine blue is shown. The visual result of the mixture is a dull olive color whose spectral reflectance curve is not predictable simply by taking into account the curves of the components. The problem of color mixing in heterogeneous pigment systems arises because the light observed from the surface is mostly reflected light and not transmitted light. The particles of the two pigments may have different light scattering power, different light absorbing power, different particle size and shape, and a different index of refraction. The individual components also may have spectral reflectance patterns which are not wholly yellow, green, blue, etc., to begin with. Under these conditions the reflected light cannot be the simple sum of the component parts.

On the positive side of the problem, we can look at what colors tend to be produced when some common pigments are mixed. When vermilion and lamp black are mixed, a chocolate brown color is obtained. Flesh tones can be obtained when vermilion is combined with lead white and

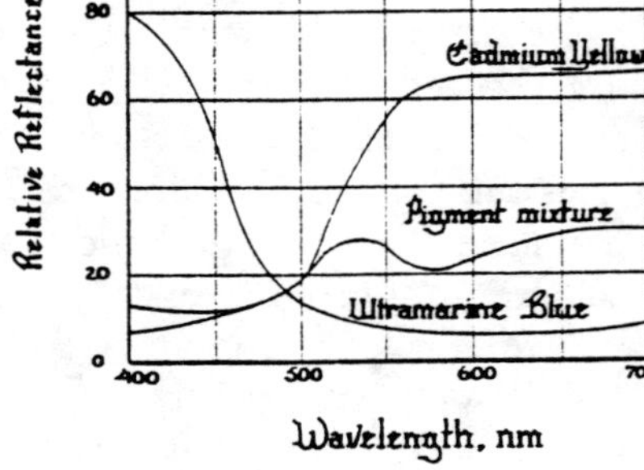

Figure XI-7. The spectral reflectance of two parts cadmium sulfide and one part ultramarine.

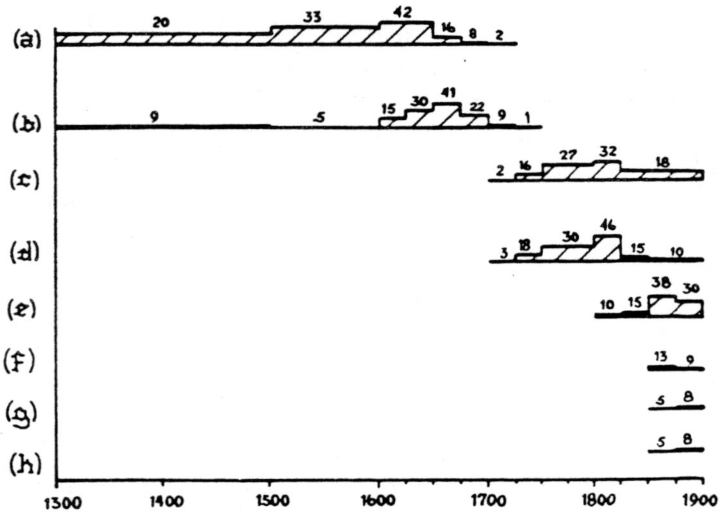

Figure XI-8. The occurrence of blue and yellow pigment mixtures to give green in European easel paintings. (a) Azurite and lead–tin yellow; (b) azurite and yellow ochre; (c) Prussian blue and Naples yellow; (d) Prussian blue and yellow ochre; (e) Prussian blue and chromium yellow (chrome green); (f) Prussian blue and cadmium yellow; (g) cobalt blue and Naples yellow; (h) cobalt blue and cadmium yellow. [Courtesy of H. Kühn, Terminal dates for paintings derived from pigment analysis, in *Application of Science in the Examination of Art Work*, Museum of Fine Arts, Boston, Massachusetts (1973), p. 199.]

yellow ochre. To make a bright orange color by mixing pigments, it is essential that the yellow pigment incline toward orange rather than green, and the red pigment tend toward orange rather than crimson or purple. Otherwise a dull, muddy tone results.

Yellow tends to be the most luminous of all colors. Hence it is difficult to obtain a satisfying yellow from a mixture of two other colored pigments. When yellow is mixed with black it gives a dull yellow-green or olive color.

Green is frequently perceived as not a bright color and it is possible to make many satisfactory greens by mixing yellow and blue pigments. The green pigments in European easel paintings are often found to be glazed or are mixtures of yellow and blue pigments. Figure XI-8 summarizes the pigment mixtures used to make green and their relative use. The mixture of colors to produce green is often a source of eventual discoloration of green areas in paintings. In time one of the components may lose its color and produce a change in tonality. Green, when mixed with gray, gives sage green colors.

Prussian blue, indigo, cobalt blue, ultramarine, and cerulean blue, when mixed with madder lake, will produce a wide variety of violet colors. However, because a great deal of light is absorbed over the visible range, the violet and purple obtained has a dull, grayish cast. Blues, when mixed with white and black, give a slate gray. Ultramarine blue laid on a white ground and glazed with transparent madder lake gives purple and violet tones.

Decomposition of Inorganic Pigments by Light

An important practical difference between organic and inorganic colorants is the fact that the inorganic materials are much less susceptible to alteration by light. Chemically the two classes are, of course, quite different. The inorganic pigments are in all cases composed of one or more metal or metalloid ions. Many are saltlike or have ionic bonds holding the crystal lattice together. A few contain organic ligands, but most do not. Inorganic salts are often simpler in composition and structure than are the organic colorants. The much greater light and thermal stabilities of inorganic pigments come in part from this simplicity. There are plainly not very many pathways for the decomposition of the excited electronic states. The simple structures and ionic interactions lead to large crystal lattice energies and further stabilization. In contrast to organic compounds, the products that might occur upon decomposition are for the most part not volatile and therefore remain behind in the pigment. Many inorganic pigments are naturally occurring minerals which have withstood the stresses of heat, light, and weathering on the earth for millions of years. Unlike commonly encountered organic materials (Chapter IX), inorganics tend to be thermodynamically very stable.

Nevertheless, there are several important pigments which are sensitive to light. Perhaps the most well-known example is that of vermilion. The brown flesh-tones in some oriental wall paintings probably result from the darkening of vermilion. Vermilion, HgS, can undergo a physical change from the red, cinnabar crystal structure to the black, *meta*-cinnabar lattice as a result of exposure to light. Vermilion has been used by restorers for retouching of paintings, and the phase transition causing discoloration has been noted in some cases to occur within five to ten years of application. The amount of *meta*-cinnabar formed is directly related to the amount of light striking the pigment; the greater the quantity of light, the more blackening that occurs. The darkening appears to be more common when vermilion is used in linseed oil than when water or emulsions are used as the vehicle. The manufacturer of vermilion appears to have some control over the potential for conversion of cinnabar to *meta*-cinnabar. The so-called wet process or German process introduced in 1687 for the synthesis of vermilion is apt to lead to light-sensitive material more often than do later synthetic processes, in which sulfur is kept in an excess relative to mercury. The light sensitivity occurs mostly on the surface and rapidly drops off with distance from the surface. This is in accordance with Beer's law [equation (IV-12)].

Pigments such as verdigris, red lead, the chromates, and lead white are somewhat chemically reactive and may undergo spontaneous change with time. These chemical changes occur in pigments as a result of moisture, oxygen, carbon dioxide, and pollutants in the atmosphere reacting at the surface. For example, red lead in glue may decompose to brown PbO_2. The chromates may decompose to green Cr_2O_3. Some of these reactions are accelerated by light. More commonly, the vehicle or binder supporting the pigment is altered by the effects of light. Hence the apparent deterioration of the paint is frequently not due to changes in the pigment but rather is caused by alterations in the vehicle or overlaying varnish in ways discussed in Chapter IX.

"Ultramarine sickness" refers to the occasional tendency of ultramarine to turn from its blue color to a gray-green color. This transformation is a true deterioration of the pigment and occurs as a result of light, but not in a direct way. Light causes an increase in the acidity of the varnish covering the ultramarine. The acidic functional groups in the varnish appear to react with the surface of the ultramarine particles and destroy the color. For this reason, ultramarine in fourteenth and fifteenth century paintings was often left unvarnished.

Chalking of zinc white in paintings occurs in the presence of light. As with ultramarine sickness, most chalking comes about because of changes occurring in the varnish and oil as discussed in Chapter IX. However, chalking in house paints composed of zinc oxide may result from chemical changes in the white pigment as a result of ultraviolet radiation and atomspheric pollutants.

Discoloration of smalt and changes in the tonality of paintings composed mainly of inorganic pigments are occasionally noted when two or more pigments have been mixed to achieve a desired color. Some of the reasons for these changes were discussed already in Chapter VI. The inorganic material generally cannot be blamed for the change. Rather, yellowing of oils and varnishes and photodeterioration of organic lakes that are mixed with the inorganics are most commonly the cause.

On the whole, if lengthy exposure time to light is necessary, most inorganic pigments and the phthalocyanine dyes can be considered to have reliable stability toward light. Modern synthetic resins are preferable as vehicles to those made from vegetable oils, such as linseed oil, to avoid yellowing.

COLOR IN GLASS AND GLAZE

For centuries it has been known that glass and ceramics could be tinted vividly using certain natural minerals. Yet the art of glassmaking evolved mostly by trial and error until quite recently. Glassmaking procedures were passed along from artisan to apprentice and so forth. Occasionally a particular technique that was necessary to obtain good-quality color and properties in a certain glass was lost with the death of a craftsman. It might then be rediscovered centuries later. The techniques needed to achieve colors of good quality in glass are not presented here. The intent of the presentation is to outline some of the important coloring agents used since antiquity and briefly discuss the origin of the colors of glass.

In Chapter VI the structure of glass materials was introduced. It was noted that the vitreous state of silica allowed for the encapsulation of ions in the matrix. There are some restrictions on the materials that are suitable for imparting color to glass. The most obvious is the requirement that the colorant be stable at the high temperatures needed during the manufacture of the object.

The ions in silicate matrices produce color by the very same mechanisms as they do in pure pigments. Cavities exist in the silicate matrix which can accommodate inorganic ions. The walls of the cavities contain oxygen atoms that act as ligands and bond to the ions. In addition, other ions such as halides may be present, and these too may furnish electron density to stabilize the metal ion. The color of the metal ions originates from electronic transitions associated with metal and ligand orbitals. In various glasses containing transition metal ions, the color is frequently about the same as that in pigments containing those ions. For example, the trend in the polarizability of the ligands present in the glass matrix correlates with the expected progressive change in the color (the shift in the absorption band to higher or lower wavelengths). Keeping in mind that the color observed is the complement of the color absorbed, a shift toward lower energy in the absorption band is observed as the polarizability of the chromophore increases in various glasses containing the Co^{2+} ion. Table XI-2 summarizes this observation. The trend in the absorption bands follows the trend in polarization of the ligands by the Co^{2+} ion, namely $F^- < O^{2-} < I^- < Se^{2-}$. As the ligand becomes more polarizable, the charge transfer band shifts to lower energy. In the case of the cobalt–selenium glass, the absorption band is so broad and intense that it blankets the entire visible range.

Polarizability in metal–ligand bonds is an important source of color in glass, but so are "*d–d*" transitions and color centers. Using these

Table XI-2. Colors Observed and Absorbed for Co^{2+} in Several Glass Matrices

Glass matrix	Chromophore	Principal color absorbed	Color observed
Fluoroberylate	Co—F	Green	Pink
Silicate	Co—O	Yellow	Blue
Silicate containing iodide ions	Co—I	Red	Green
Silicate containing selenium	Co—Se	All	Black

mechanisms of color production discussed in Chapter X, we can summarize the effects of a number of different ions on the color of glass and glaze.

Iron

Iron ions are present in virtually all glass. Techniques for purifying the sand used in glassmaking were not sophisticated before the early twentieth century, and the iron content in the silica was of great concern to the glassmaker. This was particularly true if the iron-produced color in the glass was undesirable. The color of the iron oxide depends upon the oxidation state of the iron ions. With Fe^{2+}, a blue tint is imparted to the glass because of a strong infrared absorption band whose tail extends into the visible range. Hence Fe(II) oxides in glass are very effective absorbers of infrared radiation. The origin of the blue color is actually thought to involve a resonance between the Fe(II)–Fe(III) oxidation states in the Fe^{2+}—O—Fe^{3+} bond.

Oxides containing Fe^{3+} give red-brown or weak yellow-pink color to glass, depending upon the coordination number of the iron ions. An equilibrium can exist between two coordination numbers in the glass:

$$\mathrm{Fe(III)O_4} \rightleftharpoons \mathrm{Fe(III)O_6}$$

The four-coordinate iron ions give a rather strong red-brown color characteristic of wine and beer bottles. The tetrahedral geometry allows for reasonably intense electronic transitions to occur, leading to substantial coloration. The bands giving rise to the red-brown color have maxima in the ultraviolet range but have tails extending into the visible region. The absorption of the blue and green wavelengths gives the red-brown color. In the six-coordinate octahedral geometry, the color is much weaker, in part because the "*d–d*" transitions are disallowed in that geometry (Chapter X).

Techniques to remove the frequently unwanted coloration produced by iron in glass have become well established over the centuries. The addition of phosphates (PO_4^{3-}) or manganese oxides to the glass was a popular method. Manganese was added to glass in the form of pyrolucite, MnO_2, to oxidize Fe^{2+} to Fe^{3+} and decolorize the glass. In the early twentieth century selenium, particularly in combination with cobalt, essentially replaced pyrolucite for this purpose.

Manganese

Among the oldest glass and glaze colorants are the manganese oxides. The purple color found in Egyptian glass formed in 1400 B.C. results from manganese. The purple color is due to Mn^{3+} in the oxide, Mn_2O_3. The hues from blue to red can be obtained, depending upon the alkali metal ion (Group Ia element) present in the glass. The color of the glass progressively changes from blue to red as the atomic number of the alkali metal ion decreases. For example, the replacement of K^+ with Na^+, and Na^+ with Li^+ in the glass shifts the color increasingly toward red. The Mn^{2+} ion gives a faint yellow-orange tint in visible light but greenish-orange fluorescence in ultraviolet radiation.

Copper

Copper ions have been used in glass coloration since the earliest recorded times. The dark blue Persian glazes and the blues and greens of Egyptian glass were obtained with copper oxides. The color of the Cu^{2+} ion depends upon its coordination sphere. The distance between the Cu^{2+} ion and the oxygen atoms alters the intensity as well as the hue. There may be some participation by two oxidation states of copper in producing greens and browns. This would involve a Cu^{2+}—O—Cu^+ resonance-type interaction in which the oxidation state of the copper ions rapidly changes. Intense charge transfer bands are created.

Cobalt

There is some uncertainty as to when cobalt was first used to color glass, but its intentional presence in glass before the fifteenth century appears to have been sporadic. However, the Chinese were masters of its use in porcelain. Potters during the T'ang dynasty (A.D. 618–906) used cobalt glaze, and it is found on porcelain of the Ming dynasty (1368–1644). These Chinese glazes also contain iron and manganese, which made them difficult to imitate by later craftsmen.

Red, blue, and green colors have been obtained in glass with Co^{2+} in combination with other agents. Octahedrally coordinated Co^{2+} can be obtained by mixing CoO with MgO and produces a pink to red color. The color is not strong, because the octahedral field of oxygen ligands does not give rise to allowed "*d–d*" transitions on the Co^{2+} ion. The Co^{2+} ion mixed with Al_2O_3 gives intense blue colors because the Co^{2+} ion is in a tetrahedral field of oxygen atoms. This is similar to the structure of the pigment cobalt blue. Blues are also obtained in the form of smalt in which the Co^{2+} ion is in a silicate matrix. Green colors appear when CdO and ZnO are mixed together with CoO in the glass. The pigment cobalt green has the same composition. Another mechanism for altering the color of Co^{2+} is by variation of the sodium ion concentration. The Na^+ ion appears to produce changes in the Co^{2+} coordination sphere such that an increase in sodium ion concentration causes the color to change from blue to pink.

Cobalt oxide has been used as a blueing agent to decolorize ceramics for the purpose of competing with and imitating Chinese porcelain. Josiah Wedgwood experimented with this method in his stoneware. In Pearlware, a very small amount of cobalt oxide, CoO, was added to the slightly yellow clay. With the blue reflectance increased by the presence of CoO, there is a leveling out of the absorption spectrum because blue and yellow are complementary colors. A whiter finished product resulted.

Chromium

Chromium has knowingly been used to produce green color in glass since the early nineteenth century. Before this time iron oxides were used to yield greens. In addition, PbO with copper compounds and cobalt blue also produces greens. While Cr_2O_3 gives an emerald green, it has low solubility in glass except when the alkali content is high. Large, thin plates of Cr_2O_3 can produce Aventurine glass, which has a glittering appearance (see the section on *Metals in Glass* below). On ceramics, chromium oxide mixed with other metal oxides produces pink, yellow, and brown. The use of $PbCrO_4$ gives a yellow-orange, $ZnCrO_4$ gives brown, Cr_2O_3 and Al_2O_3 together give pink. Chromium oxide and tin oxide mixtures also produce pink.

Lanthanides and Actinides

In the early nineteenth century canary glass was first made using the uranyl ion, UO_2^{2+}. Canary glass gives a strong green-yellow fluorescence. Another uranium oxide, U_3O_8, has been used as an underglaze on ceramics to produce black. The black color results from resonance between mixed oxidation states in U_3O_8. The particular ions which resonate and produce the very strong light absorption are U^{4+} and U^{6+}. The rare earth oxides cerium oxide (CeO_2), neodymium oxide (Nd_2O_3), and praseodymium oxide (Pr_2O_3) have found use as colorants in more recent times. Cerium oxide has a high index of refraction ($\eta = 2.3$) and is also used as an opacifier in glaze. When mixed with other ions, various colors can be produced. Praseodymium oxide produces a green glass. Neodymium oxide is interesting in that it shows the "alexandrite effect." Thick glass colored with Nd_2O_3 is red, while thin glass is blue. It also has the property of changing color with the type of illumination. In daylight Nd_2O_3-colored glass is blue, but in incandescent lighting, which has abundant yellow and red wavelengths, the glass appears red.

Nonmetallic Coloring Agents

It was observed long ago that the addition of organic materials to a glass melt produced an amber color. Such glasses went by the name of *carbon–amber glass*. In more recent times it was learned that the carbon is not what produces the yellow color. Instead, sulfur impurities in the organics added to the glass are responsible. The sulfur probably catenates to form polysulfides, S_n^{2-}, where n can vary in value from unit values to tens. In addition, S_n^- radicals may also be present. The length of the chain determines the shading of color, with the longer chains producing lower energy absorption bands. In addition, iron ions that might be present in the glass react to give iron sulfide, FeS, which also has a brown color. The polysulfides and FeS together give the amber color. Blue sulfur glass probably results from the presence of S_3^- radicals trapped in the glass. The S_3^- radicals have a strong absorption band centered at 575 nm and impart color to the glass in the same way that they produce color in natural ultramarine.

Polyselenides, Se_n^{2-}, in the glass matrix give red to pink colors, depending upon the concentration. One of the main uses of selenium, however, has been in the decoloration of iron-containing glass. Use of CdS and Se together in glass gives a bright, ruby red color similar to that of the pigment cadmium red. Use of Bi_2O_3 or Cu_2O in CdS–Se glass gives a dark red, while CoO, FeO, and NiO give brown to black colors.

Metals in Glass

The use of colloidal-sized metallic particles to obtain highly prized color and optical effects dates back to ancient times. However, only copper seems to have been used very early. Copper-red glazes which contain colloidally dispersed copper metal were made by Chinese ceramicists during the Sung Dynasty. This glaze has come to be known as the Sang de Boeuf glaze. Dispersions of metal particles in glass are called *pyrosols*. Great skill is needed in the production of pyrosols because the crystallites of metal must be large enough to produce color but small enough to avoid metallic reflection. The colors produced by most pyrosols result from the inherent colors of the finely divided metals and not from physical effects, such as Rayleigh scattering.

The color and appearance of the glass are very much dependent upon the size of the metallic particles. Copper metal illustrates this point clearly. *Copper ruby* is a transparent, deep red glass containing metallic copper in crystals of size considerably smaller than the wavelengths of visible light. *Hematinone* is an opaque, red glass containing copper metal particles having sizes about the same as the wavelengths of visible light. These particle sizes produce opacity but are not large enough to give metallic reflection. *Aventurine* glass forms when a relatively small number of large copper crystals precipitate in the melt. By blowing and drawing the glass, one can orient the crystal planes parallel to the face of the glass. This gives a metallic, golden spangle appearance to the glass. Aventurine glass often broadly refers to any glass with particulate luster. Crystals of Cr_2O_3 in glass produce a similar appearance, for example. *Copper stain* results when a solution of Cu^{2+} ions is painted onto a glass surface. The Cu^{2+} ions migrate into the glass and then are reduced to copper metal. They give a bright red color.

Gold ruby glass has been made since 1685, when Andreas Cassius first described its preparation. The glass is sometimes called the "purple of Cassius" for this reason and has been used as a pigment when ground finely. The use of gold in glass dates back to much earlier times, but this glass is not what is conventionally considered to be true gold-ruby. The purple color is produced by fine particles of gold. Gold is an intense chromophore. It requires only one part gold to 100,000 parts glass to produce a pink shade. The deepest colors result from about one part gold in 1000 parts glass. The particle size of the gold is very small, being in the 5–200 Å size range in many cases. If the particulate gold grows too large, then the glass takes on a muddy or "livery" appearance. The colors obtained with gold pyrosols vary widely. This probably has to do with different sizes and shapes of gold particles, the mechanical strain in them, and the presence of ions of other metallic elements which may become attached to the surface of the gold particles. Different polymorphic forms also can be present.

The Lycurgus cup belonging to the British Museum shows an effect of overheated glass containing gold pyrosol. The cup is a diatrium with decorative relief depicting Lycurgus, a mythological Spartan lawgiver, entwisted in a vine. It is a Roman work from the fourth or fifth century A.D. The thin sections of the cup through which light can pass are red in color. The thicker sections show a green color by reflected light. The colors obtained upon heating of glass containing gold are very dependent upon the temperature at which the glass is worked, and this glass was apparently overheated.

Ambernia glass from the nineteenth century shows another variation of gold-ruby. By heating levels of the glass differently, a gradation in shading can be obtained ranging from pale amber to orange, red, and ruby.

Silver has been used to give an amber tint to glass and glazes. Persian pottery was stained in this way. Griesinger (1460) stained church windows with amber silver stains. Staining of glass is achieved by painting the surface with a solution containing silver ions. The Ag^+ ions replace other unipositive ions in the silicate matrix. The ions are then reduced to silver metal to give the yellow color.

Ultraviolet Absorption by Glass

A consideration which may have importance to objects stored in colored glass containers or in churches with stained glass windows is the relative degree of ultraviolet transmission by different colors of glass. This can be briefly summarized. Yellows, oranges, and red tend to be fairly opaque to the near-ultraviolet radiation, except for ruby glass. Chromium oxide-produced green is almost opaque to ultraviolet radiation. Cobalt and manganese blues and purples are rather transparent to ultraviolet radiation. Silver, according to Figure VI-14, is very transparent to ultraviolet radiation below 320 nm. This fact makes silver-coated glass useful for making photographs of the sun in the ultraviolet range. The silver passes radiation below 320 nm, but reflects all longer wavelengths.

Photochemical Effects

The only photochemical effect of any significance in glass is that of *solarization*. Ancient glasses were not colorless, because of the presence of iron in the silicate melt. They usually took on a greenish cast. To decolorize the glass a small amount of MnO_2 was added. A reaction between Mn^{4+} and Fe^{2+} occurred which oxidized the iron to the very pale yellow Fe^{3+} ion and reduced the Mn^{4+} ion to pale pink Mn^{2+}. With exposure to ultraviolet radiation, usually from sunlight, glass treated in this manner takes on a purple color. The reason for this color change can be traced to Mn^{3+} in the glass. A quantum of light is absorbed by Mn^{2+} and an electron is lost:

$$Mn^{2+} + h\nu \rightarrow Mn^{3+} + e^-$$

The electron is trapped in the glass matrix, where it eventually finds its way to an Fe^{3+} ion and reduces Fe^{3+} to Fe^{2+}:

$$Fe^{3+} + e^- \rightarrow Fe^{2+}$$

However, the Mn^{3+} ion imparts a purple hue as a result of "*d–d*" transitions. The glass can be discolorized again by heating to about 350°C. Old window panes receiving strong sunlight and bottles exposed to the desert sun often have a pink to purple color resulting from solarization.

A few other examples of solarization in glass are known, but these are much less important than that produced by Mn^{3+}. In general, glass and its coloring agents might be expected to be, and are, very inert to changes brought on by electromagnetic radiation. This is because their high degree of thermodynamic stability essentially eliminates any sensitivity to radiation except that of the very highest energies.

BIBLIOGRAPHY

Bamford, C. R., *Color Generation and Control in Glass*, Elsevier, New York (1977).

Barnes, N. F., A Spectrophotometric Study of Artists Pigments, *Technical Studies* **7**, 120 (1938).

Brill, R. H., Ancient glass, *Sci. Am.* **209**, 120 (1963).

Cennini, C., *The Craftsman's Handbook*, Dover, New York (1933) (translated by Daniel V. Thompson, Jr.).

Chase, W. T., Egyptian Blue as a Pigment and Ceramic Material, in *Science and Archaeology*, R. H. Brill, ed., MIT Press, Cambridge, Massachusetts (1971).

Church, A. H., *Colour*, Cassell, London (1897).

De Wild, A. M., *The Scientific Examination of Pictures*, G. Bell & Sons, London (1929).

Doerner, M., *The Materials of the Artist and Their Use in Painting with Notes on the Techniques of the Old Masters*, Harcourt, Brace and World, New York (1962).

Eastlake, C. L., *Methods and Materials of Painting of the Great Schools and Masters*, Vols. I and II, Dover, New York (1960) [originally published under the title, *Materials for a History of Oil Painting* (1847)].

Fajans, K., *Chemical Forces and Optical Properties of Substances*, McGraw-Hill, New York (1931).

Feller, R. L., Scientific examination of artistic and decorative colorants, *J. Paint Technol.* **44**, 51 (1972).

Feller, R. L., Studies in the Darkening of Vermilion by Light, *National Gallery of Art Report and Studies in the History of Art*, Washington, D.C. (1967), p. 99.

Gardner, H. A., *Physical and Chemical Examination of Paints, Varnishes, Lacquers, and Colors*, 7th ed., Institute of Paint and Varnish Research, New York (1935).

Gettens, R. J., Feller, R. L., and Chase, W. T., 7. Vermilion and cinnabar, *Stud. Conserv.* **17**, 45 (1972).

Gettens, R. J., and FitzHugh, E. W., 1. Azurite and blue verditer, *Stud. Conserv.* **11**, 52 (1966).

Gettens, R J., and FitzHugh, E. W., Malachite and green verditer, *Stud. Conserv.* **19**, 2 (1974).

Gettens, R. J., FitzHugh, E. W., and Feller, R. F., Calcium carbonate whites, *Stud. Conserv.* **19**, 137 (1974).

Gettens, R. J., Kühn, H., and Chase, W. T., 3. Lead white, *Stud. Conserv.* **12**, 125 (1967).

Gettens, R. J., Mrose, M. E., Calcium sulphate minerals in the ground of Italian paintings, *Stud. Conserv.* **1**, 174 (1954).

Gettens, R. J., and Stout, G. L., *Painting Materials, A Short Encyclopedia*, Dover, New York (1942).

Harley, R. D., *Artists Pigments C. 1600–1835; A Study in English Documentary Sources*, American Elsevier, New York (1970).

Heaton, N., *Outlines of Paint Technology*, 3rd ed., Charles Griffin, London (1947).

Jones, S. R., The History of the Artist's Palette in Terms of Chromaticity, in *Scientific Methods in the Examination of Works of Art*, Museum of Fine Arts, Boston, Massachusetts (1967), p. 71.

Keisch, B., *X-Ray Investigation of Lead White*, National Gallery of Art, Washington, D.C. (1972).

Kühn, H., Terminal Dates for Paintings Derived from Pigment Analysis, in *Application of Science in the Examination of Art Works*, Museum of Fine Arts, Boston, Massachusetts (1973), p. 199.

Kühn, H., 4. Lead–tin yellow, *Stud. Conserv.* **13**, 7 (1968).

Kühn, H., 6. Verdigris and copper resinate, *Stud. Conserv.* **15**, 12 (1970).

Laurie, A. P., *The Pigments and Mediums of the Old Masters*, Macmillan, London (1914).

Levison, H. W., *Artist's Pigments*, Color Lab, Hallandale, Florida (1976).

Mayer, R., *The Artist's Handbook of Materials and Techniques*, 3rd ed., Viking Press, New York (1970).

Mayer, R., *The Painter's Craft, An Introduction to Artists Methods and Materials*, Van Nostrand, Princeton, New Jersey (1966).

Muhlethaler, B., and Thirsen, J., 5. Smalt, *Stud. Conserv.* **14**, 47 (1969).

Paper Coating Pigments, No. 30, Technical Association of the Pulp and Paper Industry, New York (1966).

Patterson, D., *Pigments, An Introduction to the Physical Chemistry*, Elsevier, London (1967).

Patton, T. C., *Pigments Handbook*, Vols. I–III, Wiley–Interscience, New York (1973).

Plesters, J., Cross sections and chemical analysis of paint samples, *Stud. Conserv.* **2**, 110 (1956).

Plesters, J., 2. Ultramarine blue, natural and artificial, *Stud. Conserv.* **11**, 62 (1966).

Preuss, H. P., *Paint Additives*, Noyes Data Corp., Park Ridge, New York (1970).

Wehlte, K., *Materials and Techniques of Painting*, Van Nostrand Rheinhold, New York (1975) (translated by Ursus Dix).

Weyl, W. A., *Coloured Glasses*, Dawsons of Pall Mall, London (1959).

XII

Light-Sensitive Materials: The Photographic Process

In the preceding chapters the interaction of light energy with matter to produce chemical changes was looked upon as a generally undesirable event. Yet in the past 150 years or so, photoreactions have been put to use in a spectacular manner to produce "light pictures" or photographs. Photographic effects capitalize on the special behavior of certain materials in the presence of light. These effects can be reduced to chemical and physical phenomena in the same way that other photoreactions have been. This chapter is directed toward the discussion of the basic events that occur in the production of the photographic image.

Photographs of historical interest may be well over one hundred years old. As techniques changed, some of the materials used in the photograph process became obsolete. However, others remained in use due to their unusual and unsurpassed qualities. Two of these materials, the silver halides and gelatin, were brought together in the infancy of the field and, almost synergistically, formed a blend giving remarkable image reproduction characteristics. In fact, the silver halides and gelatin form so perfect a mix that they are unchallenged even today in black-and-white photographic imagery. Of course, many improvements and refinements have been made in techniques since the early days. However, with as much as is known about the chemistry of the photographic process, there still remain aspects of photography which are not well understood. Active research continues, particularly in the photographic industry, in an effort to better understand and improve the details of the very complicated processes involved.

In this chapter the chemical properties of photographic components, both recent and old, are briefly set down. The light-induced events which produce the photographic image are discussed, along with a survey of the factors leading to degradation of photographic materials.

HISTORICAL MILESTONES

The progress in the area of photography as we know it today is not likely to be appreciated without a brief review of the discoveries which form the foundation of the subject.

It has been known for several hundred years that the silver halides silver chloride (AgCl), silver bromide (AgBr), and silver iodide (AgI) are sensitive to light and darken from their normal white or pale yellow color to violet-black upon exposure. In fact, Scheele in Sweden made what appears to be the first systematic photochemical investigation (rather than simply recording a phenomenon) in 1777 when he noted that the silver halides darken more in violet light than in light of longer wavelengths. In 1833 Daguerre in France began to experiment with silver salts as a light-sensitive medium for photography. The Daguerreotype process was introduced in 1839, whereby a silvered copper plate or a silver plate was exposed to iodine vapor to form a thin layer of silver iodide on the surface. After exposure, the image was revealed by exposing the plate to mercury vapor. The mercury deposited preferentially on the exposed portions to give a positive image. In retrospect, one cannot help but to wonder about the health of persons using the Daguerreotype process. Mercury vapor was used rather indiscriminately at that time for a variety of purposes. Another common use of mercury vapor was as a fungicide for felt hats and its toxicity was not fully appreciated. Mercury poisoning is now thought to have contributed to the nervousness and peculiar behavior of many hatters of this period (possibly including the "Mad Hatter" in Lewis Carroll's *Alice in Wonderland*). The Daguerreotype process was not used to any great extent after about 1870.

William Henry Fox Talbot is generally given credit for inventing the modern photographic process. Talbot introduced the concept of obtaining a negative image and producing a positive print after development. In the period from the early to mid-nineteenth century, his process consisted in soaking a piece of paper in sodium chloride solution and then in silver nitrate solution in order to precipitate a thin layer of silver chloride on the paper. After 10–60 min of exposure to the light, the image was made permanent or "fixed" with potassium iodide or sodium chloride solution, although in 1839 Talbot used sodium thiosulfate, $Na_2S_2O_3$, (hypo) for fixation. In 1840 he substituted silver iodide for silver chloride as a light-sensitive material. An important event came later when Talbot found that the silver iodide-treated paper could be rendered more sensitive to light by treating it with a solution of silver nitrate and gallic acid. After an exposure of only 30 sec duration rather than a much longer period, the very weak image could be built up or *developed* by placing the exposed paper in a developing solution consisting of the same components as the sensitizing solution, namely silver nitrate and gallic acid. This procedure led to the concept of producing a *latent image* during the short exposure to light and then later amplifying this image by a development step. The negative was then printed onto a similar light-sensitive material to produce a positive print. Talbot's process became known as the *calotype process.*

The *collodion process* was the next major development in photography. Frederick Scott-Archer is given most of the credit for the wet-collodion process, which he invented in 1851. The process was important in photography in the period 1851–1880. The *wet-collodion* method involved treating high-quality cotton or linen with nitric acid and sulfuric acid. The cellulose fiber of the cotton degrades and becomes nitrated. The product is known as *pyroxylin.* The pyroxylin was then dissolved in an organic solvent, such as a mixture of ether and alcohol. This cellulose nitrate–organic solvent mixture is a transparent, glutinous liquid called *collodion.* Potassium iodide or potassium bromide was then added to the collodion and the mixture was poured onto the surface of a clean glass plate. The word "collodion" comes from the Greek word "to stick" and perhaps refers to this step. Before the organic solvents had evaporated, the plate was dipped into a silver nitrate solution to form a layer of silver bromide or iodide on the surface of the collodion. The exposure must be made before the plate dries. If the collodion is allowed to dry before exposure to light, it produces a freckled appearance on the photograph. The photographer had to carry a small chemical laboratory with him to prepare the film immediately prior to use.

The disadvantages of the wet-collodion process led to the development of the *dry-collodion* process by J. M. Taupenot in 1855. To make a dry collodion plate, it was necessary to wash off the excess silver nitrate from the surface of the collodion layer. However, this was known to ruin the light sensitivity of the plate. Moreover, a preservative was needed for the surface. In the course of finding a preservative, tea, coffee, sugar, honey, albumin, beer, dextrine, gallic acid, and tannin were tried. All of these materials contain organic compounds which were found to react with the silver halide grains and actually increase the light sensitivity of the film again. The outcome of this work led to the practice of using an emulsion (a colloidal suspension of one material in another) for film manufacture. The silver halide grains were precipitated in the collodion layer. To do this, the collodion and zinc bromide or ammonium bromide were dissolved in alcohol and then added to an alcohol solution of silver nitrate to give a finely divided suspension of silver halide.

The disadvantages of the collodion process led to the substitution by Maddox in the 1870s of gelatin as a matrix to support the silver halide crystals. Dry gelatin plates could be stored for a considerable length of time. This important development set the stage for the advent of the commercial photographic industry. From that year onward, many discoveries and refinements in the black-and-white photographic process have led to the field as we know it today.

Color photography got its start on the foundation of experiments reported by James Clerk Maxwell in about 1860. The use of the trichromatic red, green, and blue light filters discussed in Chapter V allowed crude color photographs to be made. In 1869, Louis Ducas du Hauron suggested a number of methods which might produce color photographs, but it was not until 1895 with the work of Joby, and 1907 with the work of the Lumiere brothers, that color photography was clearly demonstrated using additive color layers. Subtractive color relationships are far more useful today and these will be discussed briefly. Inventors such as George Eastman and Edwin Land are notable for major technical innovations in the world of modern photography.

PHOTOGRAPHIC IMAGE MATERIALS

The main steps in preparing a photographic film involve producing a light-sensitive material and affixing it in a suitable form to the film base. Silver halide microcrystals are usually prepared as a colloidal suspension in a gelatin solution by slowly adding with vigorous stirring an aqueous solution of silver nitrate to a gelatin solution containing excess potassium chloride or some other water-soluble halide salt:

$$AgNO_3(aq) + KCl(aq) \xrightarrow{\text{in gelatin}} AgCl(s) + KNO_3(aq)$$

About 10% silver nitrate in water solution is used, and the vigorous stirring helps keep the silver halide crystallites small.

After heating for several hours, the mass is cooled. The cool jelly is placed on a support with a thickness of about one-tenth that of the support. Carefully prepared paper from wood pulp (cellulose) and cotton-lintels is in greatest use as a support, but the first film bases were glass plates. Untreated paper often requires that a layer of barium sulfate (blanc fixe or baryta) be placed on the paper before the emulsion is added. The emulsion will easily stick to $BaSO_4$ but not to the paper alone.

Cellulose nitrate was used as a base for motion picture film in the early years of cinema, but the fire hazard of this material was too great, and its use after about 1912 was rather limited. After 1951 it was banned completely. Cellulose acetate has been used as a replacement for cellulose nitrate in cinema film.

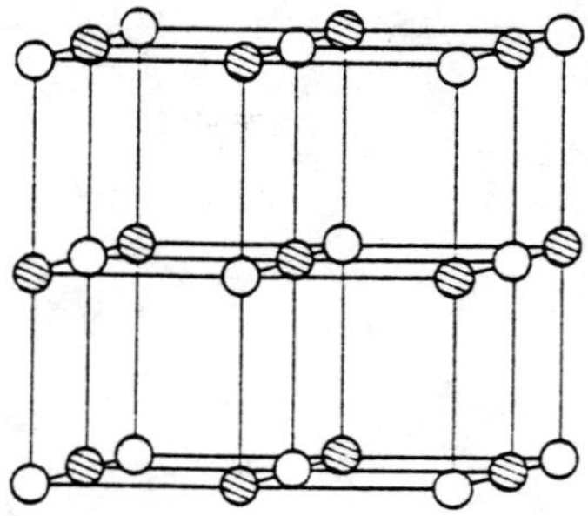

Figure XII-1. The crystal structure of AgCl and AgBr.

The *silver halides* are the most broadly useful light-sensitive materials in photography. The silver halides are simple binary salts because the most common oxidation states of silver ions and halogen ions are +1 and −1, respectively. The oxidation states follow directly from the electron configuration of the neutral elements:

$$Ag = [Kr]4d^{10}5s^1$$

$$Cl = [Ne]3s^23p^5$$

$$Br = [Zn]4s^24p^5$$

$$I = [Cd]5s^25p^5$$

The loss of the $5s^1$ electron on silver leads to a spherical electron core for Ag^+. The gain of a single electron by the halogen leads to the noble gas configuration for the halide ion.

The unit cells of silver chloride and silver bromide are face-centered cubic, as is found in the rocksalt structure of sodium chloride shown in Figure XII-1.

Silver iodide crystallizes in the hexagonal wurtzite lattice when iodide ions are in excess. Wurtzite is zinc oxide. Silver iodide in the wurtzite structure is called β-AgI (Figure XII-2a). When silver ions are in excess, γ-AgI forms in the zincblende structure of zinc sulfide (Figure XII-2b).

The morphology (the shape) of the crystals depends upon the solvent from which they are crystallized. The silver halide crystals formed in a gelatin emulsion usually have a thin, platelike appearance. The number of facets present is crudely related to the sensitivity of the crystal to light. As the number of facets increases, the light sensitivity of the crystals increases.

The solubilities of silver halide salts in aqueous solution depend

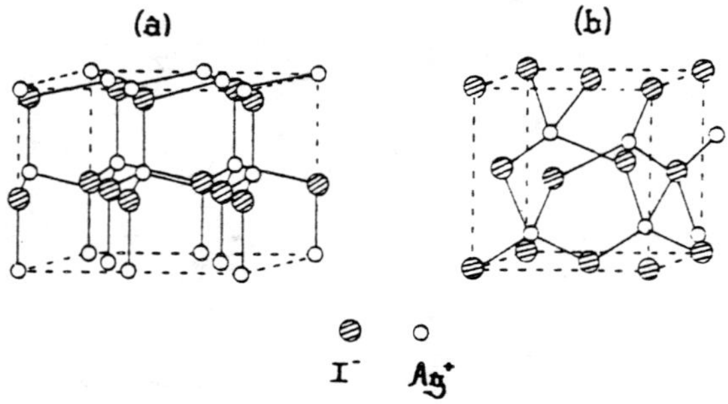

Figure XII-2. The crystal structure of (a) β-AgI and (b) γ-AgI.

markedly upon which halide anion is present. The dissolved silver halides exist in the form of ions:

$$AgX(s) \rightleftharpoons Ag^+(aq) + X^-(aq)$$

This dissociation is such that the product of the concentrations of Ag^+ and X^- at a particular temperature is equal to a constant K called the solubility product constant. If the concentrations of the ions in moles per liter are represented by placing brackets about the symbol, as in $[Ag^+]$, and K is renamed K_{sp}, then

$$K_{sp} = [Ag^+][X^-]$$

As the amount of dissociation becomes smaller, so does the value of the K_{sp}. Table XII-1 gives the K_{sp} of the silver halides at 25°C.

Silver fluoride has little utility as a light-sensitive material, partly because crystals will form only in very concentrated solutions of silver ions and fluoride ions. However, the other silver halides readily crystallize from aqueous solution. The crystals usually form with some defects because of occluded silver or halide ions in the crystal lattice. For example, in

Table XII-1. Solubility Product Constants of the Silver Halides at 25°C

Salt	K_{sp}
AgF	2×10^{2}
AgCl	1.8×10^{-10}
AgBr	4.9×10^{-13}
AgI	8.3×10^{-17}

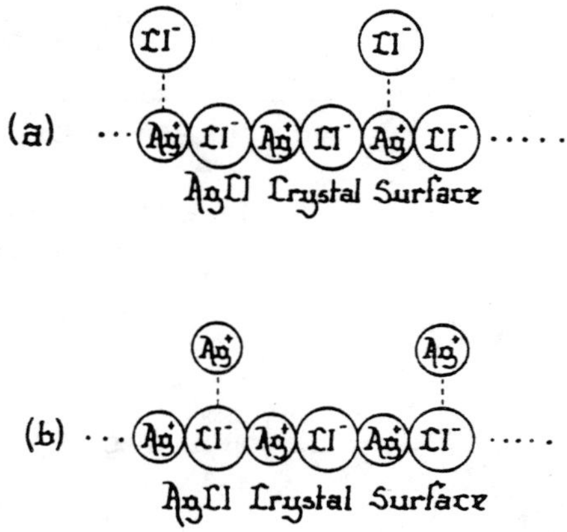

Figure XII-3. Association of ions on the surface of a silver chloride crystal. (a) A "halide body"; (b) a "silver body."

silver bromide a 0.03–0.3% excess of Ag^+ or Br^- ions may be present in 1–0.1 μm-thick crystals. Also, adsorption of ions onto the surface is common because of bumps and pockets of positive and negative charge at the surface. Figure XII-3 illustrates the electrostatic interactions that may be present on the surface.

This buildup of positive or negative charge on the surface has important physical consequences. These positive or negative charges create electrostatic repulsion and prevent the coalescence of the particles. Repulsion of the crystallites favors the formation of colloidal-sized silver halide bodies which are necessary for good image formation. If silver ions are adsorbed on the surface, then the silver halide crystal is more readily reduced in a developing solution, and is also more light-sensitive. Hence silver bodies are most useful in direct positive photographs where development of the image is not intended. In photographs intended for development, silver halide crystals with adsorbed halide are usually used. This state is obtained by using an excess of the potassium halide in the preparation of the crystals.

It is very important to control the size of silver halide grains. As a rule of thumb, the smaller the grain size is, the higher is the contrast, but the lower is its sensitivity to light. Rapid mixing of the silver nitrate and potassium halide followed by a short heating period favors the formation of small silver halide crystals. Longer heating periods tend to increase the light sensitivity and to produce crystals all having about the same size. The light sensitivity is also increased by the presence of silver iodide, which is more light-sensitive than are the other silver halides. If the crystals become too large, then their sensitivity to light decreases again and fogging in the film increases. Table XII-2 summarizes the average grain size as measured by the diameter and area in several types of film.

Table XII-2. The Diameter of Silver Halide Grains in Several Types of Films

Film type	Diameter, μm
High-resolution film	0.05
Motion picture positive film	0.3
Positive-type film	0.63
Fine-grain roll film	0.79
Portrait film	0.88
High-speed roll film	1.09
X-ray film	1.71

Gelatin replaced collodion as a matrix for the light-sensitive silver halide grains after about 1878. However, gelatin serves as more than just a support. It also is found to enhance the light sensitivity of the silver halides. Three very favorable features of gelatin make it outstanding as a photographic film component. First, gelatin tends to disperse the silver halide grains and gives high clarity. Second, cross-linking of the gelatin molecules makes the gelatin insoluble in water at temperatures well above the gelatin melting point. This allows the material to be handled in solution without damaging the image. Third, reactions occur between the silver halide and gelatin components that greatly increase the light sensitivity of the silver halide.

Photographic gelatin can be made from calf hide and ears or from pigskin and bone. Hide sources furnish collagen, while bone sources yield ossein. Collagen and ossein are fibrous proteins which are essentially insoluble in organic solvents, water, and dilute acids and bases. They have the repeating structural unit of a protein shown in Chapter VIII. The organic side chains are any of a number of organic groups characteristic of amino acids. The average molecular weight of collagen and ossein is about 360,000. They form rodlike molecules about 3000 Å in length and 13.5 Å in diameter.

To make photographic gelatin, collagen is treated with a lime slurry at 10–20°C for a period of weeks or months. The product is then neutralized with acid and extracted at elevated temperatures several times. These extractions are called "cooks." The lime has the effect of hydrolyzing some of the polypeptide bonds and breaking down the cross-linking between chains. Gelatin obtained in this process is called "lime-processed." The gelatin contains amino acids linked together in various chain lengths. As a result there is not a uniform structural formula, and the molecular weight varies between 5000 and 400,000. However, almost half of the gelatin is composed of two amino acids, proline and glycine. The sequencing of the protein chains is as follows:

—P—G—R—P—G—R—P—G—R—

About 20% of the gelatin consists of the amino acid proline (P). This is bound to glycine (G), which is 25.5% of the gelatin. Here R is any of 8–10 other amino acids.

The amino acids have acidic sites as a result of the carboxylic acid functional group, —COOH, and basic sites as a result of the amino group, $—NH_2$:

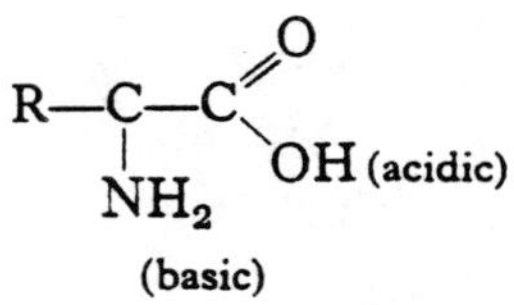

Gelatin has a characteristic *p*H in aqueous solution of 4.5–5.0.

If there is turbidity in the gelatin layer, then stray light will photoactivate silver halide particles that are not directly in the line of the light that produces the image. The net effect of this is a fuzzy picture with low resolution. This generally undesirable condition is called *fog*. Fog can also arise during development of the film.

Reactions between gelatin and silver halides are very important in the photographic process. Gelatin accumulates on the grain surface and a silver halide–gelatin complex forms as shown in Figure XII-4. This reaction protects silver halide grains and prevents them from coagulating. Gelatin reacts with Ag^+ ions to an increasing extent as the *p*H increases. This is because at lower *p*H values the $—NH_2$ groups coordinate with H^+ to give $—NH_3^+$ instead of reacting with the silver ions on the crystal surface as shown in Figure XII-4. At values greater than 8 or 9, however, silver hydroxide, AgOH, and brown-black silver oxide, Ag_2O, form. Gelatin also contains trace impurities which appear to be very important in how gelatin affects the light sensitivity of silver halide crystals. Perhaps

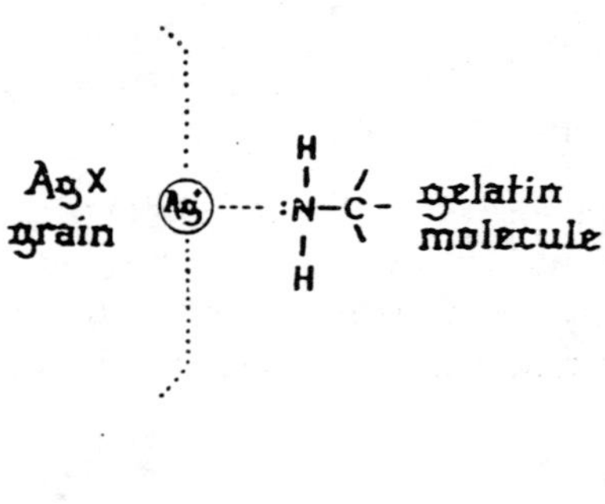

Figure XII-4. Gelatin–silver halide interactions at the surface of the grain.

the most notable impurities are those that act as sensitizers. It has been recognized for many years that sulfur-containing compounds occur naturally in gelatin in amounts of 0.001–0.1%. Many of these impurities result from the sulfur-containing amino acid cystine. Sulfur reacts very rapidly with silver to form Ag_2S. The sulfur-containing thiourea group, $-(NH)_2CS$, can act as a sensitizer even in amounts of only a few parts per million. Thiourea compounds are present in gelatin and form complexes with silver halides. In the presence of base, the complex decomposes to form silver sulfide:

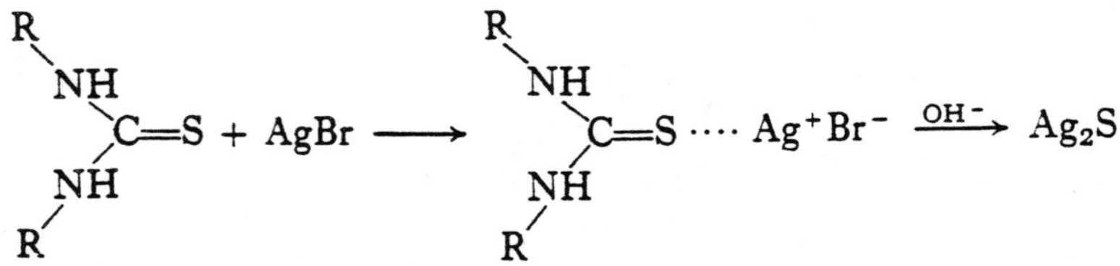

The silver sulfide "tarnish" on silver halide grains occurs mostly on the surface, but a small amount shows up in fracture lines and in internal regions of the grains. The Ag_2S specks are not uniformly spread over the surface. These specks are found to be particularly important in the photochemistry of the silver halide grains.

THE ACTION OF LIGHT ON SILVER HALIDE GRAINS

The basic film structure consists of silver halide crystallites colloidally dispersed in the gelatin matrix. The gelatin is affixed to a support, which is usually paper. Upon exposure to light some silver "metal" forms along with the release of halogen (usually as Br_2 in the case of AgBr). The intensity of the light controls the extent to which this reaction takes place. An image based on relative light intensities is thereby captured by the crystals, but it is invisible until it is developed. This invisible image is called the *latent image* in the early literature, but now the term is used to indicate any change which occurs in individual silver halide grains upon exposure whereby they become developable.

The reaction produced when light strikes the silver halide grains is a simple one,

$$AgX + h\nu \rightarrow Ag + X$$

but the process occurring in the film is an extraordinarily complex one and is still a subject under investigation. If a halogen acceptor is present,

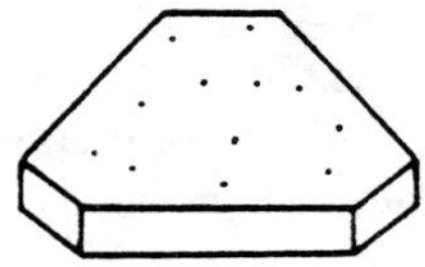

Figure XII-5. A silver bromide grain with patches of photo-produced silver metal.

then the quantum yield of the process approaches one. That is, for each photon of light absorbed, one silver atom is produced. The silver atoms aggregate in the form of patches on the surface of silver halide grains and give rise to the latent image. This ability of the silver atoms to aggregate on the silver halide surface is a unique feature of silver, and it makes silver halides particularly attractive photographic materials. Figure XII-5 illustrates the appearance of a typical silver bromide crystal which has been exposed to light.

We can represent the patches of silver metal as Ag_x, where x is an unspecified number equal to four or more. At least four silver atoms are found to be necessary to create a developable image. Most of the silver atom clusters that produce a latent image are of the Ag_4 stoichiometry, but the value of x depends in part upon the size of the grain. It is probably not correct to think of this silver as silver metal such as would be found on a silver object. Instead it appears to be more like silver ions with electrons occupying the anion vacancies of the silver halide lattice. This particular type of silver is known as *reticular silver*. The entire silver halide grain is developable if one or more patches of reticular silver are present. The patches of reticular silver are therefore called *development centers*. They help to explain why the larger silver halide grains are more sensitive to light than are smaller ones. The larger the grain size is, the greater the probability that a development center will form and thus the greater the light sensitivity of the grain becomes. Also, the larger the grain, the faster is its rate of development.

Evidence is very strong that these highly light-sensitive centers exist *before* the grain is exposed to light. The areas are called *sensitivity centers*, and are more susceptible to reduction by light than are other areas of the silver halide grain. The sensitivity centers exist both on the surface of the grain and to some extent in the interior. The nature of the exposure determines the ratio of the amount of surface image to interior grain image. Low-intensity light with a long exposure time produces mostly a surface image, while high-intensity light with a short exposure time gives a larger amount of the internal grain image along cracks and defects. The latent image of reticular silver exists on the surface and on the interior of the grain, but its intensity decreases with increasing depth into the grain. Exposure characteristics control to a great extent the depth of penetration of the latent image. Later, in discussing development processes, it will be noted that it is possible to bleach out a surface image of a photograph and then redevelop the image using an internal grain developer. The interior grain developer reacts more selectively with Ag_x patches in the interior regions of the grain than does an exterior developer.

The sensitivity centers are believed to be composed of tiny areas of silver sulfide, Ag_2S, which were formed by the reaction between the silver halide and sulfur-containing impurities in the gelatin. The amount of Ag_2S needed to produce sensitivity is very small. One part Ag_2S in 50,000 parts of silver halide gives very significantly enhanced light sensitivity to the crystal. One part Ag_2S in 5000 parts of AgX gives a tenfold increase in sensitivity. But fogging in the film is twice as great as it is with one-tenth as much Ag_2S. A lower limit of approximately 1000–50,000 molecules of Ag_2S is thought to be necessary in the patch to start a development center.

The silver sulfide increases the light sensitivity of the AgX grains by extending the long-wavelength limit of light sensitivity and offering some additional mechanisms for the elimination of the halogen atoms from the crystal. Additionally, and most importantly, the Ag_2S patches serve to concentrate the silver atoms produced when light strikes the silver halide crystal. This *concentration principle* is fundamental to latent image formation because the image could not otherwise form.

There is a need to remove the halogen formed during the exposure. Some sensitizers seem to function as absorbers of halogen, which prevents the halogen from reacting with the reticular silver and producing silver halide again. This halogen absorption may be an additional favorable function of gelatin during silver halide photolysis. As the temperature increases above 25°C, it is found that the amount of silver photolytically produced increases. This is perhaps because the gelatin reacts more rapidly with halogen atoms as the temperature increases.

The actual mechanism of photolysis of silver halide salts can be discussed on a fundamental level. Silver bromide is a good conductor of electricity in the presence of light, and this photoconductance constitutes the primary physical process in latent image formation. The theory of conduction in the solid state is complicated, and it is beyond the province of this survey to discuss the physics of conductors and insulators. However, it is straightforward to extend to the silver halide crystal the formalism described in Chapter VII with regard to energy level diagrams for diatomic molecules. If one considers a diatomic silver halide molecule in the gas phase as having a moderately covalent Ag—Br σ bond, then an energy level scheme as shown in Figure VII-6 can be used. This scheme will have empty high-energy levels lying above a set of low-energy levels filled with electron density. As the AgBr "molecules" aggregate to form an

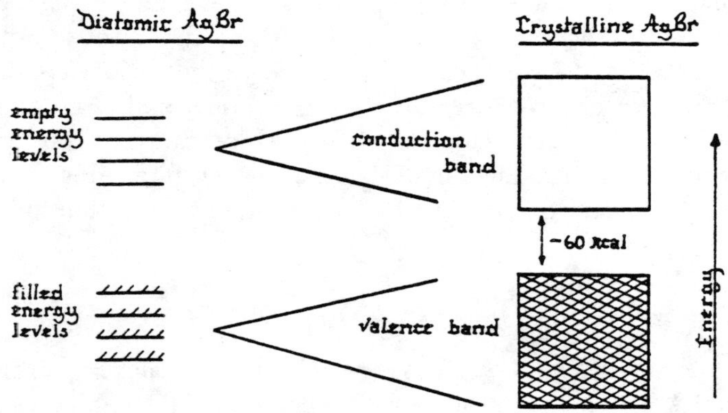

Figure XII-6. The energy levels for a diatomic silver bromide molecule and their progressive change into the valence and conduction bands in the AgBr crystal.

AgBr crystal, the electron-populated lower energy levels increase in number and pack together to form a nearly continuous set of levels called the *valence band*. The empty energy-level set lying above the valence band is called the *conduction band*. Figure XII-6 illustrates the transition of the energy level description for gaseous diatomic AgBr to the solid AgBr crystal.

The energy gap between the conduction band and valence band in AgBr is about 60 kcal, which corresponds to light energy of about 490 nm. In the dark, AgBr is an insulator; that is, no electrical current can be conducted by the crystal. If the crystal is exposed to light, then a current is conducted by the crystal. For this reason AgBr is called a photoconductor. Electrons are excited by the light from the valence band into the conduction band, where they can flow through the crystal. Figure XII-7 shows this process pictorially. The energy gap requires light of at least 490-nm wavelength to produce photoelectrons. Therefore, wavelengths of radiation

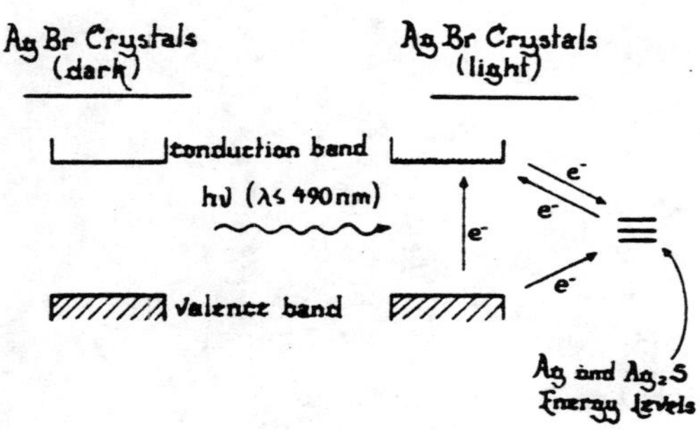

Figure XII-7. AgBr as an insulator before and AgBr as a conductor after exposure to light.

in the blue and ultraviolet regions make silver halide crystals photoconducting and produce silver "metal." It is quite important to note that the silver "metal" and Ag_2S energy levels lie below those of the conduction band, so that electrons in these two materials do not easily return to the silver bromide. Hence these two materials enhance the photoconductivity of silver bromide crystals. The removal of the halogen atom after photolysis is extremely important because the basic light-induced reaction is

$$AgBr \xrightarrow{h\nu} Ag^+ + Br + e^-$$

$$Ag^+ + e^- \longrightarrow Ag$$

Detailed analysis of the process suggests that the halogen atoms and Br_2 molecules can move through AgBr crystals. After photolysis, the bromine atom can migrate in the crystal by electron movement using the mechanism depicted in Figure XII-8. Near the surface of the crystal, the reaction of $Br + Br \rightarrow Br_2$ occurs. The Br_2 leaves the crystal, and the silver atoms are left behind to aggregate.

In 1938 Gurney and Mott considered all of the data available at that time on latent image formation, and proposed a model for the process. Their conclusions are still broadly acceptable. In the way of a summary, the important steps and hypotheses which form the basis of the Gurney–Mott theory will be given. The photochemical changes are theorized to occur in two separate stages. The first stage involves motion of electrons. The second stage involves motion of ions. Four hypotheses fit into this framework.

1. The photoconductivity and dark electrical conductance are due to the release of electrons into conduction bands because of absorption of energy by the silver halide crystal. These electrons and any excess silver ions present in the interstices of the crystal can move freely in the crystal.

2. There exist sensitivity specks on the surface of the crystal which can trap the photoelectrons and acquire a negative charge. These specks are mainly silver sulfide patches on the silver halide crystals which are created by the reaction of components of the gelatin and the Ag^+ ions.

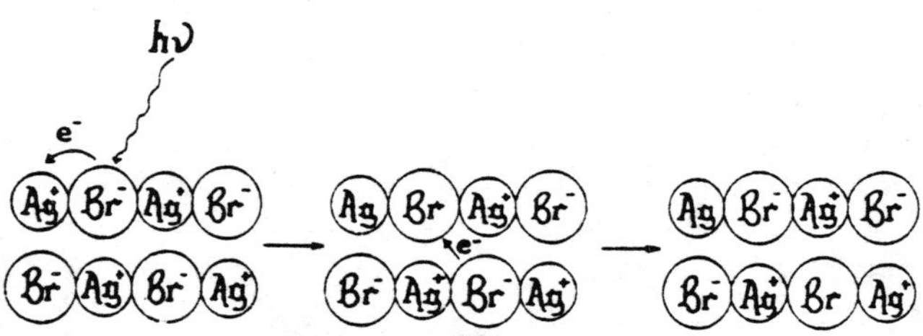

Figure XII-8. Migration of Br atoms in AgBr crystals.

3. Silver halide crystals are ionic conductors containing interstitial cations, vacant cation sites, and photoelectrons in the lattice. The electrons combine with silver ions to form silver "metal," which aggregates and grows, provided that electrons continue to be supplied by the light induced reaction:

$$Ag^+ + e^- \longrightarrow Ag \xrightarrow{+Ag} Ag_{(aggregates)}$$

4. Positive holes remain after electron movement and migrate to the surface of the crystal, where they convert bromide ions, Br^-, to bromine atoms, which escape.

There is little argument about the basic correctness of the Gurney–Mott model of latent image formation; however, some further steps occur after this point. For example, the released halogen atoms react with the gelatin and therefore do not recombine with the silver "metal" formed. When good halogen acceptors are present, the latent image occurs mostly on the surface. When poor halogen acceptors are present, the halogen atoms can react with the silver "metal" to destroy the surface image. In this case, the image predominates in the internal regions of the grain.

SPECTRAL SENSITIVITY

Silver halides in gelatin only have significant sensitivity to radiation in the ultraviolet region and the short-wavelength regions (blue–violet) of the visible spectrum. Untreated silver halide grains are not sensitive to green, yellow, orange, and red light. This behavior is unsatisfactory in photography. The composition can be made to be more sensitive to longer wavelengths by the use of spectral or optical sensitizers. Sensitizers are usually organic dyes which absorb light of longer wavelengths and transfer the energy to the crystal. It is possible to sensitize silver halide grains not only toward all of the visible wavelengths, but also toward infrared energies. Infrared sensitizers are used in infrared film.

There have been many attempts at making good sensitizing dyes since H. W. Vogel in 1873 first noted that the presence of dyes in the emulsion could alter the intrinsic wavelength sensitivity of silver halide salts. Vogel succeeded in making silver halide photographs with light wavelengths up to 600 nm by adding dyes to the silver halide–gelatin emulsion. Collodion was commonly sensitized with "silver eosine," which is silver tetrabromofluorescein (see top of page 262).

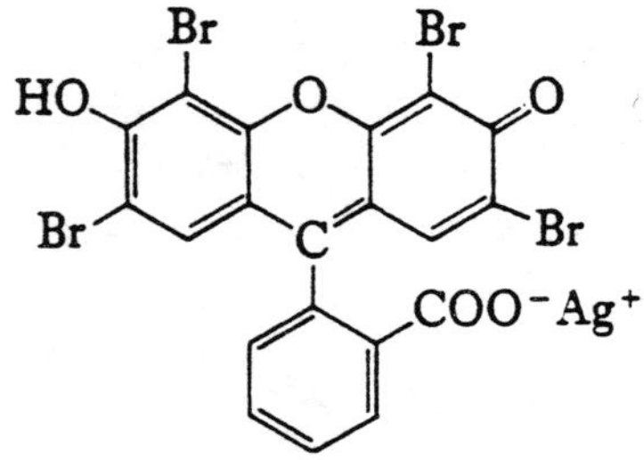

(Silver eosine)

It imparts sensitivity to silver salts in the yellow and green regions of the visible range.

By about 1904, photography with light in the full visible range had become routine. Without sensitizing dyes, it is safe to say that photography would not have found the general acceptance that it has. The terms *orthochromatic* and *panchromatic* came into use in reference to the effects of certain dye compositions which brought about wide spectral sensitivity.

The cyanine dyes discussed in Chapter VIII are of greatest interest in the photographic industry. These are flat-shaped molecules which stack against one another perpendicular to the crystal face. They are held onto the surface by van der Waals forces. The photoabsorbing properties of these dyes, which are salts and carry an overall electrostatic charge, arise from the following chromophoric resonance forms:

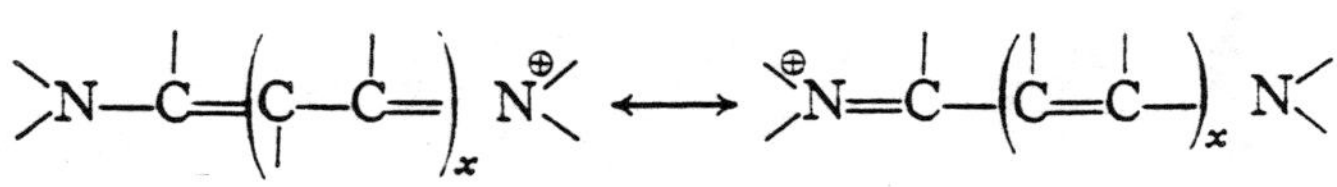

or

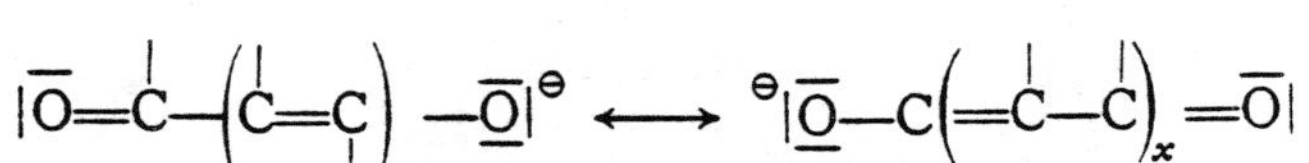

Pinacyanol was very popular for imparting spectral sensitizing effects in films after 1904 and is an example of a cyanine dye:

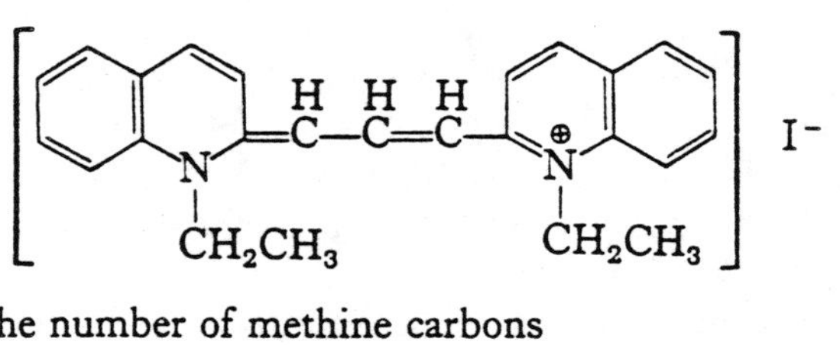

Variation in the number of methine carbons

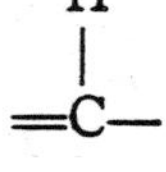

in the center of this molecule gives rise to various color changes. In general, as the number of methine groups increases, the wavelength of maximum light absorption by the dye increases. The desired result is to have dyes which absorb light over the entire visible range. To achieve this, more than one dye is often used at a time. There are thousands of structural variations that have been tried to improve upon and alter the properties of sensitizing agents.

DEVELOPMENT AND FIXATION OF THE IMAGE

After producing the latent image in the development centers of silver halide crystal grains, it is necessary to intensify this very weak image and to preserve it. Intensification of the image is the processes of development, and preservation of the image is the process of fixation. Development and fixation of the negative are not involved with the light sensitivity of the silver halides, but no discussion of photography would be complete without a brief mention of these important steps for obtaining a visible image.

The developing process amplifies the latent image about 10^9 times. Developing agents are reducing agents which effect a reduction of the light-exposed silver halides to silver metal. Reducing agents are able to supply electrons to the Ag^+ ions:

$$Ag^+ + \text{developing agent} \rightarrow Ag$$

This reaction occurs preferentially on those silver halide grains that have been exposed to light. The probability of development is near one on crystals of the silver halides that have a latent image. It is near zero on crystals where no silver metal from photolysis is present. This fact is critical in the photographic process. If reduction occurred at all grains, the latent image would be developed, but so would all other grains, and the film would show no contrast. It is clear that there must be a greatly different rate of development for those grains that already have silver "metal" specks on them resulting from the photochemical action of light. There is no simple answer as to why this is the case, although the phenomenon has been extensively investigated. The silver metal may act as a catalyst for further silver metal formation, and development may commence by catalytic action at the Ag/AgBr interface.

Gurney and Mott examined all of the facts available to them in the late 1930s and proposed that the silver metal specks act as tiny electrodes. Figure XII-9 qualitatively illustrates some of the features of their model. The first step is the approach of the developing agent to the AgBr crystal in a gelatin emulsion. An electron is released by the developing agent to

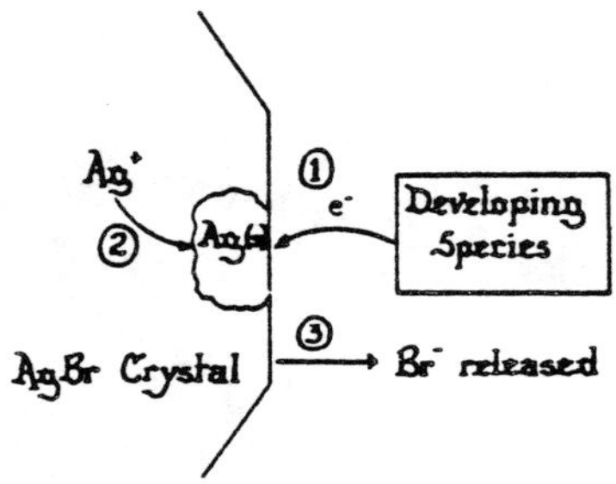

Figure XII-9. The steps involved in development of the latent image according to Gurney and Mott.

the silver metal. In the second step, an interstitial and/or lattice silver ion is attracted to the Ag/AgBr interface because of its newly acquired negative charge. Reduction of that Ag^+ ion to silver metal occurs. Because of the unfavorable electrostatic condition now present in the crystal, a Br^- ion is released from the crystal in step 3. This process is repeated many times, with the effect of building up silver metal at the AgBr/Ag interface. The details of the process are extremely difficult to investigate experimentally, and few would argue if steps one and two were reversed. That is, a silver ion could affix itself to the silver "metal" of the latent image. The positive potential produced at the silver aggregate would then attract electrons from the developer and lead to reduction.

The above description of the developing process refers to what is known as *chemical development*. That is, the film material is placed in a developing solution of controlled *p*H with the effect of reducing the AgBr grains in the film. However, another process of film development known as *physical development* exists. There is no important significance to the terms "physical" and "chemical" development. Both types of development are chemical in nature. Physical development occurs when the silver ions present in solution are reduced and then are deposited on the Ag "metal" aggregates formed by the action of light. Often this involves adjusting the *p*H of a solution of a silver salt and an organic developing agent to be favorable for precipitation of silver metal on the development centers, rather than initiating spontaneous precipitation from solution. In normal photographic development, it is believed that both physical and chemical development occur, with chemical development being most important initially and physical development becoming increasingly important with time.

Physical development is also used in *postfixation development*. In postfixation treatments, the latent image is initially fixed in the film. Development is then effected by precipitation of silver metal on these fixed silver metal specks which occurred as a result of the action of light. The exposure time necessary to give a developable image by physical means is greater than by chemical means by an order of magnitude. Also, the time needed for physical development is 2–10 times longer than that needed for chemical development.

As already noted, the latent image can be formed on both the surface of the grain and the interior of the grain. The internal and surface development centers in silver halide grains are not uniformly affected by all developers. Some agents work faster as surface developing agents than as internal developing agents, while others work equally well as both. In general, surface developers contain no silver halide solvent, so the internal region of the grain is not easily attacked by the developer. Typical surface developers are ferrous oxalate, sulfite–free catechol, and Metol–HQ and Metol–ascorbic acid mixtures. For internal grain development, silver ion complexing agents, such as SO_3^{2-}, $S_2O_3^{2-}$, and I^-, are added to Metol–HQ or Metol–ascorbic acid developing solutions.

It is often possible to bleach out a developed and fixed image from a film and then redevelop it. This principle may have application in the restoration of old photographs. The bleaching step removes the old silver metal which, in physical development, may have completely covered a silver halide grain. This layer of metal protected the grain from the fixing solution so that silver halide crystallites containing the latent image still remain in the emulsion even after fixation. After removal of this silver metal sheath, the grain can be redeveloped and fixed using an internal grain developer. Not all photographs can be treated in this manner. The procedure should be tried only with an experienced hand. A number of agents have been used to oxidize the silver in a photograph. For example, chromic acid (H_2CrO_4), copper bromide ($CuBr_2$), permanganic acid ($HMnO_4$), ammonium persulfate [$(NH_4)_2S_2O_8$], and ferric chloride ($FeCl_3$) all have found application as oxidants. The sensitivity to bleaching is lessened if the silver metal has become converted to Ag_2S or has been toned (see below) by the photographer.

Following the development step, a swollen gelatin layer containing developer oxidation products and unreacted developer is present. The next steps in the film treatment involve rinsing, fixation, hardening, washing, and drying of the photographic material.

The rinsing step employs water or an acid stop bath. The stop bath is a solution of a weak acid, most commonly acetic acid, which has the purpose of lowering the *p*H in the gelatin layer below 5.5. This neutralizes the developer and stops the developing action. The stop bath also dissolves some calcium salt scums which can occur on the film as a result of Ca^{2+} ions in natural water. The stop bath has the further effect of protecting the fixer bath and decreasing the swelling of the gelatin layer.

Fixation of the developed image into a permanent entity is the next step. In addition to developed silver metal, the emulsion contains residual silver halide which was not affected by the developer. Since these silver halide crystallites are still sensitive to light, they must be removed in order to stabilize and make permanent the image. The fixing bath serves to remove this undeveloped silver halide and therefore must contain some agent L which will complex the Ag^+ ion and form soluble AgL_n complexes. These complexes may carry electrostatic charge, depending upon the value of n. The thiosulfate anion, $S_2O_3{}^{2-}$, has been found to be a very suitable complexing agent, and has been used since the earliest days of photography. The old name for $Na_2S_2O_3$ is "hyposulfite of soda." The term "hypo" came from this and has been retained over the years to refer to the fixing bath.

Following fixation, the film must be washed to remove the reagents which would otherwise ruin the film. This step is very important to the longevity of the photographic material. With rapid agitation for 6 min, only 0.02% of the silver thiosulfate ions remain in the emulsion if glass plates or cellulose acetate is used as a support. Washing for a longer time is necessary for paper prints. The initial rate of removal of the ions is rapid, but the rate slows with time. Hypo eliminators are sometimes used to increase the rate of fixer removal. Hydrogen peroxide at $pH = 9.5$ is a common eliminator; however, it still cannot be relied upon to remove all of the hypo.

REDUCERS, INTENSIFIERS, AND TONERS

Additional treatments of photographic images can be carried out after fixing and rinsing in an effort to improve their appearance and color. Reducers are oxidizing solutions that diminish the optical density resulting from overexposure or overdevelopment. Depending upon the oxidizing agent, there can be a proportional reduction of all silver metal density, a superproportional reduction of higher density relative to lower ones, or a subproportional reduction of lower intensities relative to higher ones. Intensifiers are often used on underexposed or underdeveloped films. They convert the silver metal to a more opaque compound or use physical development to precipitate silver metal on silver grains already present in the film.

Toners can impart color to the silver image, but they can offer protection as well. Of perhaps most importance are the selenium and gold toners which protect the silver metal from degradation by environmental agents. Gold forms a protective coating on the silver, and because of the extreme inertness of gold, it offers excellent protection. Conversion of silver metal to Ag_2S by sulfur gives purplish-brown or brown-yellow images. Selenium and tellurium give Ag_2Se and Ag_2Te and create dark brown to black images. Dyes can also be used as toners if the silver metal can absorb them.

DEGRADATION OF PHOTOGRAPHIC MATERIALS

Light-induced reactions in silver halides produce the latent image in a photographic material. As long as no light-sensitive inorganic material remains in the film, further light exposure causes little change in the silver. However, the organic materials present, such as gelatin, collodion, albumin, and the paper support, are somewhat sensitive to light. As a matter of course, direct sunlight and some types of fluorescent lamps should be avoided because of their relatively high ultraviolet content. This is especially important for color prints and slides which are colored by organic dyes.

Two of the greatest concerns in the maintenance of photographs for permanent record are not related to light. These are (a) improper fixation and (b) atmospheric contamination. First, improper fixation is probably the most important cause of deterioration of film material. No more than 0.001 mg/in.^{2} of residual $S_2O_3^{2-}$ can be tolerated for permanence. The common washing procedures with water can leave as much as 0.1 mg/in.^{2} of $S_2O_3^{2-}$ in the film. Consequently, the original washing procedures are as crucial a step as any to a conservator or curator of photographs. Failure to remove the silver salts causes stain in the highlights and unexposed areas. Where $S_2O_3^{2-}$ remains in the film, gradual sulfiding of the silver image to Ag_2S will be observed. In many cases the film progressively turns to a brownish-yellow color and fades. Coarse-grain films are less sensitive to residual hypo than are fine-grain films.

Second, air pollutants generally work havoc with art materials and historically interesting objects and structures. In the case of photographic images, sulfur dioxide (SO_2) and hydrogen sulfide (H_2S) are the most villainous pollutants. They convert the silver image to a sulfide. Selenium and gold toning offer some protection against this form of degradation.

Two additional factors of concern in the stability of photographs are temperature and humidity. Increases in either or both of these will increase the rate of degradation. It is advisable to store photographic materials at as low a temperature as is reasonable, such as near 0°C or lower, and at controlled humidity of 15–30%.

Early paper photographs (1835–1855) usually fade because of imperfect fixing and washing. Daguerreotypes (1839–1855) show discoloration and imperfections in the plate as a result of airborne sulfur compounds. Difficulties with hypo are, of course, not a problem with Daguerreotypes, because this method of fixation was not used. Early plate negatives contained albumin (1848–1855) and collodion (1851–1880) as a support for the silver halides. These often show some discoloration and flaking from the glass plate. The cellulose nitrate (1858–1951) used in motion picture film and in collodion is subject to spontaneous and inevitable degradation. It first yellows, then turns sticky, and finally becomes a powder. The material can spontaneously ignite above about 50°C. Gelatin plates (1878–present) are more stable than collodion or albumin if they are stored properly.

Colored motion picture film and slides suffer from photochemical fading or color changes as a result of projection. Projection bulbs are high-intensity incandescent light sources which produce infrared and some ultraviolet energy. Ultraviolet and infrared filters should be used to protect colored film from the projection bulb. In practice, it is the cyan dyes that are often found to be the most susceptible to photochemical change.

COLOR PHOTOGRAPHY

Reproduction of both image and color together involves, as might be expected, a very much more complex problem than does image reproduction alone. The earliest color tinting of black-and-white photographs involved the use of pyrogallol stains of the gelatin during development of the image with pyrogallol. The pyrogallol stained the gelatin in and around the exposed silver halide grains to a much greater extent than the other areas. Hence it gave a brownish cast to the image. However, the basis of true color photography is the trichroma system discussed in Chapter V. There it was noted that an independent recording of the red, green, and blue constituents of any color composition followed by a subsequent recombination can reproduce the original color. Using the eye as the detector, a perfect match of the original color to the synthesized color need not be made, although the better the match is, the more impressive is the image.

Historically, the additive tricolor scheme is of interest. This mechanism of color reproduction is conceptually easy to grasp, but it is difficult to apply in practice. If two or three transparent films of the additive primary colors red, green, or blue are superimposed, then black is produced. If other colors are to be produced by the additive primaries, then they can be done so only by side-by-side spatial location of the primary elements. Depthwise superimposition will not give satisfactory results. Instead, if the colors are viewed on a white surface by *transmitting* three white light

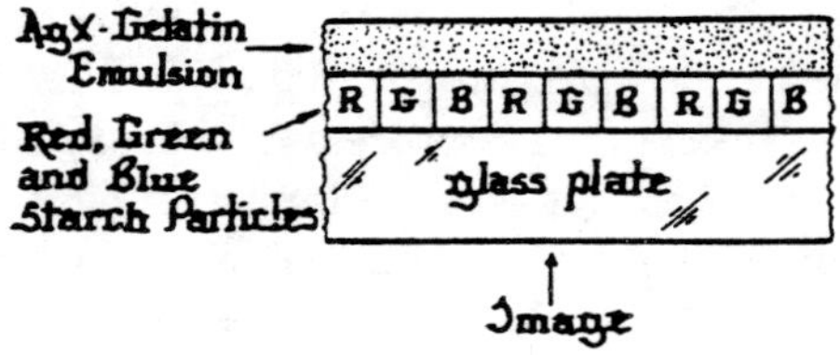

Figure XII-10. A cross section of the basic design of Autochrome color plates.

sources separately through three transparent films of red, green, and blue color, then total absorption of light does not occur. Where red and green light overlap, yellow is formed, and where green and blue overlap, cyan is formed, etc. However, photographic film involving transmission of light from three projectors is not realistically practical.

The Autochrome process devised by the Lumiere brothers and marketed in 1907 employed additive color relationships using the lateral spatial orientation of the coloring elements. They used microscopic-sized starch grains which were dyed red, green, and blue. Equal amounts of these batches were mixed together to give a gray powder. The powder was then made to adhere to a glass plate using varnish. The silver halide–gelatin emulsion was rolled on over this film as shown in Figure XII-10. A picture was taken through the glass support, developed by a reversal process into a positive, and fixed. Reversal development was achieved by developing the exposed silver halide and then bleaching out the developed silver. The remaining silver halide was then developed using a fogging developer. This produces the reverse of the image. If viewed from the silver halide–gelatin side, the positive silver image leaves open to view those color specks through which light had passed during exposure. The silver metal obscures those areas that did not receive exposure, as shown in Figure XII-11. If blue light struck the plate, the green and red particles absorbed the light, but the blue particles scattered and transmitted it. Hence, a similar color is obtained to that which actually existed in the photographed object. After World War I, gum Arabic was used instead of starch grains. Gum Arabic gave a more homogeneous color pattern in which the colored areas did not have a grainy appearance.

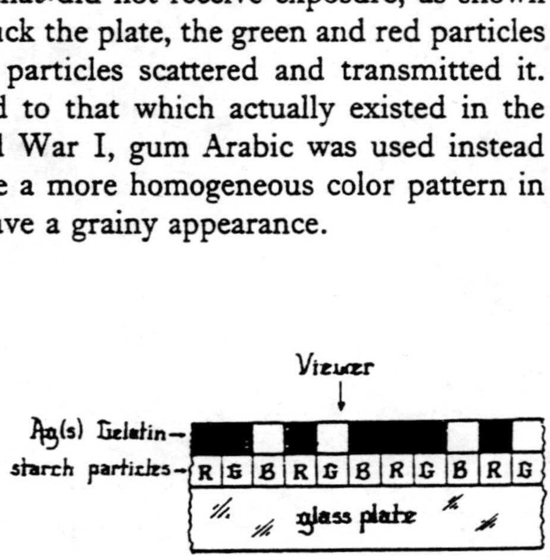

Figure XII-11. A cross section of the Autochrome plate after development.

Other early color techniques included the Dufay color film process used after 1934, in which dyed parallel lines were placed on the film, and the Keller–Dorian process marketed in 1925. The Keller–Dorian process, called Kodacolor cine film, used light filters on the camera lens which focused light on a film embossed to form minute "lenses" running the length of the film. After development, the film looked as though it were only black-and-white imagery, but by using a projector with the same lens arrangement as that on the camera, the picture could be projected in full color.

In addition to the above-mentioned disadvantages of the additive color relationships, other problems stem from the fact that the colors are often not bright, and that the film cannot be enlarged without showing some aspects of the film structure. Another technique of color photography was needed which was more routine for the photographer and more pleasing to the viewer. This came with the use of the subtractive primary methods.

In order to match colors by superimposition of the trichromatic series, it is necessary to resort to the subtractive primary scheme. This would allow covering the film with colored layers such that all colors as well as black can be observed together. Obviously, the flexibility for making color print reproductions would be much greater. The subtractive colors are yellow, cyan, and magenta; these are obtained by subtracting the additive primary colors from the visible spectrum. Two-thirds of the spectrum is transmitted and one-third is absorbed as was described in Chapter V. The basis of modern color photography is the subtractive color scheme.

The problem confronting the maker of color film was how to prepare and superimpose the colors, and how to get the right amount of the proper dye in the proper layer. A feasible arrangement for color production is shown in Figure XII-12. A photographic material is constructed consisting of a support and three light-sensitive layers each containing silver halide salts. The blue-sensitive dye lies on top, the green-sensitive layer lies in the middle, and the red-sensitive layer lies on the bottom. The silver halide

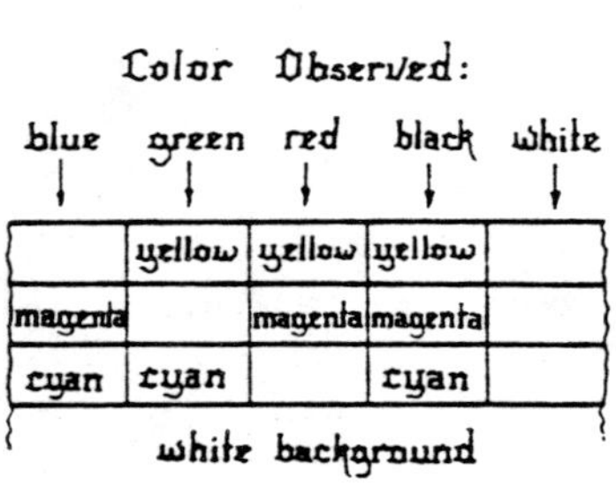

Figure XII-12. Color production by layering of the subtractive primaries (yellow, magenta, and cyan) on top of one another.

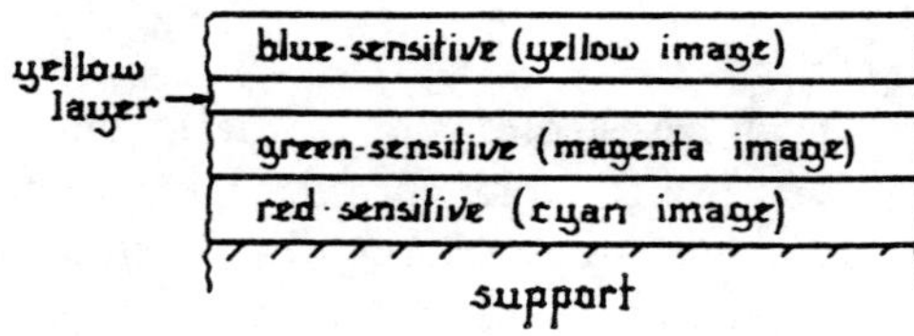

Figure XII-13. A cross section of the dye layers in typical color film.

salts are sensitive to blue and ultraviolet radiation. Therefore, sensitizing dyes must be added to the green- and red-sensitive layers. Since both the green-sensitive and red-sensitive emulsions retain blue sensitivity, the blue light must be prevented from reaching and activating the lower layers. To achieve this, an extra layer of yellow dye is placed between the top layer and the two lower layers as shown in Figure XII-13. The film is exposed to light in a camera. The yellow layer is removed by bleach in the processing of the film. The latent image is developed in a standard black-and-white developer to convert the exposed silver halide to silver metal. Then the remaining silver halide is developed in the blue-sensitive layer using a color developer to form a yellow dye. The same is done for the green-sensitive layer to form a magenta dye, and for the red-sensitive layer to form a cyan dye. Finally, all of the silver metal is removed. What remains is a silver-free image containing the colors of the original object, This development procedure is the basis of the Kodachrome, Ektachrome, and Agfacolor processes.

The azo-bleach developing process is somewhat different. Here, after the latent image has been developed into metallic silver and the residual silver halide removed, what is left is shown in Figure XII-14. This metallic silver in the layers is now used for the image-wise reductive destruction of dyes in the respective layers. Because the dyes are destroyed where the exposure occurred, a positive color image remains. The bleaching action of silver on the azo dyes used in the layers is given by the reaction

$$X—N{=}N—Y + 4Ag + 4H^+ \rightarrow X—NH_2 + NH_2—Y + 4Ag^+$$

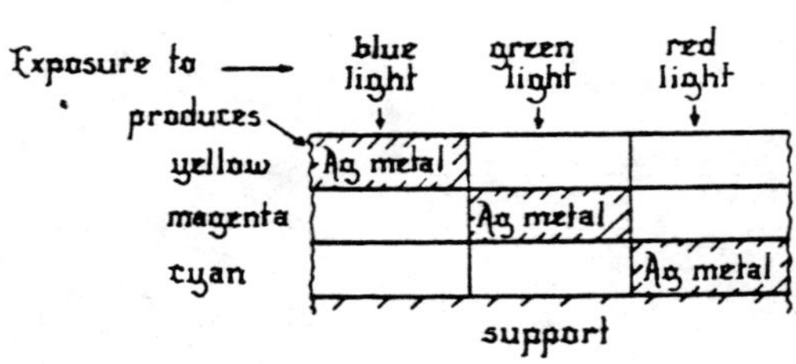

Figure XII-14. A cross section of a developed and fixed film showing the silver metal content of each layer according to the light exposure.

Figure XII-15 shows the positive image after destruction of the dye and of the silver metal.

In all cases, the color of the image determines the color of the film layer because the color forms in proportion to the exposure.

In color film processing, the development step is obviously much more complex than in black-and-white processing. It is an integral part of color formation in most cases. In the above examples of color development, a black-and-white silver halide developer was first used. At present, however, most color pictures are developed by chromogenic methods. Here the dye-forming process begins in the developing bath. The developing agent in its oxidized form condenses with certain compounds, called *couplers*, contained in the film layers to form yellow, magenta, and cyan dyes during development according to the reaction

$$\text{developing agent} + Ag^+ + \text{coupler} \rightarrow \text{dye} + Ag + H^+$$

The silver metal and acid are then removed. An example of a developing agent reacting with a coupler to form a dye is

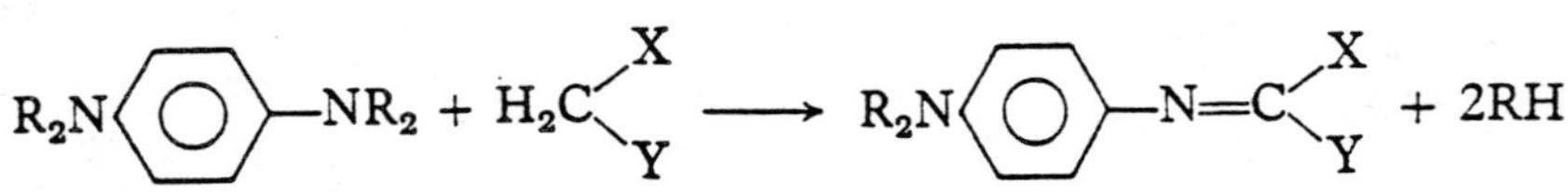

developing agent — coupler in film layer — colored dye

Some representative dyes existing in developed color film are

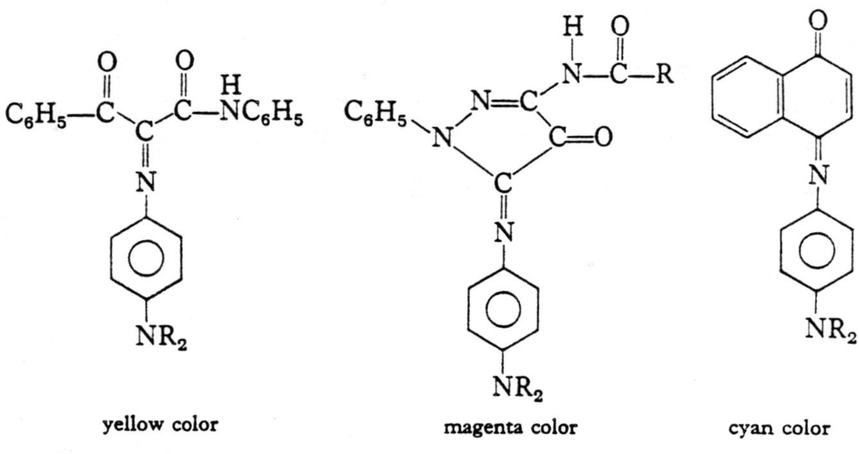

yellow color — magenta color — cyan color

The *para*-aminobenzene portion of these molecules comes from the

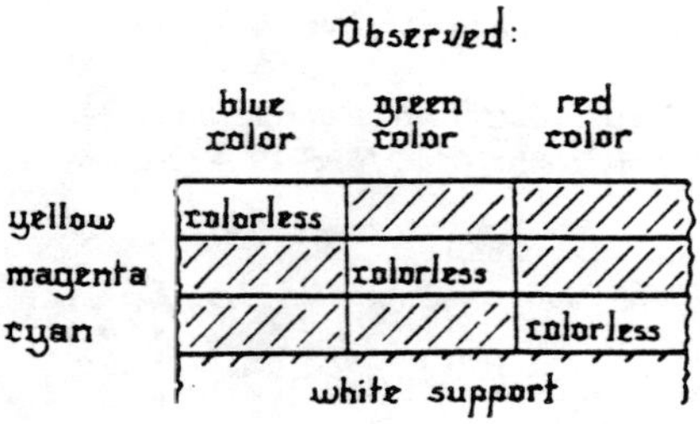

Figure XII-15. A cross section of a final transparent print after development by the azo-bleach process.

developer, and the remainder is the coupling agent originally present in the film layer.

Photographic processes have taken on a new dimension in complexity with the advent of instant color film. For example, the Eastman-Kodak process announced in 1976 uses no dye couplers in the film, but no less than nineteen layers are used to produce imaging. The various layers contain dye releasers, dyes, scavengers, uv absorbers, activators, etc. The Polacolor process developed by the Polaroid Corporation is similarly complex. Restoration of such films will be virtually impossible.

In general it is probably safe to say that conservation of color prints will take more the line of preservation rather than restoration. The composition of color film is extremely complex, and the chemical reactions and degradation patterns are difficult to establish. Color prints will unquestionably be sensitive to light, particularly to the high-energy radiation in the blue and ultraviolet. Film storage temperatures should be as low as possible to slow the diffusion of the layers into one another and to prevent swelling of the gelatin matrix by humidity.

BIBLIOGRAPHY

Adelstein, P. Z., Graham, C. L., and West, L. E., Preservation of motion picture color films having permanent value, *J. SMPTE* **79**, 1011 (1970).

Baines, H., *The Science of Photography*, Wiley, New York (1970).

Cox, R. J., *Photographic Gelatin*, Academic Press, New York (1972).

Eaton, G. T., *Photographic Chemistry*, Morgan and Morgan, Hastings-on-Hudson, New York (1965).

Eaton, G. T., Preservation, deterioration, and restoration of photographic images, *Libr. Q.* **40**, 85 (1970).

Hunt, R. W. G., *The Reproduction of Color*, 3rd ed., Wiley, New York (1975).

Katz, J., and Fogel, S. J., *Photographic Analysis*, Morgan and Morgan, Hastings-on-Hudson, New York (1971).

Klein, H. O., *Collodion Emulsion*, Penrose, London (1905).

Mees, C. E. K., and James, T. H., *The Theory of the Photographic Process*, Macmillan, New York (1966).

Mees, C. E. K., *The Theory of the Photographic Process*, Macmillan, New York (1954).

Ostroff, E., Conserving and restoring photographic collections, *Mus. News* **1974** (November), 42.

Ostroff, E., Preservation of photographs, *Photogr. J.* **1967** (October), 309.

Procedures for Processing and Storing Black and White Photographs for Maximum Possible Permanence, East Street Gallery, Grinnell, Iowa (1970).

Sheppard, S. E., *Gelatin in Photography*, Van Nostrand, New York (1923).

Tennant, J., *Developers and Development*, Tennant and Ward, New York (1912).

Index

Absorption
of gases, 59
of liquids, 59
molecular orbital theory of, 143–147
by paint, 87, 88
relation of color to, 214–215
resonance model of, 133
of solids, 59
spectra, 145
spectral, 67
Acid dyes, 156
Agate, 112
Albumin
deterioration of, 268
light sensitivity of, 267
Alexandrite effect
in alexandrite, 69, 215
definition of, 215–216
in glass, 241
Alizarin: *see also* Madder
as mordant, 158
natural, 162
as pigment, 231
structure of, 162
synthetic, 162, 170
Aluminum, 109
Ambernia, 243
Annatto, 165
Anomalous dispersion
definition of, 91
relation to hiding power, 92
relation to normal, 12, 92
Anomalous refraction, 81
Aristotle, 63
Asphaltum, 231
Atom
electron notation of, 122
structure of, 120–121
Auxochrome
definition of, 149
function of, 151–152
table of, 150
Aventurine, 240, 242
Aza[18]annulene, 167–168
Azo dye, 157
as disperse dye, 159
fading of, 177
phototendering of, 183
stability of, 166
structure of, 166
Azurite
color change in, 107
desaturation of, 96
properties of, 223

Baeyer, Adolf, 170
Barbizon School, 171
Baryta, 220
Basic dye, 157
Becquerel, Henri, 24
Beer's law, 61, 236
Beer–Lambert law, 61
Bethe, Hans, 206
Biaxial crystal, 81
Blackbody, 16–18

Blanc fixe: *see* Baryta
Blood, 167
Bloom
  in camera lens, 55
  in varnish, 104
Blue bice, 223
Blue verditer, 223
Bohr, Niels, 120
Bone black: *see* Carbon black
Bouguer, Pierre, 33
Brazilwood
  history of, 163
  as mordant, 158
  properties of, 163
Brewster, David, 50
Brewster angle, 51
Brightness, 74
Bronze powder, 232
Burnt sienna: *see also* Sienna
  as glaze, 106
Burnt umber: *see* Umber

Cadmium orange, 215
Cadmium red
  as glaze, 106
  properties of, 231
Cadmium yellow
  color in, 215
  particle size of, 96
  properties of, 229, 231
Calendering, 114
Candela, 35
Candlepower, 34
Carbon black
  properties of, 231
  with vermilion, 233
Carbonyl dye
  as acidic dye, 157
  as disperse dye, 159
  photoreduction of, 181
  phototendering in, 183
  as vat dye, 169
Carotenoid, 167
Cassius, Andreas, 242
Cellulose: *see also* Cotton, Paper
  photooxidation of, 186
  structure of, 185
  yellowing of, 186
Cellulose acetate, 250, 266
Cellulose nitrate, 250
Cennini, 108
Ceramics
  effect of radiation on, 13
  opalscence in, 95
  reflection from, 101
  repair of, 88
Cerulean
  color change in, 107
  d–d transitions in, 212
  with madder, 234–235
  properties of, 224
Chalk
  covering power, 86
  properties of, 220
  uv reflection from, 117
Chalking
  of lithopone, 221
  of titanium dioxide, 221
  of zinc white, 220–221
Charcoal: *see* Carbon black
Charge transfer transition
  color production by, 204
  in inorganic compounds, 203
  intensity of, 204–205
  ligand-to-metal, 204, 212–213
  metal-to-ligand, 204
  metal-to-metal, 205
  in organic compounds, 148
Chlorophyll, 167
Chromaticity point, 77
Chrome yellow, 229; *see also* Lead chromate
Chromium
  electron configuration of, 208
  in glass, 240
  reflection from, 109
Chromophore
  definition of, 149
  function of, 151–152
  table of, 150
Chrysocolla
  properties of, 225
  refractive index of, 82
Cinnabar: *see* Vermilion
Cobalt
  d–d transitions in, 210–211, 240
  electron configuration of, 209
  in glass, 239–240
Cobalt blue
  d–d transitions in, 212
  desaturation of, 96

Cobalt blue (*cont.*)
as glaze, 106
with madder, 234–235
properties of, 224
Cobalt green
d–d transitions in, 212
properties of, 227
Cobalt violet
d–d transitions in, 212
properties of, 231
Cochineal
history of, 160, 166
as mordant, 158
as pigment, 231
properties of, 160
Collagen, 254
Collodion
deterioration of, 268
dry, 249
light sensitivity of, 267
wet, 249
Color
additive primaries of, 71–72
in ancient glass, 55
brightness of, 74
CIE system of, 76
complementary, 64–66
defects in rosette of, 65–66
effect of light on, 70
electronic origin of, 144, 151–152
of feathers, 55, 58, 94, 105–106
in glass, 237–244
history of, 63
of landscape, 94, 107, 204–205
luminance factor of, 76
of lusterwares, 55
matching of, 77
of metal, 55, 108–111
mixing of, 74, 77
of mother-of-pearl, 55
Munsell system of, 75–76
of oil films, 55
of oyster shells, 55
in paint glaze, 106–107
of peacock feathers, 55, 58
perception of, 67–68
in photographic film, 73, 268–269
principle of, 64
from prism, 11
purity of, 76
reflectance of, 64, 66–67
Color (*cont.*)
rosette, 65, 73–74
of sky, 93–94, 107
of soap bubbles, 55
subtractive, 72–74
of tempered steel, 55
theory of, 11
Color center
definition of, 213
in glass, 237–238
Color photography
history of, 250
principles of, 268–273
Color temperature
application of, 18
definition of, 17
of fluorescent lamp, 18, 28
of incandescent lamp, 20
of paintings, 69–70
of sunlight, 20
Conduction band, 259
Coordination number, 206
Copal, 104
Copper
in glass, 205
electronic configuration of, 209–210
reflection from, 109–110
Copper resinate, 226
Corot, 69
Correlated color temperature, 17
Cotton: *see also* Cellulose, Paper, Textiles
azo dyes on, 166
basic dyes on, 157
carbonyl dyes on, 170
direct dyes on, 158
madder on, 163
photooxidation of, 186
photoreduction of dyes on, 181
phthalocyanine dyes with, 168
quercitron on, 164
rate of fading in, 177
redwood on, 163
structure of, 114
sulfur dyes in, 169
synthetic, 116
use of, 114
Covalent bond, 201
Covering power: *see* Hiding power
Creta, 220
Critical angle, 53–55
Crystal field, 208

Crystal violet
  photochromism in, 180
  structure of, 180
  use of, 169
Crystalware, 84
Cutch, 158, 165
Cyanine dyes, 167, 262

d–d transition
  in cobalt, 240
  color production by, 210
  definition of, 210
  in glass, 237–238
  intensity of, 211–212
da Vinci, Leonardo, 79, 93
Daguerreotype
  deterioration of, 267-268
  method of, 248
Damask, 114-115
Dammar: *see also* Varnish (natural, synthetic)
  color of, 102, 104
  fluorescence of, 10
  oxidation of, 188–189
Daylight, 18–20, 215–216
De Broglie, Louis, 120
Declaration of Independence, 184, 191
Democritus, 63
Desaturation, 97
Developers
  examples of, 265
  exterior, 258
  interior, 258
  pyrogallol, 268
Development, 263-265
Diamond
  color in, 47, 84
  structure of, 197–198
Dietrich of Freiburg, 63
Direct dyes, 158
Disperse dyes, 158–159
Dispersion, 11
Dragon's Blood, 108
du Hauron, Louis Ducas, 72, 250
Dyes
  definition of, 155
  degradation of, 191–193
  method of attachment, 156
  natural, 159–165
  sensitivity to light, 165
  synthetic, 165–170

Edison, Thomas, 20
Efficacy, 21
Efficiency, 20
Egyptian blue, 221-222
Einstein, Albert, 3, 4, 175
Electromagnetic wave: *see also* Light
  amplitude, 2, 5–6
  anomalous dispersion of, 12
  dispersion of, 11
  electric vector of, 4, 7
  frequency, 2-5
  intensity of, 5–6
  interaction with molecule, 7–9, 147
  interference of, 31, 55–59
  magnetic vector of, 4, 7
  spectrum of, 8–9
  theory of, 2-7
  velocity of, 2, 4, 33, 45
  wavelength of, 2, 4
  wavenumber of, 5
Electrum, 110
Emerald green, 212, 227
Enamel, 113, 221
Encaustic, 85
Energy, table of values, 6, 65
Enthalpy, 192
Entropy, 192
Epoxy
  refractive index of, 88
  as resin, 104
Exhibitions, 71, 189-191
Extinction coefficient, 60, 146
Extraordinary ray, 82
Eye
  color perceived by, 69
  spectral sensitivity of, 68

Fabrics, 114-116
Fading: *see also* Photoreactions
  bond energies in relation to, 178
  mechanisms, 178–183
  external factors, 183–184
Fajans, K., 201
Faraday, Michael, 3
Feathers, 94, 105–106
Filters
  infered, 190–191
  ultraviolet, 190
  variable density, 180
Fixation, 266–267
Flashbulbs, 29–30, 36, 39

Flax, 114
Fluorescence
  in canary glass, 240
  definition of, 24, 174
  of materials, 10
Fluorescent lamp
  color rendition by, 25, 71
  efficacy of, 26
  luminous flux of, 36
  operational characteristics of, 27, 28
  principle of operation, 24
  spectral distribution of, 25–27
  ultraviolet content of, 30, 32
Fog, 258
Footcandle, 37
Fraunhofer lines, 19
Free energy, 192
Fresco
  azurite in, 223
  composition of, 100
  lead white in, 220
  Prussian blue in, 223
  reflection from, 100–101
Fresnel, Augustin Jean, 3, 50
Fresnel's equations, 50
Fuchsin, 165
Fustic, 158, 165

Gamboge, 107
Gamma-rays, 9
Gastrofiberscope, 55
Gelatin
  complexes of, 255
  composition of, 255
  deterioration of, 268
  favorable features of, 254
  halogen absorption by, 258
  impurities in, 255
  lime-processed, 254
  pH of, 255
  sensitivity to light, 267
  sensitizers in, 256
  source of, 254
  transmission of light, 9
Gesso, 220
Glass
  additive colors in, 72
  alexandrite effect in, 216
  Beer–Lambert law in, 62
  canary, 240
  carbon amber, 213, 241
Glass (*cont.*)
  chromium in, 240
  cobalt in, 239–240
  color in, 237–244
  decolorization of, 238–239, 243–244
  effect of ir on, 13
  "frosted," 99
  iron in, 238–239
  manganese in, 239
  metals in, 241–243
  Millifiori, 112
  nonreflecting, 58–59
  opalescence of, 95
  rare earths in, 240–241
  repair of, 88
  staining of, 242–243
  structure of, 111–112
  sulfur in, 241
  transmission of light through, 10
  uranium in, 240–241
Glaze
  additive colors in, 72
  Beer–Lambert law in, 62
  on metal, 108, 113
  opacity in, 111–112
  Sang de Boeuf, 242
  structure of, 111–112
  titanium dioxide in, 221
Glazing
  of ceramics, 111–112
  of metals, 108, 113
  of paints, 106–107, 234
Gloss
  in fabrics, 115
  in paints, 98, 100
Glue, 86
Gold
  glazing of, 108, 113
  reflection from, 109–110
Gold leaf, 232
Gold ruby, 242, 243
Graphite, 109
Green earth, 225
Grimaldi, Francesco, 3
Ground state, 144
Gypsum, 220

Haematein, 163
Heating effects, 7, 13, 88, 190–191
Heisenberg, Werner, 121
Hematinone, 242

Hertz, Heinrich, 3
Hiding powder
  change with age, 89–90
  of colored paints, 87
  definition of, 85
  of white paints, 86-87
Hooke, Robert, 3
Hue, 74
Hund's rule, 125, 207-208
Huygens, Christian, 3
Hydrogen bonding, 155, 158

Iceland spar, 81, 82
Illumination: *see also* Light
  effect on color, 215-216
  raking, 101
  of textiles, 113
Impressionists, 71, 72
Incandescence, 20
Incandescent lamp
  color rendering by, 21, 71, 215
  efficacy of, 22
  efficiency of, 20
  illumination by, 190–191
  luminous flux of, 36
  quantity of light from, 39
  spectral energy distribution of, 20–21, 31, 69
  table of characteristics of, 22
  ultraviolet content of, 31, 32
Indigo
  history of, 160
  with madder, 234–235
  particle size of, 96
  photooxidation of, 181
  as a pigment, 107, 160
  structure of, 161
  synthetic, 170
Infrared radiation
  degradation by, 190–191
  heating effects by, 13, 190–191
  in photography, 14, 98, 261
  properties of, 9, 12–14
  transmission by opaque films, 14, 98
Inorganic compounds, 197
Intensifiers, 266
Intensity
  of absorption bands, 145–146
  of charge transfer bands, 204–205
  of *d*-*d* transitions, 211-212
  recommended values for light, 189–190
Interference
  in Moiré, 115
  principles of, 55–59
Inverse square law, 37
Ionic bond, 201
Ionic forces, 155
Ionic potential, 202-203
Iridescence, 55
Iron
  decolorization of, in glass, 238-239
  electron configuration of, 209
  in glass, 238–239
  glazing of, 108
Isotropic crystals, 82
Ivory black, 106

Jasper, 112

Kamal al-din al-Farisi, 63
Kermes, 160
Kirchhoff's law, 43

La Grange, J. L., 3
Lake, 158
Lambert, Johann, 33
Lambert's law, 59
Lamp black: *see also* Carbon black
  oil absorption by, 98
  particle size of, 96
Lapis Lazuli: *see* Ultramarine (natural, synthetic)
Latent image
  bleaching of, 265
  concentration principle in, 258
  definition of, 256
  development of, 264–265
  exterior, 258
  Gurney-Mott theory of, 260-261
  interior, 257-258
Lead chromate: *see also* Chrome yellow
  color in, 212-213
  crystal forms of, 214
Lead oxide, 214
Lead-tin yellow, 228
Lead white
  discoloration of, 236
  hiding power of, 87
  oil absorption value of, 98
  particle size of, 96

Lead white (*cont.*)
properties of, 219-220
soap formation by, 90
ultraviolet reflection from, 117
with vermilion, 233
Lewis, G. N., 128, 206
Lewis structures
in absorption, 148
in atoms, 128-129
limitations of, 129, 131-132
of molecules, 129-132
Ligands
definition of, 198
effect on transition metal, 206-210
Light: *see also* Electromagnetic wave
distribution in space, 39-40
electromagnetic wave theory of, 2-7
intensity of, 5-6
interference of, 55-59
luminosity factor of, 34
measurement of, 34, 38
photon theory of, 3-7
polarization of, 47-52
reflected, intensity of, 52-53
table of conversions, 37
units of, 34, 37
Litmus paper, 163
Logwood, 158, 162
Linseed oil
chalk in, 86
chrysocolla in, 225
crosslinking in, 189
degradation of, 191-194
lead white in, 219
malachite in, 225
oxidation of, 10
refractive index change in, 89
solubility changes of, 188
transparency of, 188
Turner's patent yellow in, 229
Van Dyke brown in, 229
with whites, 87
yellowing of, 98, 107, 188
Litharge: *see* Massicot
Lithopone, 221
Lord Kelvin, 3
Lorentz, Ludwig V., 3
Lumen, 36
Luminance factor, 76
Luminous efficacy, 21
Luminous flux, 36
Luminous intensity, 34-37
Luster, 100
Lusterware, 55, 105
Lux, 37
Lycurgus Cup, 242

Madder: *see also* Alizarin
as glaze, 106
as mordant, 158, 161-162
origin of, 161
as pigment, 231
synthetic, 170
Main group elements, 198-200
Malachite
desaturation of, 96
infrared transparency of, 98
properties of, 225
Malus, Etienne, 3
Manganese, 239
Martin's yellow, 167
Massicot, 82, 228
Mastic: *see also* Varnish
color of, 102, 104
fluorescence in, 10
oxidation of, 188
Matte surface
in glaze, 112
in paint, 98
reflection from, 110
Mauve, 159, 165
Maxwell, James Clerk, 2, 4, 250
Memling, Hans, 94
Mendeleev, Dmitri, 126
Mercurochrome, 169
Mercury, 29, 248
Mercury lamps, 22-23
Metals
absorption by, 108-109
electronic structure of, 108
reflection from, 108-109
Microwaves, 9
Minium, 230, 236
Moiré, 115
Moisture, 184
Molecular orbital theory
of diatomic molecules, 135-137
definition of, 133
description of, 133-135
nonbonding electrons in, 141
orbital energies, 135-136
of polyatomic molecules, 137-143
relation to absorption, 143-147

Molybdenum blue, 205
Moonlight
  intensity, 37
  spectral characteristics of, 24
Mordant dyes, 157
Mulliken, R. S., 133
Mummy, 231

Naples yellow, 106, 228
Neodymium oxide, 216, 241
Newton, Isaac, 3, 11, 45, 63, 64, 93
Nitro dyes
  as acid dye, 157
  fading mechanism of, 167, 181
  structure of, 167
Nitrocellulose, 232
Nitroso dyes: *see* Nitro dyes
Nonreflecting glass, 58–59
Nylon, 157

Octet rule, 129, 131–133
Oil: *see* Linseed oil
Oil absorption
  definition of, 98
  dependence on pigment, 98
  of zinc white, 220
Onyx, 112
Opacifiers, 112
Opacity
  definition of, 45
  index of refraction in relation to, 53–5 99
  particle size in relation to, 95–96, 99, 112
Optic axis, 82
Orbital
  antibonding, 133–135
  bonding, 133–135
  definition of, 122
  energy of, 124
  three-dimensional picture of, 123
Ordinary ray, 82
Organic compounds, 197
Orpiment, 227–228
Orthochromatic, 262
Ossein, 254
Oxygen
  bonding in, 131, 136
  degradation in presence of, 192–193
  effect on fading, 181–187
Ozone, 19

Paint
  chalking of, 188
  color temperatures of, 69–70
  enamel, 98
  fluorescence of vehicle, 10
  of industrial era, 218–219
  oil absorption by, 98
  of pre-industrial era, 218
  preparation of, 97
  quick dry, 220
  satin, 98
  subtractive colors in, 73, 233–235
Palatine orange, 167
Panchromatic, 262
Paper: *see also* Cellulose, Cotton
  acidity of, 186–187
  bleaching of, 187
  degradation of, 191–194
  foxing in, 10
  illumination of, 189–190
  oxidation of, 186
  photoactivation of, 187
  viscosity of, 186
Particle size, 214–216
Pastels, 100
Pauli exclusion principle, 122
Pauli, Wolfgang, 122
Pauling, Linus, 138, 206
Pearlware, 240
Pentimenti, 89–90
Periodic table
  electronic description of, 125–128
  pigment classification in, 198
  table, 127
Perkin, William Henry, 159, 165
Persian berries
  as mordant dye, 159, 165
  structure of, 164
Phenolphthalein, 169
Phosphorescence, 24, 174
Phosphors, 25
Phot, 37
Photochemistry, 174–184
Photochromism, 180
Photographic film
  effect of air pollution on, 267
  effect of humidity on, 267
  effect of temperature on, 267
  infrared, 261

Photographic film (*cont.*)
preparation of, 250
preservation of, 267-268, 273
silver halides in, 250
subtractive colors in, 73, 268-273
support for, 250
Photography, infrared, 14, 98
Photon, 3, 4
Photooxidation
in cellulose, 186
mechanism of, 181-183
Photoreactions
activation energy of, 184
application for Beer-Lambert law to, 62
external factors in, 183-184
Gurney-Mott theory of, 260-261
intramolecular, 179
photochromism, 180
photooxidation, 181-183
photoreduction, 180-181
phototendering, 183
rates of, 176-177
in silver halides, 258-261
ultraviolet radiation in, 10
wavelength of light in, 184
Photoreduction, 180-181
Phototendering, 183
Phthalocyanine dyes, 167-168, 213-214
Pigments
characteristics of, 96
composition of, 198
definition of, 155
light scattering by, 95-96
particle size of, 96
reflectance of, 217-218
stability of, 235-236
Planck, Max, 3
Planck's constant, 4-5
Pointillism, 72
Polarizability, 237
circular, 48
definition of, 48-50
elliptical, 48
filter, 52
Fresnel's equation for, 50-51
of ions, 201-203, 237
mechanism of, 49-50
plane, 48
of sunlight, 51
uses of, 52
"Polaroid," 152
Polyene, 149
Polyester, 104
Ponceaux, 166
Porcelain, 239
Protein
in gelatin
photosensitivity of, 187-188
of silk, 156
of wool, 156
Prussian blue
characteristics of, 223
color of, 205
lattice type, 198
with madder, 234-235
particle size of, 96
Purity, 76
Purpurin, 162
Pyrolucite, 239
Pyrosol, 242
Pyroxylin, 249

Quantity of light, 38-39
Quantum, 3
Quantum efficiency: *see* Quantum yield
Quantum numbers, 121-122
Quantum yield, 175, 257
Quartz, 9, 22
Quartz lamp, 22
Quercitron
as mordant, 158, 164
as pigment, 229
structure of, 164

RAT law, 44, 66, 99
Radiowave, 8
Raw sienna: *see also* Sienna
oil absorption of, 98
Raw umber: *see* Umber
Rayleigh scattering
by atmosphere, 93-94
by bluejay feathers, 94
by colophony, 94
by glaze, 112
laws of, 87
by milk, 94
polarization of light, 51
by varnish, 94
Realgar, 231
Red lead: *see* Minium
Red ochre, 230-231
Reducer, 266

Redwood, 163
Reflection
  critical angle of, 104
  diffuse, 44, 100
  of fabrics, 115
  laws of, 44
  of metals, 109–111
  of paint, 100
  particle size in relation to, 97
  spectral patterns, 66–67
  specular, 44, 100
  from transparent surfaces, 102–105
  of ultraviolet radiation, 116–117
Refraction
  definition of, 45
  laws of, 82
Refractive index
  of adhesives, 88
  of air, 46
  of biaxial crystals, 81
  change with age, 89–90
  colorless materials, 46
  definition of, 45
  dependence on wavelength of, 46
  effect of density on, 46
  in inorganic materials, 84
  of isotropic crystals; 82
  of liquids, 84
  major value of, 82
  mean value of, 83
  of metals, 80
  minor value of, 82
  of opacifiers, 112
  of organic materials, 84
  in relation to scattering, 85–87
  table of values, 83
  temperature dependence of, 80
  in uniaxial crystals, 81
  of vacuum, 46
  wavelength dependence of, 80
Rembrandt, 106
Resin, natural: *see* Varnish, natural
Resin, synthetic: *see* Varnish, synthetic
Resonance
  in cyanine dyes, 167
  definition of, 132
  examples of, 132
  in triarylcarbonium dyes, 168
Retinal, 179
Roseaniline blue, 165
Royal blue: *see* Tyrian purple
Rubens, Peter Paul, 218
Rutherford, Ernest, 120
Rutile: *see* Titanium dioxide

Safflower, 165
Saturation, 74
Scheele's green
  characteristics of, 226
  d–d transitions in, 212
Schrödinger, equation, 121
Schrödinger, Erwin, 120
Scott-Archer, Frederick, 249
Scumbling, 107
Selenium
  in glass, 241
  decolorization in glass by, 239
Sepia, 160
Sheen, 100
Shellac
  fluorescence of, 10
  on metal, 108
Sidgwick, N. V., 206
Sienna, 227
Silk
  acid dyes on, 156
  color of, 115
  direct dyes on, 158
  madder with, 163
  Moiré fabric, 115
  photosensitivity of, 187–188
  phototendering of, 183
  phthalocyanine dyes with, 168
  structure of, 114, 156
  synthetic, 116
  Tyrian purple on, 159
  use of, 114
  velvet, 115
  "watered," 115
Silver
  complexes of, 261–262, 266
  conduction by, 109
  glazing of, 108, 113
  reflection from, 109–111
  reticular, 257
  ultraviolet transparency of, 243
Silver halide
  action of light on, 256–261
  bonding in, 258–259
  complexes with gelatin, 255
  crystal structure of, 251–252

Silver halide (*cont.*)
defects in, 252
electron configuration of, 251
grain size of, 253–254
latent image in, 256
light sensitivity of, 253, 258
morphology of, 251
photoconduction in, 259
photolysis mechanism of, 258–261
photosensitivity of, 10
quantum yield of, 257
sensitivity centers in, 257
sensitization of, 261–263
solubility of, 251–252
Silver leaf, 232
Smalt
characteristics of, 223
discoloration of, 107, 236
Snell, Willebrord, 47
Snell's law, 47
"Soap," formation with lead salts, 90, 220, 228
Sodium lamp, 22–23
Solarization, 243–244
Solid angle, 35–36
Steradian, 35
Stick lac
history, 160
as mordant, 158
properties of, 160
Stone, 95
Stop bath, 265
Subtractive color: *see also* Color
in paint, 233–235
in photography, 270–271
Sulfur dyes, 169
Sunlight
color rendering by, 20
color temperature of, 20
energy distribution of, 18–19, 68–69
intensity of, 19, 37
ultraviolet content, 30, 32

Talbot, William Henry Fox, 91, 248
Taupenot, J. M., 249
Tempera
aging of, 90
azurite in, 223
degradation of, 191–194
gold leaf in, 232
Tempera (*cont.*)
lead white in, 219
malachite in, 225
ultramarine in, 223
Terre verte: *see* Green earth
Textiles: *see also* particular fiber type
degradation of, 190–194
illumination of, 189–190
viscosity of, 186
Theophilus, 108
Thermodynamics, 191–192
Tiffany, Louis C., 55
Tin, 108
Titanium dioxide
characteristics of, 221
color in, 214
hiding power of, 87
refractive index of, 83
ultraviolet reflection from, 117
Titian, 106
Toners, 266–267
Transition elements
color due to, 206
definition of, 128
presence in pigments, 198–200
Translucency, 45
Transmission of infrared radiation, 12–14
Transparency
definition of, 45
perception of, 99
Triarylmethane dyes, 157, 168
Tumeric
light fastness of, 158
use of, 165
Tungsten bronze, 205
Turkey red: *see* Madder
Turner's patent yellow, 229
Tyndall scattering, 93
Tyrian purple, 159–60

Ultramarine, natural
characteristics of, 222
color change in, 107
color in, 213
as glaze, 106
with madder, 234–235
reaction with varnish, 213
"sickness" in, 223, 236
Ultramarine, synthetic, 224
Ultraviolet content, 32

Ultraviolet radiation
  chalking due to, 221
  effect on chrome yellow, 229
  effect on massicot, 228
  effect on minium, 231
  effect on nitrocellulose, 232
  effect on red lake, 231
  effect on Turner's patent yellow, 229
  effect on Van Dyke brown, 229
  in fading, 184–185
  general, 9–11
  reflection of, 116–117
  transmission of, 243
Umber, 212, 227
Uniaxial, 81
Uranium, 205, 240–241

Valence band, 259
Valence bond theory
  applications of, 138–141
  definition of, 138
  of transition metals, 206–207
Valence electrons, 128
Value, 74
van der Waals forces, 155, 262
Van Dyke brown, 229
Van Eyck, Jan, 94, 108
Van Vleck, John, 206
Varnish, natural
  bleaching of, 188
  color of, 104
  effect on color, 101–104
  fluorescence of, 10
  on metal, 105, 108
  Rayleigh scattering in, 94
  reflection from, 101–105
  solubility of, 188
  surface characteristics of, 104
  yellowing of, 13–14, 107, 188
Varnish, synthetic
  acrylics, 104–105
  characteristics of, 104–105
  disadvantages, 105
  fluorescence of, 10
  stability of, 189
  transparency in infrared, 14
Vat dye: *see* Carbonyl dye
Velvet, 115
Verdigris
  characteristics of, 226
Verdigris (*cont.*)
  discoloration of, 236
  as glaze, 106
  infrared transparency of, 98
  particle size of, 96
Vermeer, 106, 229
Vermilion
  characteristics of, 230
  darkening of, 235–236
  glazing of, 106
  infrared transparency of, 98
  with lamp black, 233
  with lead white, 233
  particle size of, 96
  refractive index of, 82
Viridian
  characteristics of, 227
  "*d–d*" transitions in, 212
  as glaze, 106
Visible radiation, 9, 11–12
Vision
  color, 71
  human, 11, 68, 179
  insect, 11, 62
  perception of reflection, 101
Vogel, H. W., 261

Watercolor
  baryta in, 220
  chrysocolla in, 225
  composition of, 100
  covering power of, 86
  decomposition of, 87
  illumination of, 189–190
  lead white in, 219
  reflection from, 100
  Van Dyke brown in, 229
Wavelength
  definition, 2
  table of conversions, 7
Wax
  effect of infrared radiation on, 13
  fluorescence of, 10
  reflection from, 101
Weave, 114
Wedgwood, Josiah, 240
Weld
  as mordant, 158, 164
  structure of, 164
Werner, Alfred, 206

Vhiting: *see* Chalk
Voad, 160–161
Vood: *see also* Cellulose, Cotton
  bleaching of, 187
  darkening of, 187
  degradation of, 191–194
  effect of infrared radiation on, 13
  oxidation of, 187
Vool
  direct dyeing of, 158
  madder on, 163
  photoreduction of dyes on, 181
  photosensitivity of, 187–188
  phthalocyanine dyes on, 168
  quercitron on, 164
  structure of, 114, 156
  synthetic, 116
Wool (*cont.*)
  Tyrian purple on, 159
  use of, 114

X-rays, 9

Young, Thomas, 3, 71

Zinc white
  chalking of, 188–189, 236
  characteristics of, 220
  hiding power of, 87
  ultraviolet reflection from, 117